*Comparative Mechanisms
of Cold Adaptation*

Academic Press Rapid Manuscript Reproduction

*The Proceedings of
the 28th Annual Meeting of
the American Institute of Biological Sciences
Held in East Lansing, Michigan on August 24, 1977*

Comparative Mechanisms of Cold Adaptation

edited by

Larry S. Underwood
*University of Alaska
Artic Environmental Information
and Data Center
Anchorage, Alaska*

Larry L. Tieszen
*Division of International Programs
National Science Foundation
Washington, D.C.*

Arthur B. Callahan
*Office of Naval Research
Arlington, Virginia*

G. Edgar Folk
*Department of Physiology
University of Iowa
Iowa City, Iowa*

1979

ACADEMIC PRESS

A Subsidiary of Harcourt Brace Jovanovich, Publishers

New York London Toronto Sydney San Francisco

ACADEMIC PRESS, INC.
111 Fifth Avenue, New York, New York 10003

United Kingdom Edition published by
ACADEMIC PRESS, INC. (LONDON) LTD.
24/28 Oval Road, London NW1   7DX

Library of Congress Cataloging in Publication Data
Main entry under title:

Comparative mechanisms of cold adaptation.

Includes updated and edited papers from a
symposium sponsored by the Office of Naval Research, the
American Institute of Biological Sciences, and the
Ecological Society of America, which was held at the
28th annual meeting of the AIBS in East Lansing, Mich.,
Aug. 24, 1977.
Includes index.
1. Cold adaptation—Congresses.   2. Physiology,
Comparative—Congresses.   I. Underwood, Lawrence S.
II. Tieszen, Larry L.   III. United States.   Office
of Naval Research. IV.   American Institute of Biolog-
ical Sciences.   V. Ecological Society of America.
QP82.2.C6C64   574.5'42   79-19119
ISBN 0-12-708750-8

Contents

Contributors

Numbers in parentheses indicate the pages on which authors' contributions begin.

W. D. Billings (181), *Department of Botany, Duke University, Durham, North Carolina 27706*

M. J. Burke (259), *Colorado State University, Ft. Collins, Colorado 80302*

Brian F. Chabot (283), *Section of Ecology and Systematics, Langmuir Laboratory, Cornell University, Ithaca, New York 14853*

F. Stuart Chapin III (215), *Institute of Arctic Biology, University of Alaska, Fairbanks, Alaska 99701*

Keith E. Cooper (75), *Faculty of Medicine, University of Calgary, 2920 24th Avenue, Northwest Calgary, Canada T2N 1N4*

A. Ian de la Roche (235), *Ottawa Research Station, 1045 K. W. Neatby Building, Ottawa, Canada K1A 0C6*

D.R. Deavers (51), *Department of Physiology, University of Louisville Medical School, Louisville, Kentucky 40208*

Melvin J. Fregly (159), *Department of Physiology, College of Medicine, University of Florida, Gainesville, Florida 32601*

James A. Gessaman (1), *Department of Biology, Utah State University, Loga, Utah 84322*

John J. Kelley (323), *Naval Artic Research Laboratory, Barrow, Alaska 99723*

O. Heroux (169), *Division of Biological Services, National Research Council of Canada, Ottawa, Ontario, Canada*

D. R. Hettinger (91), *Department of Biochemistry, Health Science Center at San Antonio, University of Texas, San Antonio, Texas 78284*

Barbara A. Horwitz (91), *Department of Animal Physiology, University of California at Davis, Davis, California 95616*

D. W. Larson (303), *University of Guelph, Guelph, Ontario, Canada*

Gary A. Laursen (323), *Naval Arctic Research Laboratory, Barrow, Alaska 99723*

P. C. Miller (181), *Systems Ecology–Research Group, San Diego State University, San Diego, California 92182*

X. J. Musacchia (51), *Graduate School, University of Louisville, Louisville, Kentucky 40208*

W. C. Oechel (181), *Department of Biology, McGill University, P.O. Box 6070, Montreal 101, Quebec, Canada*

Jane C. Roberts (129), *Department of Biology, Creighton University, Omaha, Nebraska 68178*

George Guy Spomer (311), *Department of Biological Sciences, University of Idaho, Moscow, Idaho 83843*

Larry L. Tieszen (343), *Room 1208A, Division of International Programs, National Science Foundation, Washington, D.C. 20050*

Larry S. Underwood (343), *University of Alaska, Arctic Environmental Information and Data Center, Anchorage, Alaska 99501*

Robert G. White (13), *Institute of Arctic Biology, Irving Building, University of Alaska, Fairbanks, Alaska 99701*

Bruce A. Wunder (143), *Department of Zoology, Colorado State University, Fort Collins, Colorado 80521*

Mohamed K. Yousef (81), *Department of Biological Sciences, University of Nevada, Las Vegas, Nevada 89154*

Preface

The idea of conducting a symposium on the mechanisms of cold adaptation in the Arctic grew out of a series of informal conversations among the editors at the Naval Arctic Research Laboratory, Barrow, Alaska in 1975 and 1976. Each was involved in some phase of conducting, funding, or coordinating cold adaptation research, and each was experiencing some degree of frustration. We recognized that these studies play a significant role in man's understanding of the adaptation phenomenon in general, thus indicating a bright future for this field of inquiry. However, as interest blossomed and studies multiplied, communication among widely disparate researchers floundered. How could we best ensure adequate communication among colleagues in Florida, Alaska, and Canada? How could we cross disciplinary lines and learn what endocrinologists and biochemists, botanists and zoologists, laboratory scientists and arctic field scientists, and those interested in humans and those working with other species are discovering about cold adaptation? Our goal, therefore, was to stimulate such communication. We felt that a good start would be to bring together experts in the various areas of cold adaptation mechanisms research to discuss their work and to explore the best ways to continue information exchange in the future.

A symposium was planned under the joint sponsorship of the Office of Naval Research, the American Institute of Biological Sciences, and the Ecological Society of America, which was held at the 28th Annual Meeting of the AIBS in East Lansing, Michigan on August 24, 1977. Ten major papers and several shorter ones were presented that day, and a second day was devoted to a workshop on the question of where similar research should go in the future. A series of recommendations was presented to the cosponsors at the close of the symposium.

Participants were encouraged by what they heard from their colleagues and by the audience's response. The large lecture hall was consistently filled, and discussions were stimulating and often spirited. The idea for this book grew from the realization that interest in the subject of cold adaptation was obviously high; however, the amount of information presented could not be instantly digested. We hope this book will not only summarize our current knowledge, but will also be useful to scientists conducting research in this area, to students and others beginning their careers, and to funding agencies considering support for such research.

Investigators who want to do field research in the Arctic often have difficulty in locating where such work is being conducted. Thus, a chapter describing research opportunities for arctic field work has been included in this publication. The last chapter summarizes the symposium's recommendations for future research directions.

The editors wish to thank the Office of Naval Research, the American Institute of Biological Sciences, and the Ecological Society of America for supporting the symposium. Special thanks go to the staff of AIBS in planning and conducting the symposium. We also gratefully acknowledge Mrs. Peggy Hood and Mrs. Linda Murray, who assisted in the typing of the manuscript, Ms. Judy Brogan who assisted in editing, and Mrs. Shirley A. Zimmerman, who coordinated correspondence between authors and editors and prepared the camera-ready copy for publication by Academic Press, Inc. Most of all, the editors wish to recognize the efforts of the authors and to thank them for their excellent presentations and for meeting most, if not all, of their deadlines.

Larry S. Underwood
Larry L. Tieszen

I. ENERGY ACQUISITION AND UTILIZATION

James A. Gessaman

Department of Biology
Utah State University
Logan, Utah

Although many of the physiological adaptations among homeothermic residents of the Arctic are not unique, the following characteristics seem to be more unique to arctic species than to those of the temperate or tropical regions: 1) white plumage or pelage, which may be important in absorbing radiant heat in the spring, summer, and fall; 2) thick insulation in the form of fat, feathers, and/or fur; 3) thermolability of young, active growing sandpipers; and 4) high fat content of caribou and polar bear milk. These and other adaptations are discussed.

ENERGY ACQUISITION

Biochemical Energy

Energy Content of the Diet. The adaptive value of an energy-rich diet for birds and mammals living in the Arctic is quite clear. In winter the energy cost of thermoregulation may be high, and some animals have only a few hours in which to feed each day since their foraging is restricted to daylight hours. Along the arctic coast in summer, air temperatures usually average below 5°C. The cost of thermoregulation may be especially high for a young homeotherm before the insulative layer of their fur or feathers develops.

Birch seeds (Brooks 1968) make up 80 percent of the diet of redpolls *(Acanthis flammea)* in northern Finland, and birch and alder seeds make up 88 percent of their diet in the vicinity of Fairbanks, Alaska. The seeds of birch are substantially higher in caloric value than most seed types which have been measured.

In summer polar bears *(Ursus maritimus)* living on land along the Hudson Bay are omnivorous, feeding on a variety of grasses, berries, flightless birds, small mammals, and carrion. In winter these same bears move onto the ice pack of the bay and may feed exclusively on ringed seals *(Phoca hispida)*. They often preferentially eat the blubber (Stirling and McEwan 1975) which has the highest energy content of any tissue of their prey.

In the Arctic young homeotherms must cope with the high energy demands of growth plus those of thermoregulation before their insulative layer of fur or feathers is fully developed. The diet of young precocial rock ptarmigan *(Lagopus lagopus)* reflects their greater-than-adult rate of energy utilization. The chicks' diet consists of twenty-six percent by weight of invertebrates (Theberge and West 1973), with the remainder composed of birch and willow catkins. In contrast, their parents' diet approximates 86 percent birch and 6 percent willow, with the remainder made up of other plant species. The energy per unit weight of the invertebrates is greater than that of birch and willow.

Young caribou exist exclusively on energy-rich milk until the rumen becomes inoculated with bacteria and protozoa so they can digest lower-in-energy plant material. Caribou calves in Alaska are born during a two- to three-week period from late April to early May, a time when environmental factors can severely stress the early postpartum calf. These factors include snow-covered calving grounds, low ambient temperatures, scarcity of food (especially when vegetation is encased in hard-crusted snow), and harassment by predators. Newborn caribou calves are almost entirely dependent upon an adequate supply of maternal milk during this period. Caribou milk contains more total solids and fat than does the milk of any other species of wild or domesticated ungulates that has been studied. Mean values for Alaskan caribou milk (Luick 1974) at mid-lactation are 31.6 percent dry matter and 15.5 percent fat. The concentration of these constituents increases markedly throughout the lactation cycle. During early lactation when the herds are migrating in search of adequate food and coping with predators, such a highly-concentrated milk could have considerable survival value for the young calves.

The high fat content of the milk of marine mammals is well known. In the Arctic for example, Alaskan fur seal milk is 52.2 percent fat (Ashworth, Romaiah, and Keyes 1966). The milk of polar bears also has a high energy content; 30.6 percent fats and 43.5 percent total solids (Baker, Harington, and Symes 1963).

Quantity of the Diet. Redpolls, in comparison to non-arctic passerines, have a relatively higher rate and quantity of gross energy intake at low temperatures (Brooks 1968). To facilitate the process, the redpolls have a croplike esophageal diverticulum that they filled with "extra" food just prior to the onset of darkness.

The quantity of food eaten daily by the Arctic fox *(Alopex lagopus)* and caribou *(Rangifer arcticus)* varies seasonally e.g., the Arctic fox consumes more in summer than winter (Underwood 1971) and the food intake of caribou (White 1974) increases with the availability of live green biomass, which peaks in July. In contrast, the daily food consumption of snowy owls *(Nyctea scandiaca)* caged outdoors at Barrow, Alaska was three lemmings (60 g each)/day in October and six lemmings/day in January (Gessaman 1972). A free-flying snowy owl, however, unlike the caribou and Arctic fox, will emigrate to the lower latitudes of southern Canada and northern United States when a maintenance diet is not available on the arctic tundra.

Digestive efficiency. The digestive efficiencies (i.e., energy assimilated/energy in food eaten x 100) of arctic homeotherms are no better than those of their temperate zone counterparts. Among herbivores, the lemming (Melchoir 1972) *(Lemmus trimucronatus)*, willow ptarmigan (West 1968; Moss 1973) *(Lagopus leucurus)* and caribou (Luick and White 1971) have digestive efficiencies that average about 36 percent, 45 percent, and 56 percent, respectively. The increase in efficiency from lemming to ptarmigan to caribou may reflect the increasing complexity of the digestive systems. Cellulose and hemicellulose, which are only partially digested by gastric microflora in the lemmings, may be more fully processed by cecal fermentation in the ptarmigan. The reticulo-rumen fermentation of cellulose and hemicellulose by the caribou probably results in the most complete digestion among the three species.

The effect of low temperature on digestive efficiency has been reported for only one arctic homeotherm, the redpoll (Brooks 1968). The efficiency increased at temperatures below -30°C.

Absorption of Solar Radiation

In spring, summer, and fall, arctic homeotherms may
gain enough energy from direct solar radiation to reduce
the amount of metabolic heat they require to maintain
homeothermy. The white winter pelage of the tundra hare,
the varying lemming, weasels, the Arctic fox and the polar
bears as well as the white winter plumage of the snowy owl
and the rock and willow ptarmigans are well known. There
is no evidence that such lack of color helps to balance
heat loss with heat gain in the arctic winter when the
natural photoperiod is so short and the sun's altitude so
low. The whiteness, however, may be advantageous as a
solar energy absorber in spring and fall when: 1) the
length of the photoperiod exceeds 12 hours (e.g., at Barrow,
Alaska--71° 20' N lat.) the photoperiod increases from
nine to 20 hours from the first of March to the last of
April), and 2) mean air temperatures are well below zero.
It is generally assumed that heat gain from solar radiation
is substantially greater in birds with dark-colored plumages than in birds with light-colored plumages. Recent
studies of heat flux (Walsberg and King 1977) through black
and white pigeon plumages showed, however, that this is
true only under limited conditions. At very low wind
speeds black plumages acquired a greater radiative heat
load than did white plumages, but the heat loads of black
and white plumages rapidly converged as wind speed increased. This phenomenon was most dramatically seen in
erected plumages, in which (at wind speeds above 3m/s) the
generally accepted relation of coat color to solar heat
load reversed, i.e., white plumages acquired a greater heat
load than did black plumages. The effect was caused by
short-wave radiation penetrating further into light than
dark plumages. The implication is that the white color of
an arctic homeotherm may be energetically advantageous.

ENERGY UTILIZATION

Heat Production - Homeothermy

Basal Metabolic Rate (BMR). In 1950 Scholander, et.
al. reported that the basal metabolic rate (per unit body
weight) of arctic birds and mammals did not differ from
that of temperate and tropical species. In recent years
that contention has been disproved. For example, the

tundra hare (Wang 1973) *(Lepus arcticus)* has a BMR amounting to only 62-83 percent of the values predicted from its
weight. Similarly, the BMR of the snowy owl (Gessaman
1972) is 42 percent lower than predicted for nonpasserines
but only 6 percent less than that predicted for owls. On
the other hand, the BMR's of the willow and rock ptarmigan
(West 1972a) are 14 and 44 percent higher, respectively,
than predicted, and that of the redpoll (West 1972b) also
exceeds the predicted value. These data certainly discourage any generalizations about BMR levels of arctic species.

The BMR's of birds vary seasonally in a manner dependent upon body size. Analysis of data obtained on arctic
and temperate species acclimatized out of doors demonstrated (Weathers 1977) that, with a mass less than 150 g,
metabolism tends to be higher in winter than summer, while
the reverse holds for larger forms.

Thermoregulatory Metabolism. A resting homeotherm
(whether bird or mammal) produces heat above its basal
level (BMR) by shivering. Mammals also accomplish this end
through high rates of oxidative phosphorylation in brown
fat (commonly called non-shivering thermogenesis). Non-
shivering thermogenesis will be discussed in detail in
later chapters of this symposium publication.

The maximum rate of heat production by a resting homeotherm is commonly called summit metabolism. Summit metabolism varies seasonally (higher in winter than summer) and
varies from 3.0 to 6.0 times BMR among individuals. There
is no evidence, however, that arctic species have a greater
capacity to mobilize energy (i.e., a higher summit metabolism) than do inhabitants of lower latitudes. The summit
metabolism of cold-acclimatized redpolls has been reported
as 5.6 times their BMR. Rosenmann recently showed that the
higher winter BMR of the red-backed vole (1975), an arctic
inhabitant, is associated with a higher summit metabolism.

Wunder, et al. pointed out that microtines (1977)
(which compose the major small mammal species in boreal and
arctic regions) can combat problems of winter cold by
increasing their weight-specific metabolic rates and therefore thermogenesis. They further suggested that a drop in
body weight would compensate for the potentially higher
total energy needs and would decrease an animal's need to
accumulate calories while operating at higher metabolic
turnover rates. A weight drop in winter has been demonstrated in red-backed voles *(Clethrionomys rutilus)* (Rosenmann, Morrison and Feist 1975).

Little is known about the energy cost of thermoregulation during exercise in arctic homeotherms. For example, does the heat produced by an animal during locomotory activity reduce the energy cost of thermoregulation? In other words, is energy metabolism during exercise at cold temperatures less than or equal to the sum of the energy cost of the activity and the cost of thermoregulation when the animal is resting under the same environmental conditions?

At air temperatures from 0 to -30°C, the energy cost of exercise in redpolls (Pohl and West 1973) equalled the sum of these two energy costs. From -30 to -42°C the metabolic rate during exercise remained the same and was therefore less than the sum of activity metabolism and thermoregulatory metabolism (at rest). Comparable information is not available on other arctic species.

Activity Metabolism. Within the past 10 years, physiologists have learned much about the energy cost of flight in birds (Tucker 1975) and bats and of running in bipedal and quadripedal mammals (Taylor 1973). Almost none of this work, however, was done on arctic animals. Energy utilization associated with activity has been measured on two arctic species: the polar bear (Oritsland 1976) and the caribou (White and Yousef 1974).

Polar bears were trained to walk on a treadmill while their oxygen consumption was measured. Oritsland, et al. reported higher energy cost of walking (1976) in the polar bear than in other quadripeds which had been studied. The values for caribou activity metabolism did not differ significantly from those reported for red deer (Brockway and Gessaman 1977) and similar-sized quadripeds.

Heat Production - Hypothermy

Torpor. Hypothermy is a well-known strategy for conserving energy in a cold environment. However, it has neither been demonstrated in the laboratory nor in the wild in arctic birds that are either cold acclimated for winter or acclimatized, respectively. On the other hand, thermolability during the development of young sandpipers (Norton 1973) of the genus *Calidris* seems to be the most striking metabolic adaptation among breeding birds of the tundra. The free-living chicks consistently reduced the gradient between their core and ambient temperatures by allowing body temperatures to drop to $30-35^{\circ}$C while remaining functional, alert, and active. "This hypothermia differs from all other cases so far described among birds. First, it is characteristic of active birds. Second, in contrast to torpor, chick hypothermia coincides with the period of rapid growth and maximum rates of biosynthesis. Third, other studies of exothermy or thermolability in growing young birds have generally determined that endothermy develops gradually before fledging or independence from the nest is attained, but these sandpiper chicks showed no clear trend toward higher body temperatures during feeding periods as the fledgings approached 15-20 days of age (West and Norton 1975).

It has been shown that polar bears (Folk, Brewer and Sanders 1970) in captivity at Barrow, Alaska, entered torpor in January and February. Winter hypothermia has not been reported among free-roaming polar bears.

Hibernation. The Arctic ground squirrel *(Spermophilus undulatus)* is the only hibernator indigenous to the treeless. Arctic. This paucity of hibernators in the Arctic probably reflects the scarcity of soils suitable for a hibernaculum (Hoffman 1974), i.e., where soil temperatures remain above freezing throughout the winter. Arctic ground squirrels stay within their hibernaculums for about 220 days per each year.

Reduction of Heat Loss

Fur and Feather Thickness. In a cold environment any mechanism that reduces heat loss is certainly adaptive. In 1950, Scholander, et al. showed that arctic species are better insuthan those in the tropics. Other investigators have shown that insulation among arctic species is greater in winter than in summer. For example, the lower critical temperature of willow ptarmigans is 7.7°C in summer but drops to -6.3°C in winter (West 1972a). The insulation (Frisch, Oritsland and Krog 1974; Hart 1956) of the polar bear is 30 percent greater in winter than summer. The layering of fur on the bear is not uniform; instead, the areas of the body surface that contact the substratum are especially well insulated. Wild redpolls (Brooks 1968) have 31 percent heavier plumage in November than in July.

Behavior. Behavioral thermoregulation may be divided into two categories: 1) the selection of a less thermally stressful environment and 2) changes in the surface area-to-volume relationship, which may be used simultaneously.

The ptarmigan, much like grouse in alpine habitats, finds
shelter from severe storms and the heat sink of the arctic
sky at night by burrowing under the snow mantle. When
willow or birch are available within the cavity space the
bird may remain covered for one to two days. Cade (1953)
has observed redpolls entering and feeding in holes in the
snow formed either by protruding vegetation or by birds
themselves, and Irving reported that Eskimos at Anaktuvuk
Pass have also seen this behavior (1960).

Polar bears confronted by a cold windy environment
seek the lee of a natural wind break such as a pressure
ridge as a resting place; if this is not available the
animal will lie with its well-insulated rump oriented into
the wind. The postures of a polar bear (Oritsland 1970) at
different levels of thermal stress have been described by
Oritsland (Figure 1).

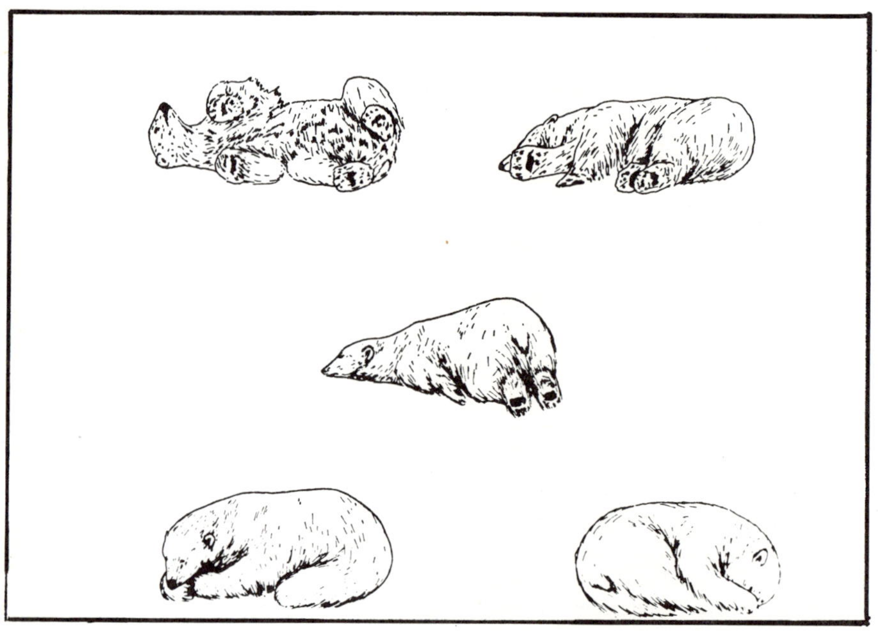

FIGURE 1. *Polar bears' postures at mean windchills
830 W/m^2 (I), 1410 W/m^2 (II) and 1910 W/m^2 (III).*

Control of Peripheral Circulation. Birds and mammals
both have certain body surfaces that are poorly insulated.
In mammals these include bare nostrils, toe pads of arctic
foxes and wolves *(Canus lupus)*, and palms and soles of
polar bears. The feet and tarsi in many arctic birds are
bare (e.g., raven, *Corvus corax*; redpoll). The tarsi and
upper surface of the foot of ptarmigan and snowy owls are
feathered but the undersurface of the toes is bare. Blood
flow through these surface tissues is precisely regulated
to maintain the temperature at or slightly above freezing
and to minimize the heat loss from the extremities.

Tissue Production

In addition to the energy devoted to heat production,
a significant amount of the energy acquired from the diet
may be shunted into processes involving biosynthesis such
as fat storage, growth, egg production, molt, fetal development, and lactation. The efficiency with which energy is
used in these six productive processes has not been examined in any species of arctic homeotherm. But then, very
few efficiencies have been measured on any non-domestic
birds and mammals. This area has been neglected for too
long by researchers.

In summary, most of the adaptations discussed in this
chapter are not unique to homeothermic residents of the
Arctic. The following characteristics, however, seem to be
more unique to arctic species than to those of the temperate or tropical regions:

1.   White plumage or pelage, which may be important
      in absorbing radiant heat in the spring, summer,
      and fall;
2.   Thick insulation in the form of fat, feathers,
      and/or fur;
3.   Thermolability of young, active, growing sand-
      pipers; and
4.   High fat content of caribou and polar bear milk.

REFERENCES

Ashworth, V.S., G.D. Ramaiah, and M.C. Keyes. 1966. Species difference in the composition of milk with special reference to the northern fur seal. *J. Dairy Sci.* *49:*1206.

Baker, B.E., C.R. Harington, and A.L. Symes. 1963. Polar bear milk. I. Gross composition and fat constitution. *Can. J. Zool. 41:*1035.

Brockway, J.M. and J.A. Gessaman. 1977. The energy cost of locomotion on the level and on gradients for the red deer *(Cervus elaphus)* *Quart. J. Exptl. Physiol. 62:*333.

Brooks, W.S. 1968. Comparative adaptations of the Alaskan redpolls to the arctic environment. *Wilson Bull. 80:*253.

Cade, T.J. 1953. Sub-nival feeding of the redpoll in interior Alaska: a possible adaptation to the northern winter. *Condor. 55:*43.

Folk, G.E., M.C. Brewer, and D. Sanders. 1970. Cardiac physiology of polar bears in winter dens. *Arctic. 23:*130.

Frisch, J., N.A. Oritsland, and J. Krog. 1974. Insulation of furs in water. *Comp. Biochem. Physiol. 47A:*403.

Gessaman, J.A. 1972. Bioenergetics of the snowy owl *(Nyctea scandiaca)* *Arctic Alp. Res. 4:*223.

Hart, J.S. 1956. Seasonal changes in insulation of fur. *Can. J. Zool. 34:*53.

Hoffmann, R.S. 1974. Terrestrial vertebrates. Pages 475–568 in Arctic and Alpine Environments, Methuen, London.

Irving, L. 1960. Birds of Anaktuvuk Pass, Kobuk, and Old Crow: A study in arctic adaptation. *U.S. Natl. Mus. Bull. 217.*

Luick, J.R. 1974. Nutrition and metabolism of reindeer and caribou in Alaska. Progress Report 1973/1974, University of Alaska.

Luick, J.R., and R.G. White. 1971. Food Intake and energy expenditure of grazing reindeer. In The structure and function of the tundra ecosystem. U.S. Tundra Biome Program.

Melchior, H.R. 1972. Summer herbivory by the brown lemming at Barrow, Alaska. Page 136 in Proceedings of 1972 U.S. Tundra Biome Symposium, Lake Wilderness Center, University of Washington.

Moss, R. 1973. The digestion and intake of winter foods by wild ptarmigan in Alaska. *Condor. 75:*293.

Norton, D.W. 1973. Ecological energetics of calidridine
       sandpipers breeding in northern Alaska. Ph.D. Thesis.
       University of Alaska, Fairbanks.
Oritsland, N.A. 1970. Temperature regulation of the polar
       bear *(Thalarctos maritiums)*.   *Comp. Biochem. Phy-*
       *siol. 37:*225.
Oritsland, N.A., et al. 1976. Physiological studies of
       polar bears at Churchill, Manitoba. In Proceedings of
       the 4th working meeting of the polar bear specialists
       group. Int. Union Conserv., Morges, Switzerland.
Pohl, H., and G.C. West. 1973. Daily and seasonal variation
       in metabolic response to cold during rest and forced
       exercise in the common redpoll. *Comp. Biochem. Phy-*
       *siol. 45A:*851.
Rosemann, M., P. Morrison, and D. Feist. 1975. Seasonal
       changes in the metabolic capacity of red-backed voles.
       *Physiol. Zool. 48:*303.
Scholander, P.F., et al. 1950. Heat regulation in some
       arctic and tropical mammals and birds. *Biol. Bull.*
       *99:*237.
Stirling, I., and E.II. McEwan. 1975. The caloric value of
       whole ringed seals *(Phoca hispida)* in relation to
       polar bear *(Ursus maritimus)* ecology and hunting
       behavior. *Can. J. Zool. 53:*1021.
Taylor, C.R. 1973. Energy cost of animal locomotion. Pages
       23-42 in Comparative Physiology. North-Holland Pub-
       lishing Co. London.
Theberge, J.B., and G.C. West. 1973. Significance of brood-
       ing to the energy demands of Alaskan rock ptarmigan
       chicks. *Arctic. 26:*138.
Tucker, V.A. 1975. The energetic cost of moving about. *Am.*
       *Sci. 63:*713.
Underwood, L.S. 1971. The bioenergetics of the Arctic fox
       (Alopex lagopus), Ph.D. Thesis. Pennsylvania State
       University, University Park, PA.
Walsberg, G.E., and J.R. King. 1977. Plumage color and
       solar heat gain: a dogma revisited. (Abstr.) 47th
       Annual Meeting of Cooper Ornithological Society.
Wang, L.C.H., et al. 1973. Adaptation to cold: energy meta-
       bolism in an atypical lagomorph, the arctic hare
       (Lepus arcticus).   *Can. J. Zool. 51:*841.
Weathers, W.W. 1977. Climatic and seasonal correlations in
       avian basal metabolism. (Abstr.) 47th Annual Meeting
       of Cooper Ornithological Society.
West, G.C. 1068. Bioenergetics of captive willow ptarmigan
       under natural conditions. *Ecology. 49:*1035.

West, G.C. 1972a. Seasonal differences in resting metabolic rate of Alaskan ptarmigan. *Comp. Biochem. Physiol. 42A:857.*

______. 1972b. Effect of acclimation and acclimatization on the resting metabolic rate of the common redpoll. *Comp. Biochem. Physiol. 43A:293.*

White, R.G., and M.K. Yousef. 1974. Energy cost of locomotion in reindeer in nutrition and metabolism of reindeer and caribou in Alaska. Progress Report 1973/1974. University of Alaska.

White, R.G., et al. 1974. Caribou-reindeer ecology in selected areas in the Prudhoe Bay region. In The Prudhoe Bay region: selected environmental reports. U.S. Tundra Biome Program.

Wunder, B.A., D.S. Dobkin, and R.D. Hettinger. 1977. Shifts of thermogenesis in the prairie vole *(Microtus ochrogaster). Oecologia. 29:*11.

II. NUTRIENT ACQUISITION AND UTILIZATION IN ARCTIC HERBIVORES[1]

Robert G. White

Institute of Arctic Biology
University of Alaska
Fairbanks, Alaska

Processes of adaptation to cold in arctic herbivores are generally interpreted in relation to body size. Small mammals and birds must rely on adequate means of increasing thermogenesis to counter the effects of frequent exposure to temperatures below their thermo-neutral range. In large mammals, coat thickness results in a decrease in the lower critical temperature so that environmental temperature is rarely below the lower critical temperature. When body temperature is lower than environmental temperature larger mammals shiver. Nutrients and water are involved intimately in the mechanisms and control of thermogenesis. Further, nutritents are involved in structural components and hence morphological and insulative adaptations to cold. For example, many macronutrients are involved in shivering and non-shivering thermogenesis through their involvement in energy stores, as messengers, in pumps, in specific chemiosmotic energy transfers, and as cofactors and modulators of enzyme reaction. Most nutrients involved in cold thermogenesis are recycled as AMP, ATP, ADP, and ITPase regenerated (the irreversible phosphate loss is presumably small). Classical laboratory studies indicate food intake is determined by the nutritive value of the food. As nutritive value increases, food intake is regulated so that intake of the most important dietary component, generally energy, remains constant. Small birds and mammals generally employ different strategies of nutrient acquisition and conservation than larger mammals.

[1]*This research has been supported by the Division of Polar Programs of the National Science Foundation (NSF Grant No. 7512945 - RATE and NSF Grant No. DPP 77-18384) and the National Institute of Health (Grant No. GM-10402).*

INTRODUCTION

The objective of this chapter is to discuss the role
of nutrients in comparative cold adaptation in northern
herbivores and to deduce where nutrient acquisition and
utilization processes may be specifically adapted for cold
resistance.

The components of the ingesta which may be involved in
such adaptations are:

1. Water,
2. Protein and N,
3. The macro-elements Ca, K, Mg, Na, P, S,
4. The trace elements, and
5. The vitamins and cofactors.

Discussions will be limited to the first three groups.
The omission of the last two groups is partly due to space
limitation but also reflects the general level of our
knowledge of involvement of nutrients in cold adaptation.

The argument is developed that nutrient acquisition
and utilization processes must provide a framework for the
physiology and biochemistry of cold adapation: at the very
least they play a permissive role. This argument is based
on the fact that nutrients and water are involved intimately in the mechanisms and the control of thermogenesis.
This involvement is outlined in detail elsewhere in this
volume, and is summarized briefly below.

Nutrients are also involved in structural components
and hence morphological and insulative adaptations to cold.
The involvement of some nutrients, mainly as cations, in
thermogenic mechanisms leads to the hypothesis that the
requirements of these nutrients may be linked to energy
expenditure; and thus, also to body size.

Finally, nutrient acquisition and utilization processes are discussed and evidence for the specific adaptation of the acquisition or utilization processes in response to cold is outlined.

INVOLVEMENT OF NUTRIENTS IN THERMOGENIC PROCESSES

Many of the macronutrients, mainly as cations, are
involved in shivering and non-shivering thermogenesis
through their involvement in energy stores (the adenosine,
guanosine, and inosine phosphates), as messengers (3'5'

cyclic adenosine monophosphate), in pumps, in specific
chemiosmotic energy transfers and as cofactors and modulators of enzyme reactions. Details of all of these modes
of action can be found in recent texts and reviews in
biochemistry, physiology and environmental physiology.
Electron probe X-ray microanalysis of cells shows that the
actual concentration of cations within and between cells
varies according to routes of movement and location of
pumps (Gupta and Hall 1979). Special alterations in the
relative, as well as absolute concentrations of cations may
alter the function of metabolic pathways as some cations
are antagonistic or exchange pairs with respect to their
effects on enzyme modulation (Behrisch 1979). Since the
amount of nutrients involved in these reactions are often
small, it has been assumed that the dietary intake readily
meets requirements. However, the effect of the level of
nutrition on thermogenic processes requires more study befor this assumption can be substantiated. In particular,
phosphorus might be a prime candidate for further study
since it is involved in many energy transactions in the
cell as well as being a cation cofactor in enzyme reactions.

Phosphate, calcium, magnesium and sulfur are involved
in the splitting of ATP using the actin-myosin ATPase
system in muscles and are therefore important in both
locomotion and shivering. Myosin-ATPase is stimulated by
Ca^{2+} and the myosin-ATPase is kept in an inhibited state by
Mg-ATP which interacts with the sulfhydryl groups of the
enzyme.

The involvement of minerals in non-shivering thermogenesis is variable as non-shivering thermogenesis can
involve:

1. Thyroxine activation of the Na^+/K^+ ATPase system;
2. Loose coupling or uncoupling of oxidative phosphorylation; and
3. Specific dynamic effect of food protein and fats.

The synthesis of thryoxine requires the micro-nu-
trient, iodine, and iodine deficiency could therefore limit
cold adaptation. However, evidence of iodine deficiencies
in northern animals is lacking. When stimulated by the
thyroid hormones (T3, T4), high energy bonds are split by
the Na^+/K^+ activated ATPase system in cell membranes of
liver, skeletal muscle and kidney. Although concentrations
of the cations Na^+ and K^+ are low, it is conceivable that a
large dietary imbalance induced by a high potassium intake
could interfere with the intracellular $Na^+:K^+$ ratio and
affect this mechanism.

The loose coupling of oxidative phosphorylation in mitochondria of thermogenic tissues such as brown adipose tissue of newborn arctic species (Blix and Steen 1979) and cold-acclimatized small mammals (Himms-Hagen 1976, 1978) and skeletal muscle of newborn harp seals (*Phoca groenlandica*) (Grav and Blix 1979) is a well regulated adaptive mechanism for thermogenesis. Phosphorus has several roles including one in the secondary messenger cAMP which transfers rate control from epinephrine induced adenyl cyclase to substrate mobilization through lipase activity in the cytoplasm of the thermogenic cell. The mitochondrial oxidative phosphorylation system is composed of iron-sulfur (Fe.S) containing cytochromes and the vitamin ubiquinone or cytochrome Q. In the process of loose coupling or uncoupling of oxidative phosphorylation, calcium and phosphate cations are accumulated by the mitochondrion against a concentration gradient. Whether this process is affected by the calcium and phosphorus reserves of animals is not documented. Manganese and ferrous ions, but not magnesium ions, may also accumulate during loose coupled metabolism (Lehninger 1970). In tightly coupled oxidative phsophorylation the availability of ions for translocation may influence the rate of mitochondrial respiration (Lehninger 1974). Potassium and sodium transport also occurs as a product of the mitochondrion K^+ pump which balances the transport of anions (fatty acids) from the cytoplasm to the mitochondrion. A deficiency in potassium can cause inhibition of cell and mitochondrial respiration (Blond and Whittam 1964; Ozawa et al. 1967); this possibility seems remote in herbivores as their diet is generally high in potassium.

Heat increment of feeding which may continue for several hours involves two processes; the work of eating (Young 1966) and the specific dynamic effect (SDE) (Rubner 1902). The SDE is thought to involve the use of ATP for the storage and utilization of ingested food; this amounts to approximately 2-4 moles ATP per mole of glucose, 2 moles ATP per mole of fatty acid hydrolyzed and reincorporated into triglyceride and approximately 4 mole of ATP per mole of amino acid incorporated into protein (or 4 moles of ATP for conversion of NH_4^+ to urea). Also, heat may be produced in the transport of ions across cell membranes. The heat increment of feeding contributes to total metabolic heat production. Thus, when the protein intake is high this contribution to the balancing of heat loss with heat production can be significant. When small meals are fed, a significant amount of the increase in heat production is stored rather than being lost (Pittet, Gygax, and Jéquier 1974) which suggests a role for SDE in substitution for thermogenic heat production. In ruminants about half of the heat increment of feeding occurs in tissues outside the digestive tract; and the heat of rumen fermentation is 22–76 kJ/MJ digestible energy intake (Webster, et al. 1975). It is argued but not proven, that in the ruminant this net exothermy of fermentation could make an important substitution for thermogenic heat production and therefore to thermoregulation (Griffin et al. 1951; Hammel, et al. 1963; White 1975).

INVOLVEMENT OF NUTRIENTS IN INSULATION AND PROTECTION OF EXTREMITIES

The formation of skin, hair, fur, wool, and feathers require the availability of the S-containing amino acids cysteine and methionine which provide the disulfide bridges necessary for protein structure. When the supply of S-amino acids is restricted there is a weakness in the structure of the wool fibers which may cause the whole fleece to be shed in domestic sheep. In wild animals moulting is precisely timed and regrowth of fur, hair, wool, or feathers takes place when temperatures are mild and food is most abundant. Whether the food supply (quantity and/or quality) could affect insulation has not been documented. However, a low plane of nutrition causes a less complete moulting in the Wiltshire sheep (Slee 1965). In marine mammals the subcutaneous fatty tissue (the blubber) serves both as thermal insulation and as an energy reserve and moulting commences in spring to early summer (Ling 1965).

Tissues which are subject to extreme cold (e.g., foot pads) contain large quantities of phospholipids, and there are specific enzyme modifications appropriate to heterothermous tissues. It is unlikely that the nutrient acquisition process could affect the formation of these tissues; however, the biochemistry may be affected through supply of cations to tissues.

BODY SIZE CONSIDERATIONS

Species variations in mechanisms of cold adaptation can be related to body size. Cold adaptations in small and large mammals and birds have been reviewed by Whittow

(1971), Himms-Hagen (1976), West and Norton (1975) West
(1976) and Miller (1978), as well as in a volume on natural
torpidity and thermogenesis edited by Wang and Hudson
(1978). The special significance of the mechanisms of cold
adaptation in the newborn has been reviewed by Blix and
Steen (1979). In summary, small mammals and birds must
rely on adequate means of increasing thermogenesis to
counter the effects of frequent exposure to temperatures
below their thermoneutral ranges. Exposure to cold is
minimized by niche exploitation and, in some species, by
such behavioral modifications such as huddling. Nonshivering thermogenesis is usually well developed in small mammals and takes place in both brown adipose tissue and
skeletal muscle. A constant high food intake is required
to meet these requirements and food intake increases as
temperature declines below the thermoneutral zone (see
reviews by Hart 1971; Yousef; this volume). Food quality
could conceivably affect body reserves as muscle glycogen
reserves are high in arctic-alpine rodents (Galster and
Morrison 1975) and these may play an important role in cold
stress. By contrast in large animals the increase in coat
thickness results in a decrease in the lower critical
temperature; such that environmental temperature is rarely
below the lower critical temperature. When it is, these
larger mammals then shiver to maintain body temperature
(Adams 1971; Whittow 1971). There may be a reduced metabolism of large herbivores in winter which involves thyroxine
function (Ringberg et al. 1978) and this lowers the energy
requirements. In an extreme instance, the Svalbard reindeer (*Rangifer tarandus platyrhyncus*) which has no natural
predators, becomes lethargic, hypothyroidic in winter and
has high levels of growth hormone which allows good control
over lipid mobilization (Ringberg 1978).

Clearly increased need for nutrients which may be
utilized in thermogenic processes must be matched by their
increased ingestion (also see Gessaman; this volume).
Water metabolism could be linked to energy metabolism in
cold acclimation (Deavers et al. 1978; Yousef and Johnson
1978), but the role of water metabolism in acclimatization
to cold is equivocal since Holleman et al. (1978) have
shown that water turnover in winter is considerably lower
than summer in red-backed voles (*Clethrionomys rutilis*).
Reindeer (*R. t. tarandus*) also exhibit lowered water flux
in winter (Cameron and Luick 1972) and the lowering is
associated with lowered metabolic rate and food and protein
intake (McEwan and Whitehead 1970; Cameron 1972). Thus, in
spite of a large difference in body size, both the red-
backed vole and the reindeer share a common lowering of

water metabolism in winter. The results could be explained
if the vole reduced its metabolism by hypothermia and
torpor: but these phenomena have not been reported for
this species. Alternately, water flux and energy metabolism may not be coupled under all conditions. Cameron
(1972) has shown that winter water flux correlates with
protein rather than energy intake in reindeer during the
winter. Also water flux increases in response to requirements for evaporative cooling or excretion of a salt or
toxin load and is not always linked to energy metabolism.

Protein turnover correlates to basal metabolic rate
(BMR) and accounts for 10 to 15 percent of it (Waterlow
1968). In young red deer (*Cervus elaphus*) in poor body
condition (~4% body fat), 30 percent of the heat produced
in response to cold stress, i.e., cold thermogenesis, is
attributable to protein oxidation (Simpson et al. 1978).
Thus, for large arctic herbivores in which a lowered BMR is
known or inferred--reindeer/caribou (*Rangifer taranous*),
moose (*Alces alces*), muskoxen (*Ovibos moschatus*), a lowered
protein requirement might be expected. In reindeer and
caribou this may be so since protein intake is extremely
low due to the low protein content of their preferred food
(lichens). Seasonal patterns of protein turnover have not
been measured. In contrast, the small mammals and birds
have a high protein intake throughout the year as would be
expected.

Except for iodine, which is excreted in urine as
thyroid hormones, most nutrients involved in cold thermogenesis are re-cycled. AMP, ATP, ADP, and ITP are regenerated and the irreversible phosphate loss is presumably
small. Few studies have involved the comparison of the
stoichiometry of mineral and energy metabolism, hence it
cannot be stated unequivocally that the requirement for
macro-elements does not rise with increasing metabolism.

An important corrolary to this strategy of survival in
the large animals, the build up of energy reserves for use
in winter, is that nutrients involved in thermogenesis must
also be stored: food acquisition in winter is not depen-
dable (see control of food intake). Evidence for storage
is shown briefly in Table 1.

TABLE I. Summary of General Macronutrient Disposition, Aquisition and Utilization in Animals

Nutrient	Location in Plant	Location in Animals	Importance in Animal Life Processes	Site of Absorption
Nitrogen	Mainly cellular proteins and structural protein in cell organeles	Proteins; free amino acids; nucleotides; NH_4+	Amino acids and proteins; structural proteins; enzymes	Amino acids absorbed in small intestine; NH_4+ uptake by rumen and omasum
Calcium	Cell walls, mainly calcium pectate; intracellular level is low	99% in bone, 1% in soft tissues	Structural component of bone and teeth; innocular irritability; blood clotting; enzyme function	Absorption in distal small intestine; secretion into upper alimentary tract
Phosphorus	Soluble fractions in cell contents; lipids; nucleic acids	Bone; organic phosphorus in all cells	Organic phosphorus compounds important in hard tissues (e.g. bone)	Small intestine and cecum; secretion into rumen and proximas small intestine
Magnesium	Chlorophyll, chelator of various organic acids; porphyrin group	a) Intracellular of soft tissues; b) bone (0.5-07% ash)	Activation of enzymes; intracellular catolysis; central and peripheral nerve function	Rumen; cecum and large intestine
Sodium	Cation in cell contents; mostly in roots	Mainly extracellular plus bone	Primary determinant of osmolarity; carrier mediated transport processes; enzyme function; nerve function	Most of the alimentary tract; undergoes rapid flux in both direction; net absorption in large intestine
Potassium	Cations in cell contents	Mainly intracellular-56% in muscle; bone	Important cytoplasmic organic anion balance; enzyme function; nerve function	Rapid flux with net passive absorption from ileum and large intestine
Sulfur	Proteins and coenzymes, vitamins (i.e. mainly organically bound); intracellular salts	Mainly in amino acids; inorganic and ester sulfates	Disulfide bridges in enzymes and structural proteins (skin, wool, hair, feathers); ossification processes	Amino acids absorbed from small intestine

Table 1. (cont.)

Mechanism of Absorption	Storage Site	Retention and Elimiration	Adaptations
Active transport for amino acids	Proteins	Conservation of ammonia at the kidney and recycled to the cecum and rumen when N levels are low.	Kidney function; aprophagy
Secreted and reabsorbed. Vitamin D has important effect	Bone	Retention at kidney is regulated; vitamin D important	Absorption/digestibility of calcium is linked to requirements
Secreted and reabsorbed; active transport requiring the presence of Ca and K	Bone; organic P(?)	Excretion through kidney is an important control mechanism; parathyroid hormone is important	High food intake may optimize phosphorus intake in the brown lemming
Absorption rate is low; could interact with K, N and volatile fatty acid absorption	Bone	Endogenous loss is low, excretion is mainly through the kidney	None documented
Active transport	Bone	Retention in the large intestine and at the kidney is very well developed	Well documented physiciological and morphological adaptations in the kidney and alimentary tract
Passive diffusion in relation to electrochemical gradient and water movement	Muscle	Retained in muscle and red cells; eliminate at the kidney during stress	Not documented
Carrier mediated uptake from small intestine (active transport)	S-containing amino acids	Recycling of inorganic S to rumen and cecum and re-incorporated into bacterial and protozoal amino acids	No specific documentation

COMPARISON OF HERBIVORES WITH CARNIVORES AND OMNIVORES

A dietary deficiency in minerals may be more rare in carnivores than in herbivores as their prey species are generally high in nutrients. On the other hand, carnivores eat intermittently and bouts of starvation normally occur in winter. The effect of starvation on cold adaptation in carnivores is largely unstudied. Periodic feeding is associated with a frequent turnover body reserves as these are stored and reused and carries with it a metabolic cost. Also during feeding bouts following a fast, the heat increment of feeding would be high and could conceivably substitute for cold thermogenesis in the winter (see above). In ominvorous mammals such as the bear and red fox (*Vulpes vulpes*), short term bouts of starvation may be minimized by resorting to herbivory.

In this chapter, more attention has been given to the control of food and nutrient intake in the mammalian herbivores since their diet contains nutrients in very low concentrations.

ACQUISITION AND STORAGE OF NUTRIENTS

Water

Water is ingested as free water in food and drink and is a product of substrate metabolism. During the oxidation of 1 g of carbohydrate, fat, or protein, the amount of water produced is approximately 0.556, 1.065, and 0.420 g respectively. The exact values depend on the degree of combustion and the hydrogen content of the substrate (van Es 1969). Metabolic water is of greatest importance to fasting animals and to those which consume a dry diet such as seeds and take in little free water.

The water content of the lean body mass in most mammalian species is 71 to 73 percent (Pace and Rathbun 1945) but may be over 80 percent in the lean body mass of the fetus and young animal. The relative water content of the body decreases as the fat content increases; body water content can be used as an index of body composition (Panaretto 1963; Searle 1970). The water content of lean body tissue and the fluid spaces is strongly controlled by antidiuretic hormone.

TABLE II. Water and Macro-nutrient Content of Summer Vegetation and Lichens
In Relation to that in Herbivorous Mammals (Brown Lemming and Caribou).

	Monocots[1]	Dicots[1]	Lichens[2]	Mammals[3]
Gross Energy[3]	4.5 – 4.9	4.2 – 5.0	4.4 – 4.7	5.2 – 6.9
Water[4]	1.0 – 5.8	0.6 – 2.3	0.3 – 5.8	2 – 5
N	2.5 – 3.5	2.5 – 3.8	0.4 – 0.8	7.2 – 10.3
P	0.15 – 0.25	0.2 – 0.45	0.01 – 0.03	1.5 – 2.7
Ca	0.05 – 0.20	0.2 – 0.5	0.04 – 0.11	3.0 – 3.1
Na	0.002 – .01	0.002 – 0.004	0.002 – 0.01	5
K	1.0 – 1.5	1.0 – 1.8	0.003 – 0.015	0.6 – 0.8
S	–	–	0.001 – 0.01	–

Units are in kcal/g for energy and g/100 g dry matter and refer to the green leaf material of vascular plants and the live thallus of lichens.
[1]Chapin (this volume); [2]Luick (1977); [3]Batzli et al. (1978); [4]R.G. White and J. Trudell (unpublished observations).

Body water is mobile and in a dynamic state of flux.
When tritiated water is injected intravenously into reindeer it can be detected in the alimentary tract and urine
within minutes (White et al. 1979). Flux of water across
the alimentary tract is of a high order (von Englhardt
1970; White et al. 1979) even though the turnover of the
whole body water may take days. The large water pool of
the rumen-reticulum and the water flux between alimentary
and extracellular pools serves to equilibrate temperature
differences due to ingestion of cold water, snow, ice, and
food. High rates of flux may also serve to redistribute
small molecules which move by solvent drag.

The possibility that water metabolism is modified as
part of the process of cold adaptation has not been demonstrated. The low water flux associated with field studies
of both reindeer and red-backed voles (see body size considerations), however, suggests at least a permissive action
in the energy savings due to the ingestion of free water.
However, in both species the energy saving due to reduced
water intake amounts to less than 10 percent of the saving
in energy due to a lowering of the metabolic rate (Holleman
et al. 1978).

Water may be stored in the rumen (Hecker et al. 1964)
and the body fluid pools of animals. This appears to be a
short-term form of storage only and most species must
aestivate or hibernate to avoid an extreme and chronic
water shortage.

Protein, N and the Macronutrients

Body tissues of animals maintain large concentrations
(mg to g percent; Table 2) of macronutrients (protein and
macro-minerals) whereas, the trace-elements and vitamins,
the micro-nutrients, are in very small amounts and the
concentrations in body tissues are usually only a few
p.p.m.

It has been suggested that arctic herbivores have the
ability to select for specific nutrients (specific euphagia)
either by selecting for plant species or parts high in
nutrient concentrations (Klein 1970) or by selecting available supplements such as mineral licks. However, unequivocal evidence for specific euphagia in arctic herbivores
is lacking. One problem in demonstrating specific euphagia
in field situations is that the concentrations of some
macronutrients (e.g., N and P) tend to be highly correlated,
and selection for one necessarily means high intake of the
other. Also some overt signs of apparent mineral deficien-

cy, such as seeking mineral licks may be difficult to study since the response may be related to an imbalance of nutrient intake rather than a deficiency of any single nutrient in forage. On ranges marginal in sodium (less than 10 mEq/kg), the high intake of potassium in new growth vegetation in spring can bring about excessive sodium loss which leads to visiting mineral licks high in sodium. This behavior has been noted with white-tailed deer (*Odocoileus virginianus*) (Weeks and Kirkpatrick 1976), fox squirrels (*Sciurus niger*) and woodchucks (*Marmota monax*) (Weeks and Kirkpatrick 1978). A similar phenomenon could be implicated in visits to licks by other ungulates including caribou, mountain goat (*Oreamnos montanus*) and Dall sheep (*Ovis dalli*) (Cowan and Brink 1949; Hebert and Cowan 1971; Calef and Lortie 1975). In reindeer a low Na:K ratio in saliva (Staaland et al. 1979) or high aldosterone levels (Ringberg et al. 1978) during summer are indicative of sodium deficiency. These symptoms could relate to a very high potassium intake coupled with a marginally low sodium intake (Staaland et al. 1979). Similar problems in sodium metabolism have been noted in wild rabbits (*Oryctolagus cuniculus*) (Myers 1967; Blair-West et al. 1968), horses (*Equus spp.*) (Clarke et al. 1978), and snowshoe hares (*Lepus americanus*) (Smith et al. 1978). Excessive potassium intake in spring herbage may also result in hypomagnesemia (Sjollema 1932, in Suttle and Field 1967).

Table 2 lists the macronutrient contents of summer tundra vegetation and indicates the range in macro-nutrients available to arctic herbivores. The green leaf portion of some monocots (grasses and sedges) are compared with leaves of willows and with the live thallus of lichens (a winter food of caribou and reindeer). In general, the range in nitrogen in dicots overlap that in monocots; however, the dicots are frequently higher in nitrogen than monocots. The same trend is shown for phosphorus, but the calcium content of dicots is considerably higher than that of monocots. It is noteworthy that the sodium level of monocots, dicots and lichens is low (1 to 1.5 g%) while that in lichens is very low. Lichens are characterized as having a very low mineral content in general. No sulfur concentrations are available for arctic plants.

Thus, although the actual concentration of macro-nutrients in the summer diet depends on dietary composition, provided live portions of vascular plants are eaten, the intake of nutrients must be reasonably high irrespective of the botanical composition of the ingesta. The nutrient content of the winter diet is subject to more

variation as less green leaf is available and, in the case
of caribou and reindeer, is largely replaced by dead and
leached material of low mineral content; or it is replaced
by lichens.

Since both small and large herbivores tend to harvest
whole leaves, it is also instructive to consider the
amount of nutrient contained in an individual leaf.
Figure 1 shows the nitrogen content of a whole leaf compared with the concentration of leaf-nitrogen. In comparison with nitrogen concentration, the peak amount of nitrogen present in a single leaf is delayed by one to several
weeks. Therefore, although nitrogen concentration of the
leaf declines in late July, the amount of nitrogen still
available to an animal who harvests whole leaves remains
high and partially compensates by an increase in food
intake. This effect could prolong the availability of
nutrients in the fall. Any mechanism which prolongs the
availability of nutrients may be important to large herbivores which must replace a depleted nutrient pool following
lactation and rutting activity, and for smaller herbivores,
it could mean the successful rearing of a late summer
litter. Productivity in northern environments is limited
by the duration of the summer period. The period during
which plants of high nutrient concentration are available
in summer depends on climatic variables and thus upon
altitude and latitude.

Finally, nutrient content of a preferred habitat
is also related to plant phenology. Large herbivores
may maximize nutrient intake by following phenological
progressions associated with latitudinal and elevational
gradients, which are often associated with a receding snow-
line and by changing vegetation types as summer progresses
(Klein 1970; Skogland 1975). Many of these phenological
progressions are predictable and result in generalized movement patterns based on indicator plants.

Small herbivores have the ability to select the most
nutrient rich parts of plants such as leaf buds, seeds, and
rhizomes. The same parts can also be taken by large
herbivores but would be mixed with surrounding material of
lower nutrient quality as selection is somewhat dependent
on the size of mouth parts.

Although it may be supposed that herbivores select
plant species and parts rich in nutrients, it can also be
argued that the selection process in herbivores is not for
nutrients but against toxic substances produced by plants
and which function as a chemical defense against the gra-
zer or browser. This hypothesis is currently being tested

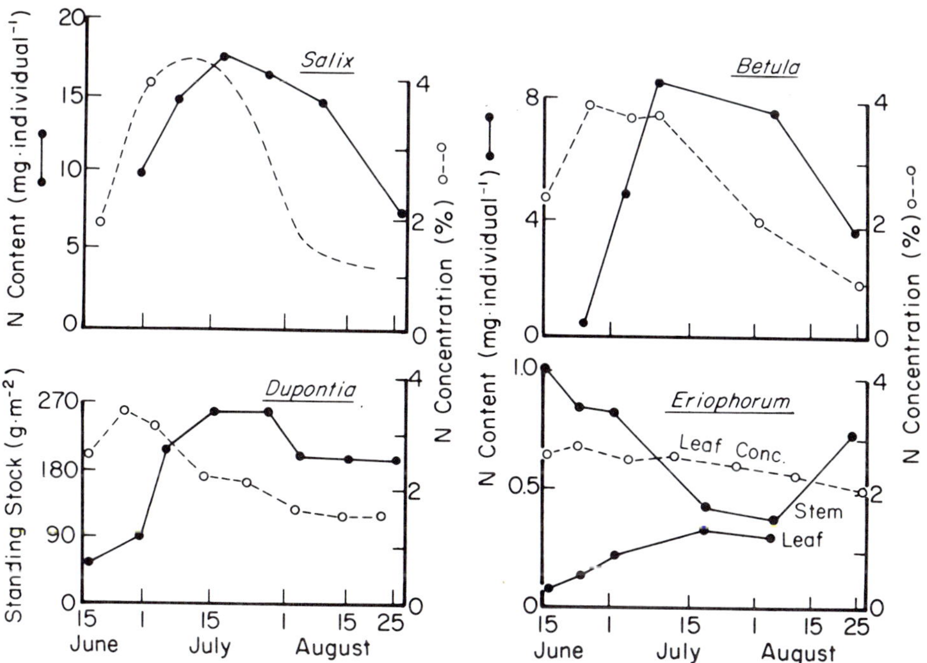

FIGURE 1. *Comparison of the seasonal trend in the N content of a whole leaf (o-o, mg.individual⁻¹) with the N concentration: o---o, g.(100 g)⁻¹ in the leaf. For both dicots (S. pulchra, B. nana) and monocots (D. fischeri, E. vaginatum) the N content of individual leaves trained 1 to 3 weeks behind N concentration. The seasonal trend in N content of individual leaves also represents the trend in standing crop of N. Data are from Chapin (this volume).*

for northern herbivores. If vindicated it would explain paradoxical instances in which unselected plants are of high nutrient content: the plants may be unpalatable because they contain substances which are unpleasant to taste, toxic, or contain digestive inhibitors. Many of these substances are extracted with the lipid fraction of plants and may make up a major proportion of them. If

secondary compounds are absorbed, they must be metabolized
or excreted and the latter way result in a lowering of the
metabolizability of digested energy. This could explain
the observation that urine energy loss is highly correlated
with fat intake (r=0.89) in white-tailed deer fed browse
(Mautz et al. 1975). Thus, northern herbivores are often
required to limit the ingestion of a highly nutritious
plant because of the presence of a harmful substance,
especially during winter. Small mammals can selectively
feed on parts of plants such as frozen buds, green vascular
structures, stem bases, and rhizomes; and, would seemingly
have a better ability to select against toxic, deterrant,
or inhibitory substances.

CONTROL OF NUTRIENT INTAKE

Clearly nutrient intake is not determined simply by
availability. A conceptual diagram of factors influencing
nutrient acquisition and utilization by herbivores is shown
in Figure 2. In the figure, mass flow is shown as intact
arrows while feedback relationships are shown as broken
lines. Nutrient intake of the animal is a product of the
nutrient concentration in forage and its bulk.
On the lower half of Figure 2, some of the factors
that influence the nutrient intake of animals are given.
By selecting the right habitat, and by selecting specific
food types from the habitat, the animal can control the
nutrient content of the diet throughout the year. Specific
nutrient requirements could be met by selecting species or
parts of plants that are high in the nutrient or by sup-
plementation, possibly from mineral licks.
The second factor in sustaining the level of nutrient
acquisition is the amount of food ingested. Food intake is
affected by both extrinsic (environmental) and intrinsic
(physiological) factors. The extrinsic factors, include
food accessibility, (it may be covered by snow or inundated
by water) and harassment of the animals by insects, preventing feeding at any particular time, and may result in
spacial displacement. The structure of the food resource
itself may also limit accessibility. The intrinsic factors
include the physiological condition of the animal and its
adaptation to cold, which may affect appetite and therefore
food intake.

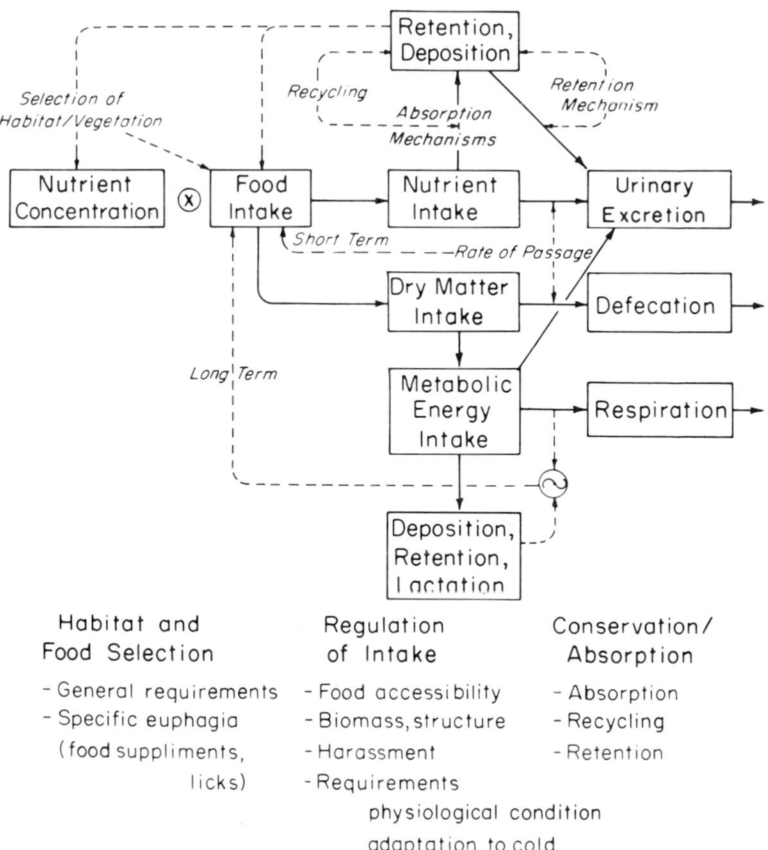

FIGURE 2. Conceptual model of factors affecting the food, energy, and nutrient intake of herbivores. Absolute ingestion (acquisition) is controlled by the vegetation, amount of food intake, and supplements while further control is then exercised at absorption, storage, and retention processes. Thus, behavioral responses involved in selection from within habitats must be considered in relation to physiological adaptation of the absorption, metabilism, and retention mechanisms.

Maximization of nutrient retention involves not only nutrient selection and level of intake but also absorption and conservation of nutrients. Since the nutrient content of animal tissues is between two and ten times that of a plant material, most macro-nutrients are absorbed against a concentration gradient (Table 2). Active transport mechanisms, and factors (inhibitors and cofactors) which regulate

active absorption of the macro-nutrients and micro-nutrients,
are important and may be under strong selective pressure
where the nutrient content of forage is low. Nutrients are
retained by mechanisms which optimize retention at the
kidney, however, alimentary losses are frequently high due
to secretion of some minerals into the alimentary tract and
also through additions as cell sloughings. For ruminants,
conservation of nutrients may involve recycling mineral
nutrients to the rumen to maintain a well developed microflora which aids in digestion, as well as adptations in
kidney function.

CONTROL OF NUTRIENT INTAKE BY FACTORS INFLUENCING FOOD
INTAKE

Factors influencing food intake are frequently related
to the amount of dry matter consumed. A large proportion
of ingested dry matter is absorbed and metabolized and
energy released as heat, unoxidized material is then retained. The amount of energy retained may constitute a
negative feedback effect on food intake. If little energy
is retained, it is generally assumed that food intake must
increase wherever the work load of the animal increases.
On the other hand, the indigestibility of the diet may
restrict food intake. Impaction of the rumen is an extreme
example of physical factors limiting food intake in ruminants consuming a low quality-high roughage diet of low
digestibility.

CONTROL OF FOOD INTAKE BY FOOD FORM (STRUCTURE) AND
AVAILABILITY

Independent of the animal's needs for energy, water,
and nutrients, the amount of food consumed may be controlled by the availability (biomass) and the form or
structure of the preferred plant material. Figure 4 shows
that instantaneous food intake in reindeer is related to
above-ground biomass. Similar relations have been shown
for other ruminants (sheep and cattle) (Arnold 1964; Arnold
and Dudzinski 1967; Allden and Whittaker 1970) and for
small herbivors such as the brown lemming (*Lemmus sibiricus*)
and Arctic ground squirrel (*Spermophilus parryii*) (Batzli,
unpub. observ.).

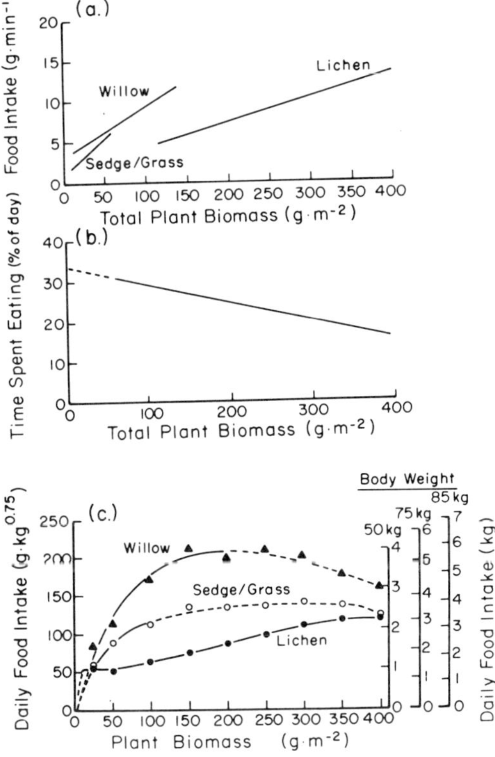

FIGURE 4. *Relationship between instantaneous food intake (a) and plant biomass in reindeer. Daily food intake (c) depends not only on biomass and plant growth form but also on time spent grazing (b). (R.G. White and J. Trudell, unpub. observ.) Plant growth form has an important control over intake at medium to high plant biomass.*

To compensate for a low instantaneous intake when food availability is low, reindeer (White 1979; White and Trudell, unpub. observ) and sheep (Young and Corbett 1972) increase time spent eating per day in order to maximize intake (Figure 4). The relation between daily food intake and biomass approaches a maximum as biomass approaches 100 to 150 g.m⁻² for vascular plants and 400 g.m⁻² for lichens (Figure 4). A similar curve has been shown for sheep

grazing mediterranean grassland systems (Allden and Whittaker 1970; Arnold and Dudzinski 1972). Unfortunately such
relations for daily food intake have not been determined
for the other important arctic herbivores such as lemmings,
ground squirrels, ptarmigan (*Lagopus spp.*), and musk-oxen.
     The influence of physiological conditon on behavior
has been shown in numerous studies: time spent grazing is
greater for lactating caribou than for non-lactating cohorts
(B.R. Thomson in White et al. 1975). From the relationships in Figure 4, it can be predicted that food intake
would increase during lactation, i.e., in response to
increased physiological demands. The mechanism of such a
response has not been determined; clearly an increase
either in rumen size and/or in turnover would have to
occur. If the digestibility and metabolizability of the
ingested food can be estimated and the energy requirements
for maintenance and for lactation is known, then the minimum plant biomass required to maintain bodyweight in reindeer and caribou can be predicted. A minimal biomass of
25-30 g.m^{-2} (dry matter basis) of this preferred forage of
high quality (digestible energy and nutrient content) could
support non-productive reindeer, and caribou (Batzli et al.
1979). These values for forage biomass are lower than that
predicted for mediterranean type of grassland system car-
rying domestic sheep.
     When preferred plant biomass is in excess of 30 g.m.$^{-2}$,
presumably food intake can be maximized and the population
density can increase. Increased plant biomass may allow
for a lower search time and could lessen animal-animal
interactions particularly between those competing for
preferred plant species and parts. That plant biomass in
summer regulates the size of wild populations of fallow
deer (*Dama dama*) in Europe, has been argued by Bobek (1977).
Similarly the average biomass of wild reindeer and caribou
populations on circumpolar tundra ranges is correlated with
above-ground biomass of vascular plants (T. Skogland,
unpub. observ.). Independent of this density effect, the
size of large arctic ungulate populations is dependent on
the availability of food in winter (Klein 1970).
     The possibility that food quality and availability is
important in controlling population levels of the brown
lemming and snowshoe hare has been emphasized in recent
ecological studies (Shultz 1964, 1969; Batzli et al. 1979;
Pease et al. 1979). In both of these species excessive
removal of preferred plant species and parts occurs during
winter grazing at peak or near-peak population numbers.
Such intensive overbrowsing by hares can lead to a decline

in winter browse for the following four years and may also
induce responses in plant secondary compounds (DeVos 1964;
Wolff 1978; Bryant, Fox and Chapin, unpub. observ.). The
immediate effect of over-grazing and over-browsing results
in a decline in the population which is exacerbated by an
extremely high predator population. The lemming model also
involves a locking-up of nutrients in lemming feces and in
dead plant material cut during the winter to line runways
and to construct nests. A number of hypotheses are being
investigated in detail as no population regulation mechanisms (e.g., intrinsic, nutritional, predation-dispersion
models) adequately explain all observed features of these
cycles.

CONTROL OF NUTRIENTS THROUGH ABSORPTION

The digestibility or availability of nutrients in food
is controlled partly by their location in plants (Table 2)
but mostly by absorption mechanisms. The digestibility of
nitrogen, calcium, phosphorus, magnesium, sodium, potassium
and sulfur in domestic mammals has been studied and Table 3
shows some values for domestic sheep. Similar values for
the availability and factors which affect digestibility of
these nutrients are unstudied in arctic mammals.

The approximate digestibility of nitrogen in forage
varies between 40 and 60 percent depending on the total
digestible protein content of the diet. Digestibility
becomes negative at total protein content less than eight
percent; in other words, there is a net loss of nitrogen
from the body into the alimentary tract. This occurs in
reindeer and caribou in winter when lichens dominate the
diet (Cameron 1972; Jacobsen and Skjenneberg 1975). Following absorption, the utilization of protein depends on
its biological value, which relates to its amino acid
composition. Milk of most species has a high biological
value; 80 percent of the absorbed protein being retained.
The biological values of proteins of other high quality
foods is in the order of 70 percent whereas for very low
quality food, it may be as low as 60 percent. Negative
synergism is indicated; in situations where protein is of
low digestibility the biological value of the digested
fractions is also low. Adaptation to poorly digestible/low
biological value protein would depend on mechanisms for
correcting these deficiencies. The return of nitrogen, as
urea, to the rumen and cecum of ruminants and cecalids
receiving low protein-high carbohydrate diets, results in

the synthesis microbial protein which has a higher biological value than the negative plant protein in low quality forage. For arctic cecalids (snowshoe hare, arctic hare-- *Lepus articus*), coprophagy is necessary to maximize this effect and may effect a more marked nitrogen economy than that of ruminants.

Table 3. Nutrient digestibility of forage.

Nutrient	Approximate Digestibility	Relationships
Nitrogen (as total crude protein)	40-60% negative for low protein diets	Biological values: milk - 80% forage: high quality 70% low quality <60%
Calcium	40-60%	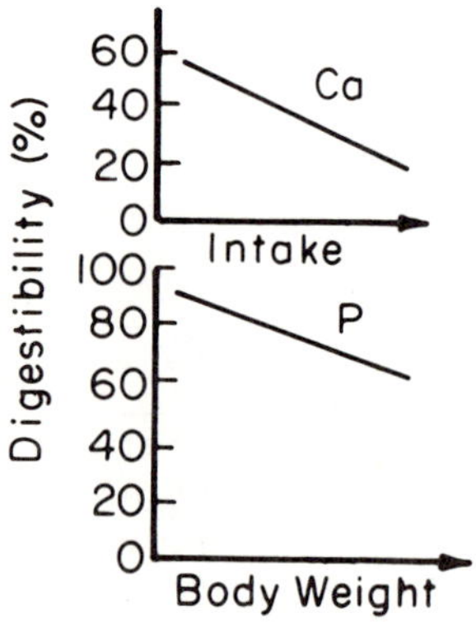
Phosphorus	60-80%	
Magnesium	variable milk - 70% forage - 20-40% (depressed when K intake is very high)	
Sodium	>90%	
Potassium	low	
Sulfur	negative to 80%	

In the urea recycling process the importance of rais-
ing the biological value of protein as opposed to increasing
protein synthesis *per se*, warrants further study in arctic
herbivores. In studies on urea cycling it is necessary to
determine that the recycled urea-N is incorporated into
microbial protein. Where dietary energy is limiting a
recycling of N may be futile (e.g., rumen--or cecal--urea
hydrolysis $\rightarrow$ absorb NH_4^+ $\rightarrow$ hepatic urea synthesis $\rightarrow$ alimentary urea hydrolysis $\rightarrow$ absorb NH^+ $\rightarrow$ etc.). This cycle uses
ATP for the synthesis of urea. On the other hand, urea
synthesis is also an exothermic reaction and heat produced
in the liver and kidney could contribute to non-shivering
thermogenesis.

The digestibility of calcium in the diet which is of
the order of 30-60 percent, is similar to that of nitrogen.
Part of the reason for the low digestibility is that much
of the plant calcium is locked up in calcium pectate and
other compounds which are poorly digested. However, in
domestic ruminants and non-ruminants the digestibility of
calcium declines markedly from 60 to 20 percent as the
intake of calcium increases. Hence, calcium is not absorbed greatly over and above its requirements. Evidence
for a specific calcium euphagia in the pectoral sandpiper
(*Calidois mylanetus*) during the time of peak egg production
has been described by MacLean (1974); sandpipers meet
calcium and phosphorus requirements by eating considerable
amounts of lemming bones. Unfortunately no estimates were
made on the digestibility of calcium in the sandpiper diet.
Calcium is absorbed by an active process and this process
is enhanced when calcium is limiting.

Phosphorus is generally available in a herbivore diet
as it is a constituent of the cytoplasm of cells (Table 3).
Little, if any, is "locked-up" in cell wall material.
Digestibility of phosphorus may be between 60 and 80 percent
in ruminants and perhaps 70 to 90 percent in non-ruminants
and it declines rapidly with increasing age or body weight
of the animal. This suggests that phsophorus digestibility
relates to phosphorus requirements of the animal. Those
animals which have a high phosphorus requirement, as during
lactation, extract more phosphorus from their diet (Luik
and Lofgreen 1957; ARC 1965). The digestibility of phosphorus in milk is again very high at 90 to 97 percent.

Antler growth of cervids may be affected by calcium
and phosphorus level in the diet; highest rates of antler
growth are noted when dietary calcium and phosphorus exceeds 0.59 and 0.54 percent respectively (Magruder et al.

1957). However, the minimum dietary level of phosphorus
which will maintain development of deer fawns is considerably lower at 0.26 percent (Ullrey et al. 1975).

Magnesium, like calcium, is an important nutrient, in
that it is involved in enzyme activation and nervous tissue
function. Magnesium has a very low digestibility (20 to 40
percent) in herbivore diets but is high (70 percent) in
milk. There is some evidence for a decreased digestibility
when potassium and/or non-protein nitrogen levels are high.
The main site of magnesium storage is in bone and no evidence for an adaptation in magnesium acquisition or utilization in arctic herbivores has been documented.

NUTRIENT REQUIREMENTS OF ARCTIC HERBIVORES

Little is known of the exact relations between nutrient requirements and intake in northern herbivores.
Figure 3 shows a highly idealized representation of nutrients in relation to three different breeding cycles in
arctic herbivores: the brown lemming, the caribou, and the
Arctic ground squirrel. The brown lemming is the only one
of these three species that has breeding cycles in winter.
Hence, the cyclic pattern may be replicated up to three
times during the winter and up to four times in the summer.
The caribou inhabits the Arctic year around, and produces
only one offspring per year, so this cycle is annual. The
Arctic ground squirrel hibernates during winter; there is
delayed implantation of the fertilized egg and winter
gestation is protracted. In studies made under the auspices of the U.S. Tundra Biome Program, George Batzli
estimated that demands made on phosphorus and calcium by
the lactating brown lemming are extremely high as are those
on protein and energy. In a model of nutrient metabolism,
it was found that the brown lemming almost completely
depleted her calcium and phosphorus reserves by the end of
lactation. These reserves must be replaced before the
female can enter the next breeding cycle. Hence, phosphorus and calcium requirements can remain high well beyond
weaning. Energy requirements during the non-lactating part
of the period are about 50 to 60 percent of peak requirements noted toward the end of lactation. Thus, Batzli and
coworkers hypothesize that the ecology of the brown lemming
involves a food acquisition process characterized by a fast
passage of food through the alimentary tract in order to

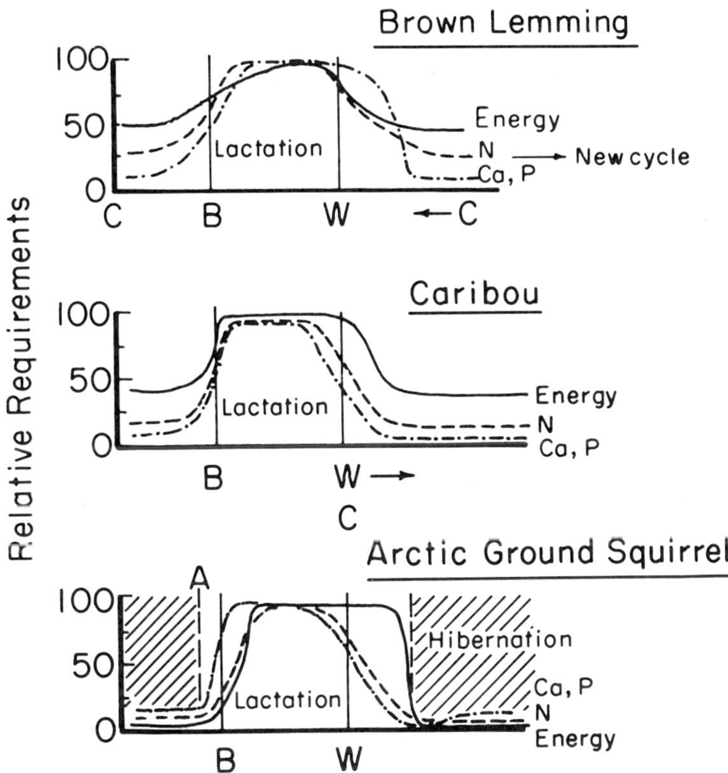

FIGURE 3. Comparison of the relative requirements for the macro-nutrients N, Ca, and P in relation to energy in the brown lemming, Arctic ground squirrel, and caribou. Lines are idealized curves. The cycle for the lemming is repeated at 20-30 d intervals in summer. The cycle for the Arctic ground squirrel includes the summer period, while the winter period invloves short-term arousal bouts. C, conception; B, birth; W, weaning; A, arousal.

maximize phosphorus and calcium intake. Also, selection of mosses which are high in calcium and phosphorus, is thought to facilitate calcium and phosphorus acquisition.

In comparison with the brown lemming, the reindeer or caribou calf is physiologically well developed at birth: it can thermoregulate despite severe climatic conditions (Hart et al. 1961; Blix and Steen 1978), and is able to walk with its mother within a few hours of birth which

permits escape from predators and the completion of the
migration to summer range. The energy cost of walking
correlates inversely with body weight (Taylor et al. 1970;
Taylor 1974), thus rapid early growth should lower the
energy cost of walking. Rapid growth involves increased
need of minerals and protein. To meet these needs of the
calf during lactation, the mother depletes her own reserves
and this continues into late lactation (Luick et al. 1974).
The young reindeer or caribou takes an increasing amount of
vegetation and can supplement a rapidly declining milk
production (Holleman et al. 1971). The energy requirement
of the mother remains high since she must now deposit a
good energy reserve before the onset of winter. The energy
requirements of the female are highest from early lactation
through the rut and until winter (November). The gradual
decline in energy requirements is thought to parallel a
lowering of resting (Segal 1962) or basal metabolism
(McEwan 1970) during winter.

The mineral requirements of the Arctic ground squirrel
are apparently high at or about arousal (Hartshorne and
Boucher 1972; Galster and Morrison 1976). This may relate
to the elevation of core temperature by shivering and non-
shivering thermogenesis: protein is mobilized for gluconeogenesis (Klain and Whitten 1968; Galster and Morrison
1970), calcium is required for muscle contraction in shivering, and magnesium is an important regulator in many biochemical pathways. In liver, K^+ and Mg^{2+} decline while the
respective exchange pairs K^+ and Ca^{2+} increase during
hibernation; during arousal the concentrations are abruptly
restored (Galster reported by Behrisch 1978), such changes
may affect the flow of substrate to thermogenic organelles
and tissues. In the female, nitrogen and energy requirements lag somewhat behind minerals since the highest demands accompany and follow arousal. Fat reserves are
utilized for thermogenesis until food is available. By
mid-lactation the nitrogen and energy requirements are
still very high, followed by a decline toward the end of
lactation. For the female the bulk of energy and nutrient
reserves are probably replenished in August and September.
Energy demand at this time is very high as the animal
fattens in readiness for hibernation. These fat reserves
are used up in the short-term arousal bouts during hibernation and for initial milk production the following spring.

EVIDENCE FOR CONTROL OF INTAKE BY NUTRIENT REQUIREMENT

In classical laboratory studies food intake is thought
to be determined by the nutritive value of the food.
Consumption of low quality food is limited by bulk. As
food quality improves food intake increases linearly until
requirements are met. As the nutritive value increases
still further, food intake is regulated so that intake of
the most important dietary component, general energy, is
constant (figure 5a). This model has been documented for
the rat, sheep, goat (Baumgardt 1970). When tested for the
brown lemming, Batzli and coworkers found that daily food
intake was more closely "regulated" to meet a requirement
for nutrients rather than energy (Figure 5b). Energy and
dry matter were taken in excess of requirements and the
lemmings became obese. However, under field conditions
lemmings rarely fatten beyond eight to nine percent body
fat. The brown lemming's requirements for nutrients rather
than energy is probably based on the problem it encounters
in lactating beneath the snow during winter. At this time,
availability of plants (particularly those of high quality)
is relatively low. The nutrient levels in the frozen
monocots is lower than in summer so that a large bulk of food
must be taken in to provide the calcium and phosphorus
requirements of the gestating and lactating animal. The
animal may supplement nutrient requirements by eating moss
which constitutes up to 40 percent of the diet during the
winter. Mosses are known to be very poorly digested, dry
matter digestibility is less than 15 percent, but are high
in calcium and phosphorus. Thus, they make a significant
contribution to the calcium and phosphorus requirements of
the lemming (Batzli et al. 1979).

EFFECT OF COLD EXPOSURE ON FOOD DIGESTION AND INTAKE

Regulation of food intake in caribou and reindeer in
relation to energy and nutrient demands is not well understood. When diets of fixed composition are fed, voluntary
food intake declines in winter compared with summer, in
parallel with a decline in resting or basal metabolism.
However, during the winter lichen intake increases in
reindeer held in outdoor pens compared with those confined
indoors in stalls and increases even farther in reindeer
which forage for lichens beneath the snow (Holleman et al.
1979). Holleman and coworkers concluded that increasing

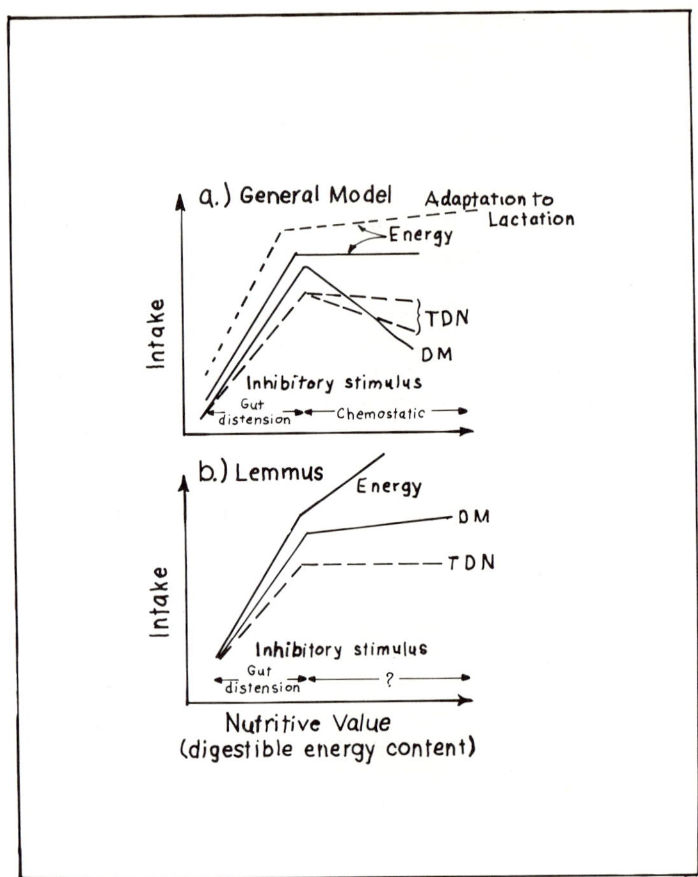

*FIGURES 5 a, b. General relationships between volun-
tary intake and the nutritive value of diets for animals
which regulate energy intake (sheep, goats, rats) (a), and
those which regulate total digestible nutrient (TDN) intake
(brown lemming) (b). Adaptation to increased energy demand for lactation is shown in (a) as described by Baumgardt
(1970). Voluntary food intake expressed as dry matter
(DM), energy and TDN, is limited by inhibitory stimuli associated with gut distension when food is of low quality.
The resultant food intake based on the regulation of TDN
intake results in a very high energy intake (G.O. Batzli,
unpub. observ.). An analagous type of food intake regulation has been suggested for fatty Zucker rats (Radcliffe
and Webster 1974).*

energy demands of foraging and outdoor activity lead to an
increase in food intake. The observed increase in lichen
intake was unlikely to be due to cold exposure *per se*
since environmental conditions were mild and reindeer were
probably within their thermoneutral range of environmental
temperature. Recent evidence suggests that with sheep and
cattle, cold exposure sufficient to cause a rise in oxygen
consumption, also causes increases in peristalsis, an
increase in the rate of passage of food, and decreases in
rumen retention time, dry matter digestibility (Christopherson 1976), and apparent N digestibility (Kennedy and
milligan 1978). In cattle and sheep dry matter digestibility declined by 0.18 to 0.31 digestibility units per degree
of C decline in ambient temperature. Effects on food
intake were less clear, but a tendency for this to increase
as temperature declined was noted. In subsequent work, it
was shown that thyroid hormones are involved in the responses to cold exposure (Westra and Christopherson 1976;
Kennedy et al. 1978) presumably by decreasing the rumen
turnover time. In spite of the decline in apparent N
digestibility, the passage of plasma urea-N to the rumen
was increased as was the efficiency of microbial synthesis
in the rumen (Kennedy and Milligan 1978).

Thyroxine is known to affect peristalsis in monogastric animals (Levin 1969), suggesting that an increased
intake of macro-nutrients could occur upon cold exposure of
monogastric herbivores of the Arctic (e.g., lemmings,
hares). Thyroxine injections also result in increased food
intake and decreased rumen turnover time in reindeer in
late winter (Ryg; Korg; Jacobsen; unpub. observ.). However, the effect should be treated with caution as T_4
levels of free-grazing domestic reindeer arc lower in
winter than summer (Ringberg et al. 1978); a finding in
agreement with a general energy-sparing metabolism predicted for winter from laboratory studies. Obviously, food
and nutrient intake would also be lower in winter than
summer. Any stimulating effect of cold *per se* on food
intake would occur only if temperatures declined below the
lower critical temperature and food availability was not
limiting. For reindeer and caribou in winter pelt, the
lower critical temperature is thought to be less than $-50^{\circ}C$
in still air (Hart et al. 1961).

SUMMARY AND CONCLUSIONS

Processes of adaptation to cold in arctic herbivores are generally interpreted in relation to body size; it follows that for a nutrient whose utilization or metabolism is linked to energy metabolism a medium to high requirement would be predicted: i.e., for the small herbivores (microtine rodents, hares) during winter. An added nutrient requirement for lactation would exist for those microtine rodents which reproduce beneath the snow. Evidence is presented that food intake in at least one of these species, the brown lemming, is controlled by requirements for nutrients rather than for energy. In large herbivores low demand for nutrients in winter would be expected because resting metabolism is lower in winter than summer. Activation of nutrient conservation mechanisms in the large herbivores is necessary to minimize the loss of nutrient reserves particularly when food availability limits nutrient intake in winter and early spring.

No adaptations of the nutrient acquisition and utilization processes to cold *per se* could be documented. However, compensation for the high energy expenditure in cold stressed ruminants results in an increased rate of passage of food and an increased efficiency of microbial protein synthesis in the rumen. It is suggested that nutrient acquisition and utilization processes play a permissive role in that they ensure a ready availability of nutrients for specific adaptations to cold centered on aspects of energy metabolism, including cold thermogenesis, and insulation. This suggestion is speculative in that before specific adaptations in nutrient acquisition and utilization processes can be ruled out, considerable work needs to be done in relating whole animal nutrient metabolism to thermogenic processes and to determining the actual seasonal requirements for nutrients of arctic species.

ACKNOWLEDGMENTS

I would like to thank my colleagues at the Institute of Arctic Biology, including H.W. Behrisch, J. Bligh, F.S. Chapin, III, D.R. Klein, and P. Tallas, who took part in critical discussions and the review of this chapter.

REFERENCES

Adam, T. 1971. Carnivores. In Comparative physiology of
    thermoregulation. G.C. Whittow, ed. Academic
    Press, NY. 151-189 pp.
Allden, W.G., and I.A. McD. Whittaker. 1970. The determin-
    ation of herbage intake by grazing sheep: the inter-
    relationship of factors influencing herbage intake and
    availability. *J. Agric. Res. 21:*755-766.
ARC. 1965. The nutrient requirements of farm livestock,
    No.2 Ruminants. Agricultural Research Council. Her
    Magesty's Stationary Office, London, ENG. 264 pp.
Arnold, G.W. 1964. Factors within plant associations af-
    fecting the behavior and performance of grazing ani-
    mals. In Grazing in terrestrial and marine environ-
    ments. D.J. Crisp, ed. Blackwell Scientific Publica-
    tions, Oxford. 133-154 pp.
Arnold, G.W., and M.L. Dudzinski. 1967. Studies on the diet
    of the grazing animals. III. The effect of pasture
    species and pasture structure on the herbage intake of
    sheep. *J. Agric. Res. 18:*657-666.
Batzli, G.O., et al. 1979. The herbivore-based trophic
    system. In an Arctic ecosystem: the coastal tundra of
    northern Alaska. J. Brown, et al., eds. (In press).
Baumgardt, B.R. 1970. Control of feed intake in the re-
    gulation of energy balance. In physiology of digestion
    and metabolism in the ruminant. A.T. Phillipson, ed.
    Oriel Press, Newcastle. 235-253 pp.
Behrisch, H.W. 1979. Metabolic economy at the biochemical
    level: the hibernator. In Strategies in cold: natural
    torpidity and thermogenesis. L.C.H. Wang, and J.W.
    Hudson, eds. Academic Press, NY. 461-497 pp.
Blair-West, J.R., et al. 1968. Physiological, morphological
    and behavioural adaptation to a sodium deficient
    environment by wild native Australian and introduced
    species of animals. *Nature. 217:*922-928.
Blix, A.S., and J.B. Steen. 1979. Temperature regulation in
    newborn polar homeotherms. *Physiol. Rev. (in press).*
Blond, D.M., and R. Whittam. 1964. Effects of Na and K on
    oxidative phosphorylation in relation to respiratory
    control by cell-membrane ATPase. *Biochem. Biophys.
    Res. Commun. 17:*120-124.
Bobek, B. 1977. Summer food as the factor limiting roe deer
    population size. *Nature. 268:*42-49.
Calef, G.W., and G.M. Lortie. 1975. A mineral lick of the
    barren ground caribou. *J. Mamm. 56:*240-242.

Cameron, R.D. 1970. Water metabolism by reindeer (*Rangifer tarandus*). Univ. of Alaska Ph.D. Thesis, Fairbanks.

Cameron, R.D., and J.R. Luick. 1972. Seasonal changes in total body water, extracellular fluid, and blood volume in grazing reindeer. *Can. J. Zool.* *50*:107-116.

Christopherson, R.J. 1976. Effects of prolonged cold and the outdoor winter environment on apparent digestibility in sheep and cattle. *Can. J. Anim. Sci.* *56*:201-202.

Clarke, L., J. Moore, and H. Garner. 1978. Diurnal variation of plasma aldosterone in horses: dependence on feeding schedule. In Proceedings of the 29h Annual Fall Meeting of American Physiological Society, St. Louis, MO.

Cowan, I. McT. and V.C. Brink. 1949. Natural game licks in the Rocky Mountain national parks of Canada. *J. Mamm.* *30*:379-387.

Deavers, D.R., J.W. Hudson, and X.J. Musacchia. 1978. Comparison of total body water, water turnover and metabolism during cold exposure in three species of rodents. *The Physiologist 90*:165(abst.).

DeVos, . 1964. Food utilization of snowshoe hares on Manitoulin Island, Ontario. *J. Fores. 62*:238-244.

Galster, W., and P. Morrison. 1976. Seasonal changes in body composition of the Arctic ground squirrel, *Citellus undulatus*. *Can. J. Zool. 54*:74-78.

Grav, H.J., and A.S. Blix. 1979. A source of non-shivering thermogenesis in fur seal skeletal muscle. *Science. (in press)*.

Griffin, D.R., et al. 1951. The comparative physiology of temperature regulation and thermal insulation. Arctic Aeromedi. Lab. Tech. Doc. Rep. Cont.No. AF-33(038)-12764.

Gupta, B.L., and T.A. Hall. 1979. Quantitative electron probe X-ray microanalysis of electrolyte elements within epithelial tissue compartments. *Fed. Proc. 38*:144-153.

Hammel, H.T., et al. 1962. Thermal and metabolic measurements on a reindeer at rest and in exercise. Arctic Aeromedi. Lab. Tech. Doc. Rep. Contr. AAL-TDR-61-54.

Hart, J.S. 1971. Rodents. In Comparative Physiology of Thermoregulation. G.C. Whittow, ed. Academic Press, NY. 1-149 pps.

Hart, J.S., et al. 1961. The influence of climate on metabolic and thermal responses of infant caribou. *Can. J. Zool. 39*:845-856.

Hartshorne, D.J., and L.J. Boucher. 1972. Calcium, and the control of muscle activity. F.E. South, et al., eds. Elsevier Pub. Co., NY. 357-367 pp.

Hebert, D., and I. McT. Cowan. 1971. Natural salt licks as a part of the ecology of the mountain goat. *Can. J. Zool. 49:*605-610.

Hecker, J.F., O.E. Budtz-Olsen, and M. Ostwald. 1964. The rumen as a water store in sheep. *J. Agric. Res. 15:* 961-968.

Himms-Hagen, J. 1976. Cellular thermogenesis. Annuals Reviews Inc., Palo Alto, CA. *Physiology 38:*315-351.

_____. 1978. Biochemical aspects of non-shivering thermogenesis. Pages 595-617 in Strategies in cold: natural torpidity and thermogenesis. L.C.H. Wang and J.W. Hudson, eds. Academic Press, NY.

Holleman, D.F., J.R. Luick, and R.G. White. 1971. Transfer of ragiocesium in milk from reindeer cow to calf. Pages 76-80 in Radionuclides in ecosystems. D.J. Nelson, ed. National Technical Information Service, U.S. Dept. of Commerce. Springfield, VA.

_____. 1979. Lichen intake estimates for reindeer/caribou during winter. *J. Wildl. Mgmt. (in press).*

Holleman, D.F., R.G. White, and D.D. Feist. 1978. Body water turnover and energy metabolism in free-living arctic homeotherms. Proceedings of 29h Annual Fall Meeting. American Physiological Society. St. Louis, MO.

Jacobsen, E., and S. Skjenneberg. 1975. Some results from feeding experiments with reindeer. Pages 95-107 in Proceedings of the First International Reindeer and Caribou Symposium. J.R. Luick, et al., eds. University of Alaska, Anchorage. Biological Papers, Special Report No. 1.

Kennedy, P.M., and L.P. Milligan. 1978. Effects of cold exposure on digestion, microbial synthesis and nitrogen transformations in sheep. *Br. J. Nutrition. 39:*106-117.

Kennedy, P.M., B.A. Young, and R.J. Christopherson. 1978. Studies on the relationship between thyroid function, cold acclimation and retention time of digesta in sheep. *J. Animal Science. 45:*1084-1090.

Klein, D.R. 1970. Tundra ranges north of the boreal forest. *J. Range Mgmt. 23:*8-14.

Klein, G.J., and B.K. Whitten. 1968. Free amino acids in hibernation and arousal. *Comp. Biochom. Physiol. 27:*617 619.

Lehninger, A.L. 1970. Biochemistry. 2d ed. North Publishers, Inc. NY. 1104 pp.

Lehninger, A.L. 1974. Ca^{2+} transport by mitochondria and
    its possible role in the cardiac contraction-relaxa-
    tion cycle. *Circ. Res. 35*:83-90.

Levin, R.J. 1969. Review. The effects of hormones on the
    absorptive, metabolic and digestive functions of the
    small intesting. *J. Endocr. 45*:315-348.

Ling, J.K. 1965. Hair growth and moulting in the southern
    elephant seal, *Mirounga leonina* (Linn.). Pages 525-544
    in Biology of the Skin and Hair Growth. A.G. Lyne, and
    B.F. Short, eds. American Elsevier Publishing Company,
    Inc., NY.

Luick, J.R. 1977. Diets for grazing reindeer. Pages 267-278
    in CRC Handbook Series in Nutrition and Food. Section
    G: Diets, Culture Media, Food Supplements vol. 1.
    Diets for Mammals. M. Recheigl, Jr. ed. CRC Press,
    Inc. Cleveland, OH.

Luick, J.R., G.P. Lofgreen. 1957. An improved method for
    the determination of metabolic fecal P. *J. Anim.
    Science. 16*:201-206.

Luick, J.R., et al. 1974. Compositional changes in the milk
    secreted by grazing reindeer. I. Gross composition and
    ash. *J. Dairy Science. 57*:1325-1333.

McLean, S.F., Jr. 1974. Lemming bones as a source of cal-
    cium for Arctic sandpipers *(Calidris spp.)*
    Ibis. 116:552-557.

Magruder, N.D., et al. 1957. Nutritional requirements of
    white-tailed deer for growth and antler development.
    II. Pennsylvania State University, Agriculture Exp.
    Stn. Bulletin 628. 21 pp.

Mautz, W.W., H. Silver and H.H. Hayes. 1975. Estimating
    methane, urine, and heat increment for deer consuming
    browse. *J. Wildl. Mgmt. 39*:80-86.

McEwan, E.H. 1970. Energy metabolism of barren ground
    caribou *(Rangifer tarandus).*   *Can. J. Zool.
    48*:391-392.

McEwan, E.H., and P.E. Whitehead. 1970. Seasonal changes in
    the energy and nitrogen intake in reindeer and cari-
    bou. *Can. J. Zool. 48*:905-913.

Miller, L.K. 1978. Phusiological studies in arctic animals.
    Comp. Biochem. Physiol. 59A:327-334.

Myers, K. 1967. Morphological changes in the adrenal glands
    of wild rabbits. *Nature. 213*:147-150.

Ozawa, K., et al. 1967. Rapid liberation of potassium ions
    from brain mitochondria. *J. Biochem. Tokoy. 62*:584-590.

Pace, N., and E.N. Rathbun. 1945. Studies on body composition. III. The body water and chemically combined
    nitrogen content in relation to fat content. *J. Biol.
    Chem. 158*:685-691.

Panaretto, B.A. 1963. Body composition *in vivo:* III. The
    composition of living ruminants and its relation to
    the tritiated water spaces. *J. Agric. Res.
    14*:944-952.

Pease, J.L., R.H. Vowles, and L.B. Keith. 1979. Interaction
    of snowshoe hares and woody vegetation. *J. Wildl.
    Mgmt. (in press).*

Pittet, P.H., P.H. Gygox, and E. Jéquier. 1974. Thermic
    effect of glucose and amino acids in man studied by
    direct and indirect calorimetry. *Br. J. Nutr.
    31*:343-349.

Radcliffe, J.D., and A.J.F. Webster. 1974. Regulation of
    food intake during growth in fatty and lean female
    Zucker rats given diets of different protein content.
    Br. J. Nutr. 36:457-469.

Ringbert, T. 1978. Survival strategies in Spitzbergen
    reindeer. Proceedings of 29h Annual Fall Meeting of
    the American Physiological Society. St. Louis, MO.
    October 22-27.

Ringberg, T., et al. 1978. Seasonal changes in levels of
    growth hormone, somatomedin and thyroxine in free-
    ranging, semi-domesticated Norwegian reindeer (*Rangifer tarandus tarandus L.).*   *Comp. Biochem. Physiol.
    60A*:123-126.

Rubner, M. 1902. Die Gesetze des Energieverbrauchs bei der
    Ernahrung. Franz Dauticke. Leipzig, Vienna.

Schultz, A.M. 1964. The nutrient-recovery hypothesis for
    Arctic microtine cycles. Pages 57-68 in Grazing in
    Terrestrial and Marine Environments. D.J. Crisp, ed.
    Symposium No. 4. British Ecological Society. Blackwell, Oxford, ENG.

______. 1969. A study of an ecosystem: the arctic tundra.
    Pages 77-93 in The Ecosystem Concept in Natural Resource Management. Chap.V. Academic Press, NY.

Searle, T.W. 1970. Body composition in lambs and young
    sheep and its prediction *in vivo* from tritiated water
    space and body weight. *J. Agric. Sci., Camb.
    74*:357-362.

Segal, A.N. 1962. The periodicity of pasture and physiological functions of reindeer. Pages 130-150 in Reindeer in the Karelian ASSR. (translated by Dep. Secy.
    of State Bur. Trans., Canada).

Simpson, A.M., et al. 1978. Energy and nitrogen metabolism
     of red deer *(Cervus elaphus)* in cold environments:
     comparison with cattle and sheep. *Comp. Biochem.
     Physiol. 60:*251-256.

Sjollema, B. 1932. Tijdschr. Diergenseesk. 59:57-329.

Skogland, T. 1975. Range use and food selectivity by wild
     reindeer in southern Norway. Pages 342-554 in Pro-
     ceedings of the 1t International Reindeer and Caribou
     Symposium. J.R. Luick, et al., eds. University of
     Alaska, Fairbanks. Biological Papers Special Report
     No. 1.

Slee, J. 1965. Seasonal patterns of moulting in Wiltshire
     horn sheep. Pages 545-563 in Biology of the Skin and
     Hair Growth. A.G. Lyne, and B.F. Short, eds. American
     Elsevier Publishing Company, Inc., NY.

Smith, M.C., J.F. Leatherland, and K. Myers. 1978. Effects
     of seasonal availability of sodium and potassium on
     the adrenal and cortical function of a wild population
     of snowshoe hares, *Lepus americanus.   Can. J.
     Zool. 56:*1869-1876.

Staaland, H., et al. 1979. Seasonal and dietary influences
     on sodium and potassium metabolism of reindeer. *Can.
     J. Zool. (in press).*

Suttle, N.F., and A.C. Field. 1967. Studies on magnesium in
     ruminant nutrition. 8. Effects of increased intakes of
     potassium and water on the metabolism of magnesium,
     phosphorus, sodium, potassium, and calcium in sheep.
     *Br. J. Nutr. 21:*819-831.

Taylor, C.R. 1974. Exercise and thermoregulation. Pages
     163-184 in Physiology series one. vol. 7. Environ-
     mental Physiology. D. Robertshaw, ed. Butterworths,
     London and University Park Press, Baltimore, MD.

Taylor, C.R., K. Schmidt-Nielsen, and J.L. Raab. 1970.
     Scaling of energetic cost of running to body size in
     mammals. *Am J. Physiol. 219:*1104-1107.

Ullrey, D.E., et al. 1975. Phosphorus requirements of
     weaned white-tailed deer fawns. *J. Wildl. Mgmt.
     39:*590-595.

Van Es, A.J.H. 1969. Report to the sub-committee on con-
     stants and factors regarding metabolic water. Pages
     513-514 in Energy Metabolism of Farm Animals. K.L. Bl-
     axter, et al., eds. Oriel Press, Ltd., Newcastle upon
     Tyne.

von Englehardt, W. 1970. Movement of water across the rumen
     epithelium. Pages 132-146 in Physiology of Digestion
     and Metabolism in the Ruminant. A.T. Phillipson, ed.
     Oriel Press, Newcastle upon Tyne.

Wang, L.C.H., and J.W. Hudson, eds. 1978. Strategies in cold: natural torpidity and thermogenesis. Academic Press, NY. 715 pp.

Waterlow, J.C. 1968. Observations on the mechanism of adaptation to low protein intakes. *Lancet.* 2:1091-1097.

Webster, A.J.F., et al. 1975. The influence of food intake on portal blood flow and heat production in the digestive tract of sheep. *Br. J. Nutr. 34:*125-139.

Weeks, H.P., Jr., and C.M. Kirkpatrick. 1976. Adaptations of white-tailed deer to naturally occurring sodium deficiencies. *J. Wildl. Mgmt. 49:*610-625.

______. 1978. Salt prederences and sodium drive phenology in fox squirrels and woodchucks. *J. Mammol. 59:*531-542.

West, G.C. 1976. Seasonal adaptation of birds to cold. Pages 462-490 in Progress in Biometeorology. H.D. Johnson, ed. Sweto and Zeitlinger B.V., Amsterdam.

West, G.C., and D.W. Norton. 1975. Metabolic adaptations of tundra birds. Pages 301-329 in Physiological Adaptation to the Environment. F.J. Vernberg, ed. Intext Educational Publishers, NY.

Westra, R., and R.J. Christopherson. 1976. Effects of cold on digestibility, retention time, of digesta, reticulum mobility and thyroid hormones in sheep. *Can. J. Anim. Sci. 56:*699-708.

White, R.G. 1975. Some aspects of nutritional adaptations of arctic herbivorous mammals. Pages 239-268 in Physiological Adaptation to the Environment. E.J. Vernberg, ed. Intext Educational Publishers, NY.

______. 1979. Vegetation use and forage consumption by reindeer and caribou in arctic tundra. Arctic and Alpine Research (in preparation).

White, R.G., et al. 1979. Models of sodium and water exchange in physiological pools of reindeer (manuscript prepared for American Journal of Physiology).

White, R.G. et al. 1975. Ecology of caribou at Prudhoe Bay, Alaska. Pages 151-201 in Ecological Investigation of the tundra biome in the Prudhoe Bay Region Alaska. J. Brown, ed. University of Alaska, Fairbanks, Biological Paper Special Report No. 2.

Whittow, G.C., ed. 1971. Ungulates. Pages 191-281 in Comparative Physiology of Thermoregulation. Vol. 2. Academic Press, NY.

Wolff, R.G. 1978. Food habits of snowshoe hares in interior Alaska. *J. Wildl. Mgmt. 42:*148-153.

Young, B.A. 1966. Energy expenditure and respiratory ac-
    tivity of sheep during feeding. *J. Agric. Res.*
    17:355-359.
Young, B.A., and J.L. Corbett. 1972. Maintenance energy
    requirement of grazing sheep in relation to herbage
    availability. I. Calorimetric estimates.
    J. Agric. Res. 23:57-76.
Yousef, M.K., and H.D. Johnson. 1978. Thyroid function in
    heat and cold acclimated desert wood rats. Proceedings
    of the 29h Annual Fall Meeting of the American Phy-
    siological Society. St. Louis, MO. October 22-27.

III. MECHANISMS OF THERMAL TOLERANCE

X. J. Musacchia
D. R. Deavers

Dalton Research Center
and
Department of Physiology
University of Missouri
Columbia, Missouri

Department of Physiology and Biophysics
University of Louisville
Louisville, Kentucky

Hibernation is a widely explored natural phenomenon involving responses to cold. Physiological investigations utilize field collected subjects, laboratory reared subjects and animal models which utilize hypothermia. In our laboratory, water turnover in ground squirrels, (Spermophilus tridecemlineatus) showed that total water turnover (TWT) is about 14 ml/day and evaporative water loss (EWL) about one half TWT. VO_2 falls from 400 cc O_2/hr in normothermia to 8 cc O_2/hr in hibernation. TWT and VO_2 were comparably reduced in hibernation, both about 2% of normothermic values. Helium-code hypothermia in hamsters, (Mesocricetus auratus) is utilized as a model for hibernation. Hamsters are induced into hypothermia to body temperature (T_{re}) $7^{O}C$ and, with glucose maintenance through i.v. infusion, they survive for four days, a period comparable to a bout of natural hibernation. Pretreatment with cortisone enhances gluconeogenesis and survival in hypothermia, and through endogenous thermogenesis 50 percent rewarm to normothermia, T_{re} $37^{O}C$. This review proposes a role for glucocorticoids in hibernation and suggests that animal models be used to broaden our understanding of thermogenic mechanisms.

AN ORIENTATION FOR LABORATORY AND FIELD RESEARCH IN
HIBERNATION

     The title, "Mechanisms of Thermal Tolerance", auto-
matically evokes the idea of animal subjects undergoing
exposure to environmental temperatures higher or lower than
routinely experienced and/or animal subjects which routinely exist at environmental temperatures tolerated by relatively few species. Additional considerations are the
questions about the mechanisms whereby such organisms meet
these environmental challenges. Environmental biologists
and temperature regulation physiologists, among others,
have approached these ideas in several ways. In one approach,
they have gone into the field to study the animal subject
in its natural habitat. In another approach, the animal
has been brought into the laboratory in order to examine
individual physiological, biochemical and behavorial characteristics under controlled conditions. Extending from this
latter approach has come the development of animal model
systems which mimic, simulate, or compare with the animal
under natural conditions. With investigations of specific
physiological or biological systems under laboratory
conditions, the investigator has been able to predict
events which probably take place under natural or field
conditions. Investigators such as Wang (1972, 1973),
Hudson and Deavers (1973), and Deavers and Hudson (1977),
have stressed that laboratory measurements must relate in a
meaningful way to conditions encountered by the animal in
its natural environment.
     Our interests have been chiefly concerned with animal
responses to cold and experimentally lowered environmental
temperatures. This brief review will consider some studies
of animal systems known to adapt to environmental extremes
and will emphasize laboratory approaches to investigations
of hibernators.
     One of the most widely explored natural phenomenon
involving animal responses to cold is that of hibernation.
The foremost representatives of hibernators are found among
the *Rodentia,* the *Insectivora,* and *Chiroptera*. The current
level of understanding of this phenomenon is for the most
part due to laboratory investigations where quantitative
data are obtained as the result of controlled experimentation. However, it is incumbent that we take into account
the improved experimental approaches, such as radioteleme-
try (Wang 1973; Devaney et al. 1976), computer data storage
and data analysis which will become more widely used in the

field. With the availability of such new technology the
opportunities to readily monitor several physiological
systems will expand. It is anticipated that quantitative
data will accrue in considerable measure and availability
of resources of field oriented laboratories such as the
Naval Arctic Research Laboratory (NARL) will serve to
accomodate field oriented quantitative studies.

The Arctic offers one of the most unique animal subjects for studies of thermal tolerance, namely, the Arctic
ground squirrel, *Spermophilus parryii*. This common inhabitant of the arctic slope has been the subject of physiological investigation for almost 30 years and our first
experiments were conducted in 1949 and 1950. Among these
early works was the report which showed that blood glucose
levels in the long-term hibernating Arctic ground squirrel
were only slightly less than those in normothermic subjects
(Musacchia and Wilber 1952). That report supported the
view that blood glucose reserves were not depleted during
hibernation. In the subsequent twenty-five years the
specific role of blood glucose, and the role of carbohydrate metabolism in overall maintenance of hibernators have
been investigated in a limited manner. There remains a
relatively unclear picture of a role for carbohydrates in
the cyclic events of hibernation. It is our contention
that in this area of investigation, the laboratory approach, despite the rather exaggerated and even artificial
controlled conditions, continues as a most efficient and
productive means of investigating the biology of hibernation.

IMPROVEMENTS WITHIN ANIMAL COLONIES

It is important to call attention to the fact that
laboratory experiments with hibernators are fraught with
problems relating to the physiological, biochemical, behavorial, and even the nutritional status of the animal.
For the most part, our experiments as well as those in the
majority of laboratories, are conducted with animals that
are captured in the wild and then transferred to laboratory
quarters. Whether or not caging and confinement, removal
from the need to search for food, avoidance of predators
and other natural factors, versus the regulation of nutritional status and control of other environmental variables
are of importance remain to be assessed. Answers to such
immediate questions are as yet unavailable. Under current

research conditions, one tends to accept that the laboratory measured phenomena are probably representative of that which occurs under field conditions. Another approach has been to obtain the animal under field conditions and to make measurements immediately at the time of capture and/or sacrifice of the subject. Certainly a variety of tests and measurements can be made immediately when the animal is killed and a considerable number of tissue biochemical characteristics have been obtained in this manner. However, the limitation is that these are acute or "one-time" analyses. How then can some of these limitations be overcome?

Two approaches have been found convenient in our laboratories. One practice is to capture pregnant females in the wild and transfer them to the laboratory where litters are raised under laboratory conditions. Another related approach is the establishment of breeding colonies. Ground squirrels do not ordinarily mate and breed in captivity; however, we have had limited success with *S. parryii* (Musacchia 1958), and with *S. tridecemlineatus* (Barr and Musacchia 1968). Recently, in the Dalton Research Center vivarium, some breeding trials with *S. tridecemlineatus* have been successful (the vivarium manager is confident that with additional careful planning, laboratory breeding of ground squirrels can be achieved). In both these approaches there are recognizable advantages in terms of more homogenous populations with information about age, nutritional state, parasite control and other husbandry procedures. It is of note that the capacity of hibernators to undergo hibernation both in the wild and in the laboratory is inherent, and thus they have been easily utilized in experiments. The potential that more uniform colonies of experimental animals can be easily obtained, warrants the extra effort of field capture of pregnant females and/or the establishment of breeding colonies.

LABORATORY APPROACHES TO PHYSIOLOGY OF HIBERNATION

There is another approach to the utilization of laboratory facilities and experimentation in hibernation. This is the use of model systems for the development of procedures to study metabolic cold torpor or the simulation of natural hibernation. One system which is readily suited as a model for hibernation is the helium-cold method of inducing hypothermia (Fisher and Musacchia 1968; Musacchia 1972, 1976).

Two series of experiments, currently underway in our laboratory are illustrative of research which could not be done in the field at this time. In each case there is an area of research which asks pertinent questions of the nature of specific mechanisms in response to cold exposure. The results of these experiments are illustrative of approaches to studies of cold tolerance, particularly among the hibernators.

Oxygen Consumption and Water Turnover During Cold Acclimation and Hibernation

The fact that energy savings result from natural torpidity is classic and unquestioned. However, as pointed out by Hudson and Deavers (1973) and Deavers and Hudson (1977), little quantitative work has been done on water metabolism during hibernation and cold acclimation.

Using the thirteen-lined ground squirrel, *S. tridecemlineatus*, measurements of evaporative water loss and total water turnover (with $^{3}H_2O$ as a tracer) showed that total water turnover is about 14 ml/day (Deavers and Musacchia 1976). Evaporative water loss amounts to about one-half of the total water turnover (Figure 1).

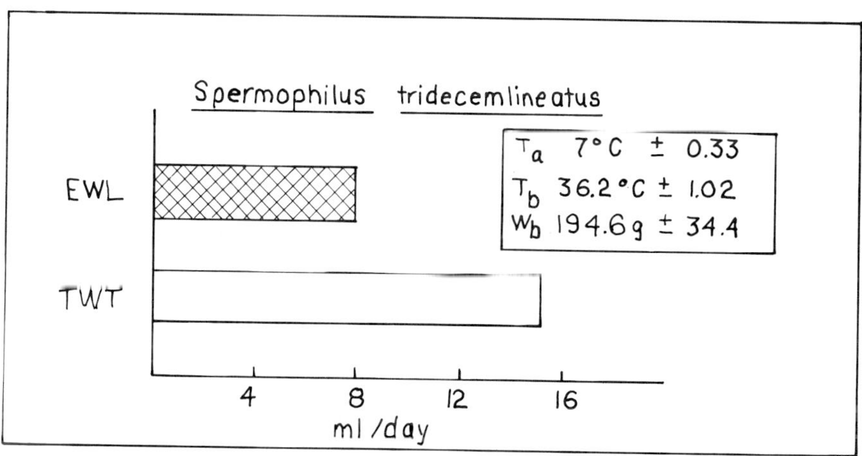

FIGURE 1. *Evaporative water loss (EWL) and total water turnover (TWT) in the thirteen-lined ground squirrel, Spermophilis tridecemlineatus; T_a= ambient temperature, T_b= body temperature and W_b= body weight.*

In 1962 Richard, Langham, and Trujillo studied water turnover of mammals of widely ranging body weights and they calculated a regression equation that allows prediction of water turnover as a function of body weight. Water turnover in the thirteen-lined ground squirrel (Figure 1) is about one-half the rate predicted from their equation. This difference may be explained in terms of evolution of efficient water conservation mechanisms. It is of interest that the thirteen-lined ground squirrel is found in mid-western latitudes in the United States where the environments are often characterized by summers with lengthy periods of limited availability of water and drought, and this is the time when these animals are normothermic and highly active. It would be interesting to determine if the Arctic ground squirrel, a northern relative, also showed an adaptive low water turnover (breeding, rearing young, etc.). Consider the fact that the arctic slope is a marsh during summer and water is continuously available.

Attempts at using the tritiated water turnover method for measuring total water turnover in hibernating squirrels were unproductive. The total water turnover was so low that changes were undetectable. In this case, pulmocutaneous water loss is probably equal to, or almost entirely, the route of total water loss during hibernation at low T_b. Evidence from Moy (1971) and from Tempel and Musacchia (1975) as well as Volkert and Musacchia (1976), suggests that renal function nearly ceases in hibernating ground squirrels and hamsters, respectively. Also, there is minimal or no fecal water loss during hibernation. Thus, it appears reasonable that evaporative water loss accounts for nearly all the water loss of hibernating squirrels.

A comparison of both evaporative water loss and oxygen consumption in normothermic and hibernating ground squirrels clearly shows the extreme differences between these two conditions (Figure 2). As is readily seen $\dot{V}O_2$ is only 8cc O_2/hr in hibernating animals as compared with over 400cc O_2/hr in normothermic animals; and in hibernating subjects, evaporative water loss has fallen to less than 0.2 g daily.

The comparative relationships between evaporative water loss and oxygen consumption in hibernating and normothermic ground squirrels are shown in Figure 3. When either evaporative water loss or oxygen consumption in hibernation are shown as a percentage of the levels in normothermic animals, there is a marked similarity. In our view, since the magnitude of decrease of both evaporative loss and metabolic rate during hibernation are similar, it is a reasonable assumption that decreased metabolism is responsible for decreased evapor-

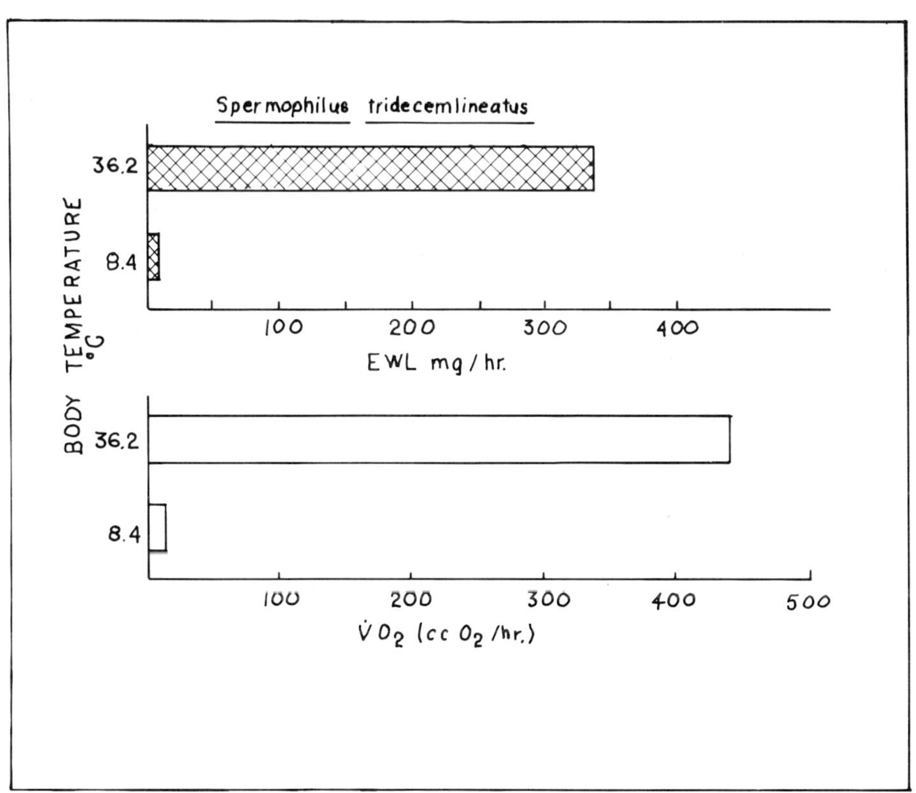

FIGURE 2. A comparison of evaporative water loss (EWL) in mg/hr and oxygen consumption in cc/O₂/hr in active nor-thermic (36.2°C) and hibernating (8.4°C) thirteen-lined ground squirrels, *Spermophilus tridecemlineatus*.

ative water loss. It is clear that considerable water, as well as energy, is saved during hibernation.

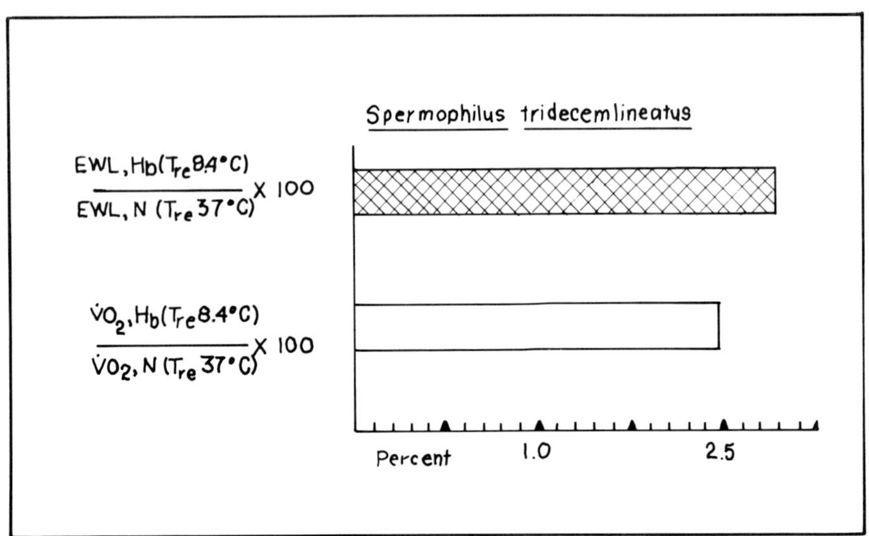

FIGURE 3. *Comparative relationships between evaporative water loss (EWL) and oxygen consumption (VO$_2$) in
active normothermic (N), T$_{re}$ 37°C, and in hibernating (HB),
T$_{re}$ 8.4°C, thirteen-lined ground squirrels, Spermophilus
tridecemlineatus.*

Helium - Cold Hypothermia, A Model for Hibernation

The hibernator as an experimental subject creates some
rather complex problems for the investigator. Not the
least of which is that in winter, the normal physiology of
the animal differs considerably from the normal physiology
of the same subject in summer. This is true even if the
animal is prevented from hibernating in winter and is maintained at usual room temperatures of 22 ± 2°C. Thus research projects must be designed with full appreciation for
a variety of limitations imposed by hibernation.

In many respects, experimental models for studies of
hibernation present a variety of intriguing problems. A
major question is: Can an animal be induced into a condition of cold torpor for a period of time which simulates
hibernation? Also, the hibernating animal is viewed as a
subject which is in physiological homeostasis at greatly
reduced body temperature and with concomitant reduced
metabolic functions. Hibernation is achieved after several
modifications involving a great number of specific physiolo-

gical systems, but in particular a reduction in cardiovascular, respiratory and temperature regulatory responses. After several days of torpor, through inherent signals (as yet unidentified), the animal awakens to a normothermic condition in a period of about two hours.

Experimental hypothermia and natural hibernation have, in common, generalized responses of metabolic depression: in the case of hypothermia, the initiating cause for this effect is cold exposure to the extent that heat production is overcome by heat loss. In our experience the hamster, made hypothermic using the helium-cold method, represents a reasonable subject in developing a hypothermic laboratory model for hibernation.

We have explored the problem by attempting to mimic hibernation through the induction of hypothermia in the golden hamster, *Mesocricetus auratus*. In the hamster, hibernation occurs after several weeks of exposure to T_a of about 4° to 7°C. In the course of hibernation, certain physiological adjustments take place, for example: a fall in venous PCO_2 and PO_2, a slight rise in pH and a marked fall in oxygen utilization (Table 1). The helium-cold hypothermic hamster, clearly parallels these changes.

Despite the marked reduction in cardiac activity, there is a redistribution in percent of cardiac output in which heart, lung and brown fat receive a greater proportion of blood flow and visceral organs such as intestine and kidney receive considerably less (Figure 4). Considering this information in a teleogical sense, it is reasonable to deduce that in hibernation the cardiovascular and respiratory systems are essential to continued life support, albiet at a low rate compared to normothermic animals and, in contrast, the intestine and kidney are functionally idle during hibernation, and blood, blood oxygen and other nutrients are of limited need in those tissues. The marked similarity of hypothermic and hibernating subjects is noteworthy in further considerations of a model.

An examination of cardiovascular responses shows that in hypothermia there is a marked decrease in cyctolic and diastolic pressures to values of about 54.6 and 43.9, mm Hg, respectively (Tempel, Musacchia, and Jones 1977). In those experiments the mean arterial pressure decreased from about 100 mm Hg to 49.9 mm Hg, a fall of about 60% in hypothermia, and the early data from Chatfield and Lyman (1950), using the same species in hibernation, likewise showed a decrease of about 55%. The effect of experimental helium-cold hypothermia on blood pressure and heart rate in

TABLE I. Comparison of Venous Blood Gasses, pH and VO_2
Normothermia (N), Hibernation (HB) and Hypothermia (HY)

Condition AND T_B *	PO_2 MM HG	PCO_2 MM HG	pH	$\dot{V}O_2$ CC/G HR AND (T_A)**
N 37-38°C	30.4+3.3[1]	55.5+4.1[1]	7.39+0.03[1]	3.80 (0°C)[2]**
N	----	60.0[3]	7.39[3]	----
HB 5°C	----	32.4[3]	7.44[3]	0.070 (4°C)[2]
HY 6°C	9.1[1]	33.3[1]	7.39[1]	0.050 (7°C)[4]

* T_b= Body Temperature
** T_a= Ambient Temperature

[1]Volkert and Musacchia
[2]Lyman
[3]Lyman and Hastings
[4]Resch and Musacchia

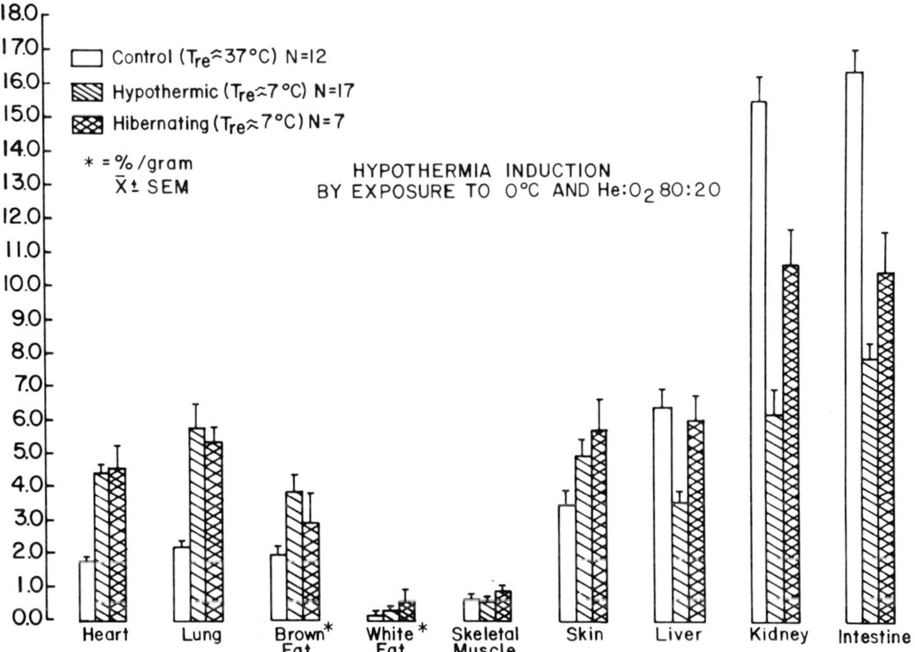

FIGURE 4. Shows a comparison of percent of cardiac output distributed to various organs, using the ^{86}Rb C1 method, in normothermic (control), hypothermic and hibernating hamsters.

hamsters shows a lowered mean pressure and lowered cardiac frequency, as shown in Figure 5a and b. The body temperature changes during rewarming show that with an increase in body temperature, there is a concomitant increase in heart rate. At first, the increased heart rate is slow in comparison to blood pressure, which rapidly increases until about T_{re} 12°C, then more slowly until about T_{re} 16 - 17°C and then there is a plateauing.

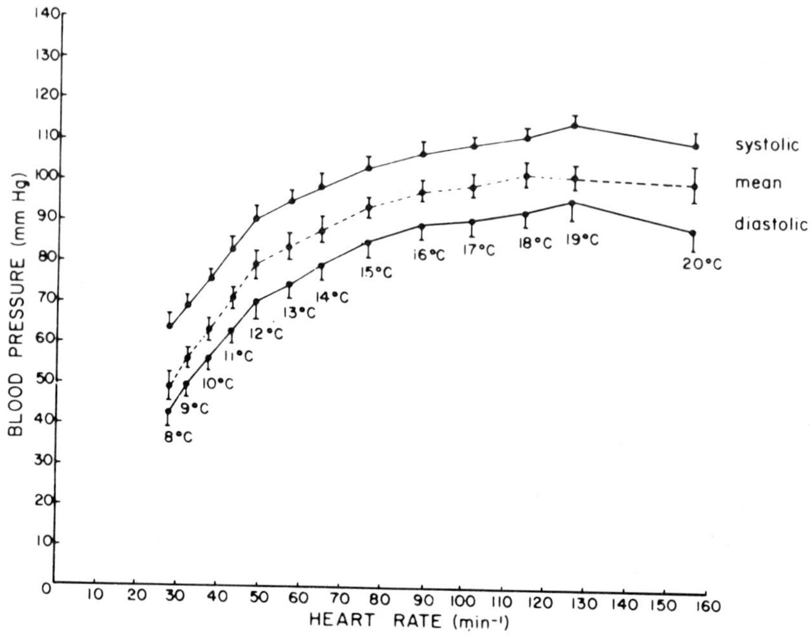

FIGURE 5a. Systolic, diastolic and mean blood pressures during rewarming of hypothermic hamsters.

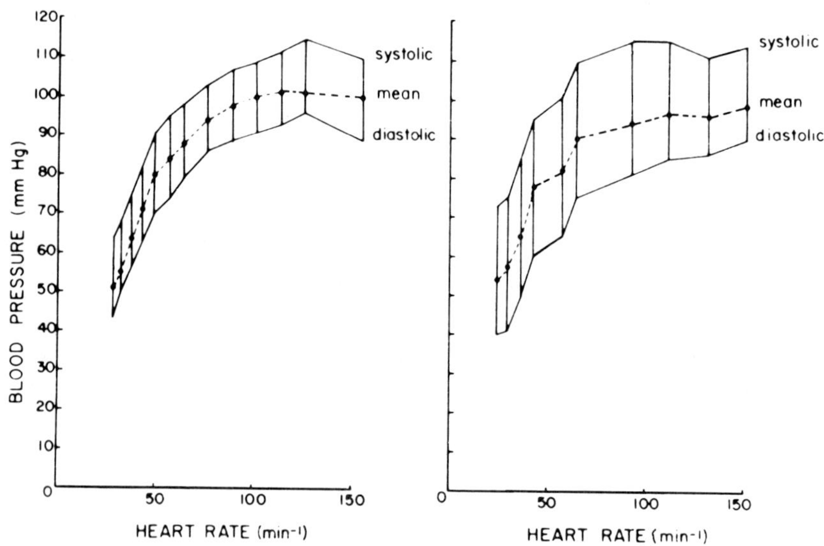

FIGURE 5b. A comparison of blood pressure and heart rate in rewarming hypothermic (lft) and HB hamsters (right)

CARBOHYDRATE REQUIREMENTS IN HIBERNATION

A great deal is known about energy requirements of
hibernating animals specifically, lipid metabolism in
hibernators. In fact, lipid metabolism in terms of sub-
cellular components and whole animal nutritive status,
has been researched for many years. In contrast, an area
which is less understood is that of carbohydrate metabolism.

Among the more general approaches to understanding the
role of carbohydrates in hibernation have been measurement
of blood glucose and tissue glycogen levels. From a host
of studies, with more or less quantitative data and with
more or less specific knowledge of the state of hibernation
for a particular subject, there has emerged a relatively
confused picture of the role of carbohydrate metabolism and
the regulation of carbohydrate metabolism under conditions
of cold acclimation and in depressed metabolism.

Carbohydrate metabolism during hibernation appears to
be in a state of relative equilibrium. For example, Lyman
and Chatfield (1955), showed that concentrations of blood
glucose in hibernating hamsters are somewhat lower than in
normothermic hamsters; work with Arctic ground squirrels
late in the winter hibernating cycle, indicate blood glucose levels comparable with those in normothermic summer
animals. Similar findings in *Spermophilus lateralis* were
reported by Twente and Twente (1967). More recently (Deavers, Musacchia, and Meininger 1977) discovered is that
sometimes blood glucose levels in hibernating hamsters are
as high as 110 mg percent, whereas sybling normothermic
controls are about 100 mgpercent. These results compare
favorably with those of Lyman and Leduc (1953), showing
that circulating glucose may be elevated in a hibernating
hamster.

The status of liver glycogen stores during hibernation
is also somewhat controversial. For example, Galster and
Morrison (1970, 1975) found that during hibernation, hepatic
carbohydrate reserves are slowly depleted and then replenished during brief periods of normothermia. In two earlier
studies, utilizing hamsters (Lyman and Leduc 1953) and Arctic ground squirrels (Musacchia and Wilber 1952), it was
reported that liver glycogen differs very little between
hibernating and normothermic animals. However, in many of
the studies cited there are some limitations due to lack of
information about the precise physiological state of the
animal. For example, length of time in hibernation, the
number of times the animal hibernated, and the nutritional

state of the animal. Many earlier studies merely select
one or two parameters in the whole animal, and leave unanswered happenings in specific tissues. More recently,
investigations have tended to examine tissue metabolism.
The general rationale is that blood glucose levels are a
reflection of both gluconeogenesis and glycogenolysis.
Burlington and Klain (1967) and Burlington (1972) support
the view that gluconeogenesis is an active mechanism in the
maintenance of blood glucose levels in hibernation. They
reported that kidney cortex from hibernating squirrels
showed enhanced ability for gluconeogenesis. Earlier,
Klain and Whitten (1968) reported that there was enhanced
incorporation of $^{14}CO_2$ into glucose and glycogen by liver
during hibernation. A role for glycogenolysis contributing
to the pool of circulating glucose was shown by Tashima et
al. (1970). They calculated that glycogenolysis could
maintain normal blood glucose levels for 4-5 days of
hibernation, but Burlinton (1972) points out that gluconeogenesis must supply some of the circulating glucose during
longer periods of torpor.

     Lastly, consider the fact that maintenance of a carbo-
hydrate reserve and blood glucose level is essential for
proper function of the central nervous system (CNS).
Without adequate supplies of carbohydrates, the respiratory
center and CNS functions are grossly impaired. The regulation of awakening from hibernation and return to hibernation may not be fully understood, but the need for proper
CNS functions must be considered to participate in an
ordered sequence of events. Thus, in order to utilize a
model for hibernation we must produce a subject in which
blood glucose levels would suffice for CNS metabolism.

A ROLE FOR CARBOHYDRATE METABOLISM IN HYPOTHERMIA

     Endocrine functions are greatly reduced in hiberna-
tion, but the animal remains in physiological and biochemical homeostasis. In contrast, in hypothermia reduced
endocrine functions can bring about a reduced regulation of
carbohydrate metabolism, and one can assume the eventuality
of pathophysiological events. Apparently the balance
between regulation and maintenance of an effective homeostatic condition in the hypothermic subject can be described
as precarious, and current understanding of the mechanisms
involved are in initial exploratory stages. Our questions

are and have been: Can we overcome the pathophysiology and can we enhance the endocrinologic regulation? In light of the foregoing, laboratory efforts have focused on an evaluation of carbohydrate metabolism in the functional maintenance of hypothermic and hibernating animals.

Maintenance of blood glucose levels and liver glycogen levels are a relative indication that an animal is capable of homeostatic regulation and is capable of arousal from a depressed metabolic state. Several years ago, it was found that with hypothermia there is a progressive decline in blood glucose to an extreme hypoglycemic level. Summarized data is presented in Figure 6. Also a hypoglycemic hamster does not arouse from hypothermia and prior to death there is clear evidence of respiratory distress. This is seen in the form of interrupted regular respiration, with repeated gasping and periods of apnea, and a marked interruption in the pattern of oxygen uptake (Anderson, Volkert, and Musacchia 1971).

Interpretation of these findings indicated that the lack or reduction of available glucose (a necessary metabolite for CNS function), resulted in a marked failure of the respiratory center and this predictably led to death. It was postulated that if glucose was essential to maintenance and survival, particularly since it was utilized during hypothermia, then replenishment of blood glucose should improve survival (Resch and Musacchia 1976). This hypothesis was readily confirmed by a series of experiments in which glucose infusion, via carotid cannula resulted in greatly improved survival in hypothermic hamsters (Figure 7). Recently questioned was the possibility that some hormonal regulation could play a role in carbohydrate maintenance. Since the earlier work (Resch and Musacchia 1976; Lyman and Leduc 1953; Burlington and Klain 1967; Burlington 1972) evidenced that carbohydrate metabolism is regulated in hibernation, a gluconeogenic hormone was utilized in order to determine if a hypothermic animal has the capacity to effectively sustain carbohydrate metabolism. With daily (5 days) administration of cortisone, 5 mg i.p. in 0.1 ml 0.9 percent saline per 100 – 110 gm hamsters, prior to induction of hypothermia, there is a marked increase in circulating glucose (Deavers and Musacchia 1977; Deavers, Musacchia, and Meininger 1977; and Musacchia and Deavers 1978). Those animals which are administered cortisone prior to induction of hypothermia and after twenty hours of hypothermia, are better able to survive at T_{re} 7°C; there was a fourfold improvement in survival (Figure 8).

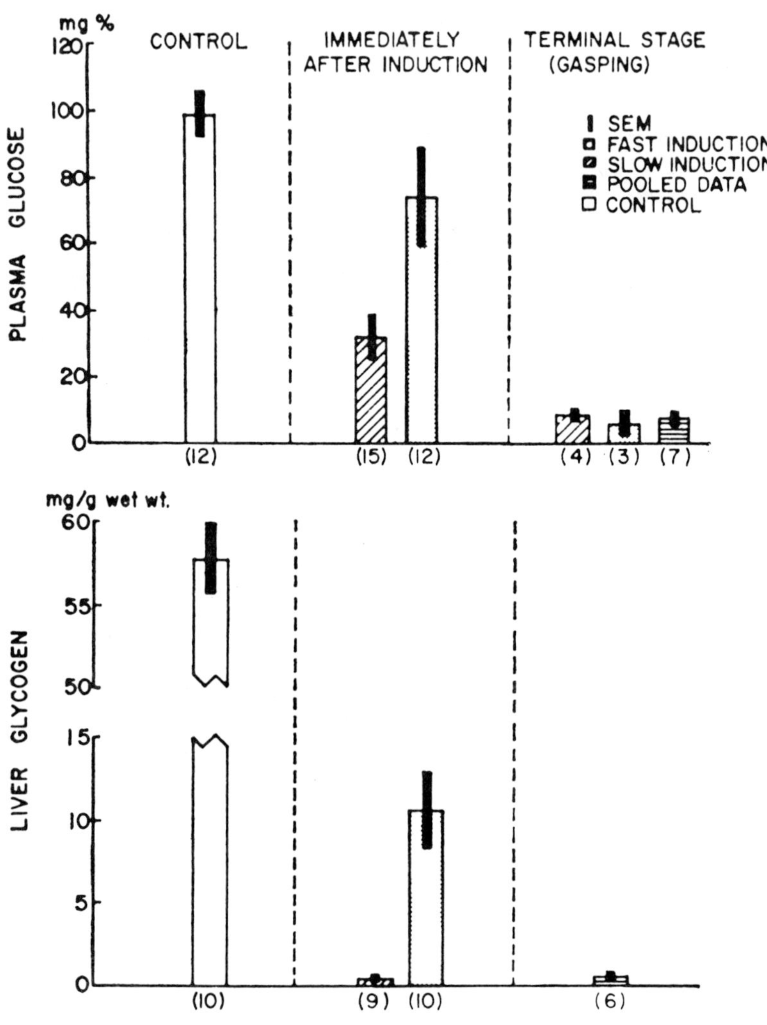

FIGURE 6. Plasma glucose and liver glycogen. A comparison of lowered plasma glucose and liver glycogen with fast and slow induction of hypothermia. At terminal stages of survival there is severe hypoglycemia. (Illustration taken from Chapter 8, "Regulation of Depressed Metabolism and Thermogenesis," L. Jansky & X.J. Musacchia, eds. 1976.)

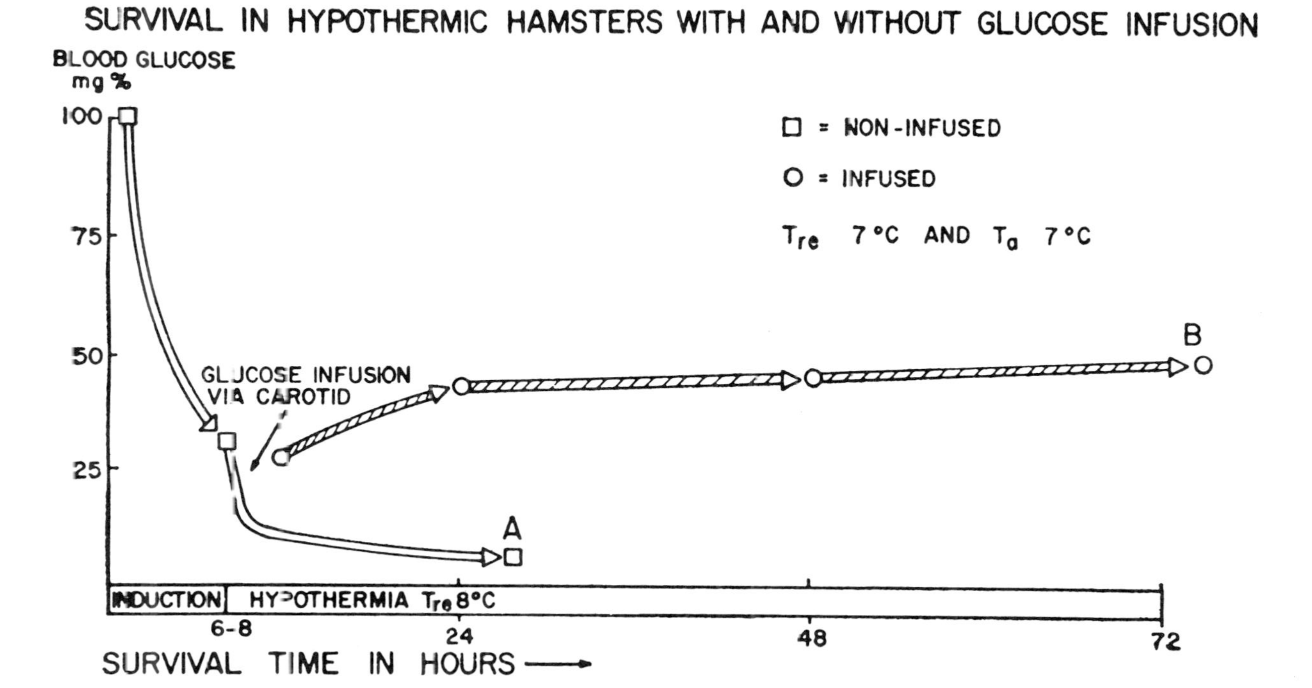

FIGURE 7. An illustration showing that with glucose infusion, survival of hypothermic hamsters is improved. (Illustration taken from Chapter 8, "Regulation of Depressed Metabolism and Thermogenesis," L. Jansky and X.J. Musacchia, eds., Charles C. Thomas, Publisher 1976.)

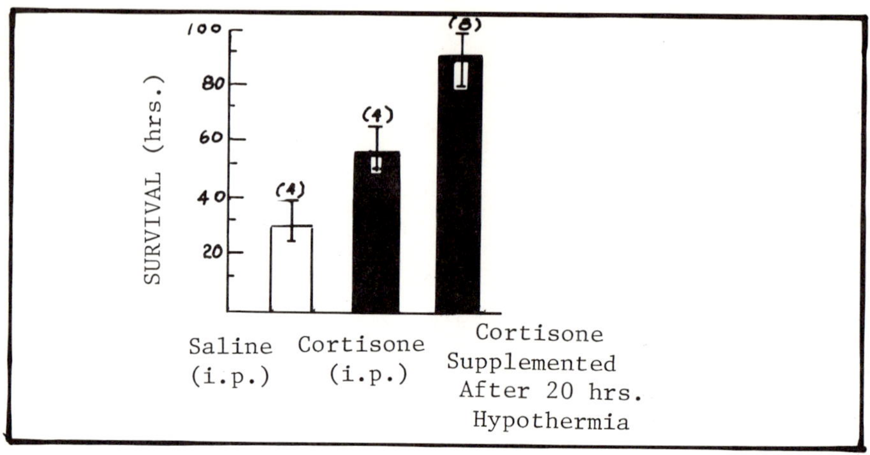

FIGURE 8. Shows the improved levels of survival in hypothermic hamsters treated with cortisone.

In addition, in cortisone treated hamsters about 50 percent aroused spontaneously from T_{re} 7°C. In these aroused animals the blood glucose levels approached normothermic, control values of about 100 mg percent (Figure 9). These findings provided initial evidence that cortisone treated hypothermic hamsters and hibernating hamsters were comparable in at least three features: maintained level of blood glucose (i.e. @ 100 mg%), a 3 - 4 day period of torpidity at T_{re}7°C, and spontaneous awakening in T_a 7°C. The pervasive effects of glucocorticoids, particularly with respect to enhancing thermogenesis, have recently been reviewed (Deavers and Musacchia 1979). The combined effects of glucocorticoids in potentiating catecholamine induced lipolysis and vasopressore effects, as well as permissive and direct effects in promoting gluconeogenesis, may permit the level of thermogenesis necessary for arousal to normothermia. It can be reasoned that with the current experimental approach, a hypothermia model can be developed (and in fact was) which can be used for experimentation leading to a better understanding of hibernation.

The adrenal cortex is playing a vital role in the maintenance of gluconeogenesis and possibly in the arousal from torpidity during hypothermia. In this respect, there is a relationship to events known in hibernation. Some time ago Vidovic and Popovic (1970) showed that the adrenal cortex is necessary for survival during hibernation. Normothermic, adrenalectomized ground squirrels do not enter hibernation, but if a fragment of the cortex is left

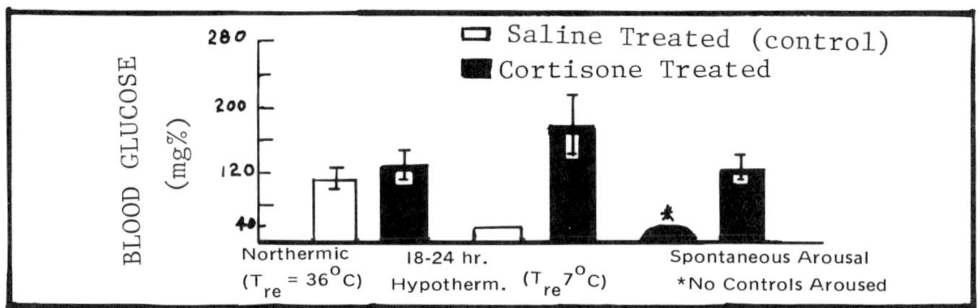

FIGURE 9. Shows the effect of cortisone treatment on blood glucose levels in hypothermic hamsters after 18-20 hours, and after spontaneous arousal.

in situ, squirrels will hibernate and survive. In addition to enhancement of gluconeogenesis, glucocorticoids may be implicated in thermogenesis. Consider, for example, treatment with glucocorticoids resulted in arousal in about half of the hypothermic hamsters (Deavers and Musacchia 1977; Deavers, Musacchia, and Mcininger 1977), whereas, with glucose infusion (Resch and Musacchia 1976), there was improved survival but no arousal. As was mentioned earlier, it is recognized that in hibernation carbohydrates may play a role in arousal. Results from other laboratories also implicate glucocorticoids in enhancing thermogenesis. Experiments by Tyslowitz and Astwood (1942), and Maickel et al. (1967a, 1967b), showed that cold resistance is reduced in adrenocortico-deficient rats, and glucocorticoid administration restores cold resistance. In it known that glucocorticoids are essential in potentiating catecholamine induced lipolysis (Maickel, et al. 1967a,b) and they also play a role in potentiating vasopressor effects (Lepri and Christiani 1964). Since non-shivering thermogenesis (NST) utilizing mainly kipid fuels is most important during the early stages of arousal (Jansky 1973), and since selective vasopressor control is equally important for rapid arousal from hibernation (Lyman 1948), glucocorticoids may be essential in initiating the arousal process through interactions with other hormones.

Lastly, the hypotheses that glucocorticoids are not produced in hypothermia resulting in a lack of gluconeogenesis, is consistent with the data of Egdahl et al. (1955). They found that in hypothermic dogs secretion of both adrenocorticotropin and cortisol are greatly reduced and rewarming resulted in a return to normal plasma levels of these hormones.

PROSPECTS FOR A MODEL FOR NATURAL HIBERNATION

The results bring us closer to a model for natural
hibernation using experimental hypothermia. It appears
that glucocorticoid treatment resulted in: 1) gluconeogenesis, and 2) thermogenic enhancement capacity, since about
50 percent of the treated hamsters were capable of arousal
to normothermia from T_{re} 7-8°C at T_{a} 7°C. In this regard
these hypothermic hamsters are similar to naturally hibernating hamsters which arouse periodically; (in contrast,
they differ from other animals experimentally induced into
"deep" hypothermia.) Also, cortisone treated hamsters which
did not arouse, survived hypothermia for an average of four
days which is comparable to periods of hibernation between
periodic awakenings. Survival in cortisone treated hypothermic hamsters is about four times the duration of survival in control (i.e., non-treated hamsters). We propose
that, during "deep" hypothermia in the hamster, glucocorticoids are not produced; or if they are produced, they are
ineffective thus resulting in lack of gluconeogenesis and a
depletion of blood and liver carbohydrates occurs. Glucocorticoid treatment apparently restores gluconeogenesis and
elevates carbohydrate reserves during hypothermia.
In brief, there is a role for glucocorticoids in
hibernation, and animal models may be utilized in broaden-
ing our understanding of thermogenic mechanisms. Also, the
helium-cold hypothermic hamster provides a model which can
be used to extend present knowledge of the role of carbohydrate metabolism in depressed metabolic states. Animal
model systems for studies of mechanisms of cold tolerance
can be extended to utilize more species. Furthermore,
field oriented laboratories which have access to "wild"
subjects and in which natural habitat conditions are closely paralleled, should be able to enhance experimental
proposals utilizing animal model systems.

COMBINED USE OF HYPOTHERMIC MODELS AND NATURAL HIBERNATORS

The Arctic ground squirrel is readily available for
detailed studies of hormonal regulation and carbohydrate
metabolism. Their distribution on the arctic slope and
their behavior is known well enough that pregnant females
could be captured and transported to the Naval Arctic Research Laboratory. Seasonal and winter hibernating studies

could be expanded to include comparisons with the hypothermic
hamster model, since it is also easily maintained in the
laboratory. In brief, questions concerning the role of
carbohydrate metabolism in hibernation can be designed,
hopefully for more efficient investigation around a laboratory model, in close proximity to field and natural conditions.

ACKNOWLEDGMENT

We wish to acknowledge Ms. Janet Burnett and Mr.
Richard Riek for their technical contributions to much of
the research in our laboratory.

REFERENCES

Anderson, C.L., W. Volkert, and X.J. Musacchia. 1971. O_2
    consumption, electrocardiogram and spontaneous re-
    spiration of hypothermic hamsters. *Am. J. Physiol.*
    221:1774-1778.
Barr, R.E., and X.J. Musacchia. 1968. Breeding among cap-
    tive *Citellus tridecemlineatus.*   *J. Mammal.*
    49(1):343-344.
Burlington, R.F. 1972. Recent advances in intermediary
    metabolism of hibernating mammals. In South, et al.,
    Hibernation and hypothermia, perspectives and challen-
    ges. Elsevier, New York.
Burlington, R.F., and G.J. Klain. 1967. Gluconeogenesis
    during hibernation and arousal from hibernation. *Comp.*
    Biochem. Physiol. 22:701-708.
Chatfield, P.O., and C.P. Lyman. 1950. Circulatory changes
    during process of arousal in the hibernating hamster.
    Am. J. Physiol. 163:566-574.
Deavers, D.R., and J.W. Hudson. 1977. Effect of cold-ex-
    posure on water requirements of three species of small
    mammals. *J. Appl. Physiol: Respir. Environ. Exercise*
    Physiol. 43:121-125.
Deavers, D.R. 1976. Water turnover, total body water, and
    metabolism of hibernating and cold-acclimated normother-
    mic 13-lined ground squirrels. *Fed. Proc. 35(3)*:639.
Deavers, D.R., and X.J. Musacchia. 1977. Cortisone and
    gluconeogenic function in hypothermia and hibernation.
    Fed. Proc. 36(3):419.

Deavers, D.R., and X.J. Musacchia. 1979. Function of gluco-
    corticoids in thermogenesis. In symposium of Metabolic
    Aspects of Thermogenesis, Nural and Hormonal Control.
    Feder. Proc. (in press).

Deavers, D.R., X.J. Musacchia, and G.A. Meininger. 1977.
    Glucocorticoids, carbohydrate metabolism, and thermo-
    genesis in hibernating and hypothermic hamsters. In
    J.W. Hudson, and L.C.H. Wang, eds. Abstracts of the
    Fifth International Hibernation Symposium: Strategies
    in cold: Natural torpidity and thermogenesis. Jasper,
    Alberta, Canada.

Devaney, M.J., et al. 1976. Telemetry of body temperature
    and heart rate of hibernating and normothermic ground
    squirrels. In T.B. Fryer, and H.A. Miller, eds. Bio-
    telemetry III. Academic Press, New York.

Egdahl, R.H., D.H. Nelson, and D.M. Hume. 1955. Adrenal
    cortical function in hypothermia. *Surg. Gynecol.
    Obstet. 100*:715-720.

Fisher, B.A., and X.J. Musacchia. 1968. Responses of ham-
    sters to He-O$_2$ at low and high temperatures: induction
    of hypothermia. *Am. J. Physiol. 215*:1130-1136.

Galster, W., and P.R. Morrison. 1970. Cyclic changes in
    carbohydrate concentrations during hibernation in the
    Arctic ground squirrel. *Am. J. Physiol. 218*:1228-1232.

______. 1975. Gluconeogenesis in Arctic ground squirrels be-
    tween periods of hibernation. *Am. J. Physiol.
    228*:325-330.

Hudson, J.W., and D.R. Deavers. 1973. Thermoregulation at
    high ambient temperatures of six species of ground
    squirrels *(Spermophilus spp.)* from different habitats.
    Physiol. Zool. 46:95-109.

Jansky, L. 1973. Non-shivering thermogenesis and its ther-
    moregulatory significance. *Biol. Rev. 48*:85-132.

Klain, G.J., and B.K. Whitten. 1968. Carbon dioxide fixa-
    tion during hibernation and arousal from hibernation.
    Comp. Biochem. Physiol. 25:363-366.

Lepri, G., and R. Christiani. 1964. Ability of certain ad-
    renocortico hormones to potentiate the vasoconstrictive
    action of noradrenolin on the conjuctival vessels in
    the rabbit and in man. *Brit. J. Opthal. 48*:205-210.

Lyman, C.P. 1948. The oxygen consumption and temperature
    regulation of hibernating hamsters. *J. Exptl. Zool.
    109*:55-78.

Lyman, C.P., and P.O. Chatfield. 1955. Physiology of hiber-
    nation in mammals. *Physiol. Rev. 35*:403-425.

Lyman, C.P., and A.B. Hastings. 1951. Total CO_2, plasma pH and PCO_2 of hamsters and ground squirrels during hibernation. *Am. J. Physiol. 167*:633-637.

Lyman, C.P., and E.H. Leduc. 1953. Changes in blood sugar and tissue glycogen in the hamster during arousal from hibernation. *J. Cell. Comp. Physiol. 41*:471-491.

Maickel, R.P., et al. 1967a. The sympathetic nervous system as a homeostatic mechanism. I. Effect of adrenocortical hormones on body temperature maintenance of cold-exposed adrenalectomized rats. *J. Pharmacol. Exptl. Therap. 157*:103-110.

Maickel, R.P., et al. 1967b. The sympathetic nervous system as a homeostatic mechanism. II. Effect of adrenocortical hormones on body temperature maintenance of cold-exposed adrenalectomized rats. *J. Pharmac. Exptl. Therap. 157*(1):111-116.

Moy, R.M. 1971. Renal function in the hibernating ground squirrel *(Spermophilus columbianus)*. *Am. J. Physiol. 220*:747-753.

Musacchia, X.J. 1958. Observations on a new born litter of *C. undulatus*. *J. Mammal. 39*:594-595.

______. 1972. The role of heat and cold acclimation in helium-cold hypothermia in the hamster, *Mesocricetus auratus*. *Am. J. Physiol. 222*:495-498.

______. 1976. Helium-cold hypothermia, an approach to depressed metabolism and thermoregulation. In L. Jansky an X.J. Musacchia, eds. Regulation of depressed metabolism and thermogenesis. C.C. Thomas, Springfield, IL.

Musacchia, X.J., and D.R. Deavers. 1978. Effectors of thermogenesis, glucocorticoids and carbohydrate metabolism in hypothermic and hibernating hamsters. *Experientia Suppl. 32*:247-259.

Musacchia, X.J., and C.G. Wilber. 1952. Studies on the biochemistry of the Arctic ground squirrel. *J. Mammal. 33(3)*:356-362.

Resch, G.E., and X.J. Musacchia. 1976. A role for glucose in hypothermic hamsters. *Am. J. Physiol. 231*:1729-1734.

Richmond, C.R., W.H. Langham, and T.T. Trujillo. 1962. Comparative metabolism of tritiated water by mammals. *J. Cell. Comp. Physiol. 59*:45-53.

Tashima, L.S., S.J. Adelstein, and C.P. Lyman. 1970. Radioglucose utilization by active, hibernating, and arousing ground squirrels. *Am. J. Physiol. 218*:303-309.

Tempel, G.E., and X.J. Musacchia. 1975. Renal function in the hibernating and hypothermic hamster, *Mesocricetus auratus*.   Am. J. Physiol. *228(2)*:602–607.

Tempel, G.E., X.J. Musacchia, and S.B. Jones. 1977. Mechanisms responsible for decreased glomerular filtration in hibernation and hypothermia. *J. Appl. Physiol: Respir. Environ. Exercise Physiol. 42(3)*:420–425.

Twente, J.W., and J.A. Twente. 1967. Concentrations of D-glucose in the blood of *Citellus lateralis* after known intervals of hibernating periods. *J. Mammal. 48*:381–386.

Tyslowitz, R., and E.B. Astwood. 1942. Pituitary and adrenal cortex in resistance to cold. *Am. J. Physiol. 136*:22–31.

Vidovic, V., and V. Popovic. 1970. Studies on the adrenal and thyroid glands of the ground squirrel during hibernation. *J. Endocrinol. 11*:125–133.

Volkert, W.A., and X.J. Musacchia. 1970. Blood gases in hamsters during hypothermia by exposure to HeO_2 mixture and cold. *Am. J. Physiol. 219*:919–922.

______. 1976. Renal function in hypothermic hamsters. In L. Jansky, and X.J. Musacchia, eds. Regulation of depressed metabolism and thermogenesis. C.C. Thomas, Springfield, IL.

Wang, L. 1972. Circadian body temperature of Richardson's ground squirrel under field and laboratory conditions: A comparative radiotelemetric study. *Comp. Biochem. Physiol 43A*:503–510.

______. 1973. Radiotelemetric study of hibernation under natural and laboratory conditions. *Am. J. Physiol. 224(3)*:673–677.

DISCUSSION I.
THE ROLE OF NEURAL INPUTS
AND THEIR PROCESSING IN COLD ADAPTATION

Keith E. Cooper

Division of Medical Physiology
Faculty of Medicine
University of Calgary
Calgary, Alterta, Canada

A plea is made for a detailed study of the transduction of cold sensation in the skin, and its central nervous system processing in arctic and subarctic mammals and during cold acclimation. In addition, the thermal properties of skin, and the relationship of skin blood flow to skin temperature need examining. The sensory aspects of cold acclimation must be determinants of the ultimate biochemical changes and this too, needs study.

While there is a great deal of evidence concerning the metabolic or biochemical responses of animals in the cold, which make them more able to withstand cold climatic conditions, one must remember that these mechanisms are the end result of a chain of events. Perception of cold is the beginning, processing of the perceived information follows, and this leads to a sequence of events which culminate in the biochemical changes. One's own recent experience of travelling to Britain in the last winter and finding the interior of houses uncomfortably cold upon arrival, but within three weeks being less cognizant of the cold surroundings; and, upon returning to Canada finding that the interior of one's own house was too hot, to which one again adjusted in a few weeks, illustrates the problems before us. The questions which have to be asked are: What is responsible for the gradual alteration in the perception of the altered thermal environment and one's response to it? We know from the work of Hensel (1970) and Iggo (1964), that there are in the skin, temperature-sensitive neuronal transducers, the firing rate of which in relation to their temperature, is described by an approximately bell-shaped curve. Some of the "receptors" respond with increasing firing frequency as the skin temperature rises, up to a maximum at some specific skin temperature – the warm receptors – and others whose firing frequency increases as the temperature is reduced – the cold receptors. One could postulate that there is a change in the behavior of these receptors or in the position of the peak of the bell-shaped curve during prolonged exposure to the cold which might account for the apparently altered sensation. Such measurements as have been made so far on only cold receptors in the nose of the cat (Hensel 1973), would suggest that prolonged exposure to cold does not change the firing patterns of these peripheral transducers. It is, however, possible that some change in the quality of the skin, i.e., thickness of the stratum corneum, or a change in the moisture content of the skin, might alter the superficial thermal conductivity of the skin, and therefore, the actual temperature of the receptor or transducer, following a period of prolonged exposure to a cold climate.

It is possible that some long-term change in peripheral blood flow could also alter the temperature at the skin neuronal transducer site. It has been shown (Cooper et al. 1949; Shepherd 1963) that a plot of skin blood flow against skin temperature reveals a gradual increase in blood flow as the skin temperature is raised up to about 30°C, and then there is an abrupt inflexion of the curve in which, for a further small increment in skin temperature, there is a very large rise in skin blood flow. No one to this author's knowledge has repeated these whole curves of skin blood flow against local temperature during the process of cold acclimation in the laboratory to see whether they are shifted so as to maintain a higher skin blood flow, and therefore a higher deep skin temperature, for a given surface skin temperature. Again, this might modify the apparent behavior of the receptor.

It is possible that some slight alteration might take place in either the hypothalamic blood flow or the hypothalamic heat production, which would lead to a slightly raised hypothalamic temperature during cold acclimation, and this in turn would result in a modification of the behavioral response to cold or to a change in perception of the information arriving from the periphery. We know from the work of Hellon (1973) that there is, between the sub-

stantia gelatinosa input of the thermal information to the
spinal cord and the ventromedial thalamus, some gating of
information related both to the rapid transient response to
sudden cold exposure, and more prolonged changes in thermal
topography in the skin. It would seem that this would be a
fruitful area to investigate to see whether the modifications of the information arriving in from the peripheral
nerves between the substantia gelatinosa and the thalamus,
could be altered by prolonged cold exposure, thereby giving
a modification to the second or processing step of the
thermal information.

Some very interesting work by Kawakami et al. (1972)
has shown that if spectral analysis is done on the frequency components of the electroencephalogram derived from
the amygdala or hippocampus in the animal, there are
certain components which alter during repeated exposure to
heat and others which alter with repeated exposure to cold.
While it is not possible at present fully to interpret
these changes in a physiological measure of activity in the
limbic system during cold/heat acclimation, it would seem
that further study of the input from the limbic system into
the sensory processing of information derived concerning
the environment would be of very great value.

There is little evidence in man concerning the metabolic responses to chronic cold exposure if the metabolic
pathways are disassociated from the thermoregulatory ap-
paratus within the hypothalamus. One patient studied by
Hockaday et al. (1962) had no thermoregulatory responses
but suffered periods of prolonged hypothermia which were at
times severe. These authors were able to make measurements
of whole body oxygen consumption over a range of temperatures during various phases of this disorder, and the
average Q_{10} for the oxygen consumption was approximately
2.3 to 2.4, which does not differ widely from any other
isolated tissue and certainly shows no dramatic deviation
from what would be expected, even though this woman had
been exposed to quite severe cold internal and external
environmental stresses.

There are some other objective measures of response to
acute cold exposure. For example, Cooper and Martin
(1976) demonstrated on a small group of subjects that the
hyperventilation response to sudden immersion in cold water
diminishes if the subject is given a 10-minute immersion
each day over a period of six to eight days. This at-
tenuation of a well-established response to a cold stimu-

lus, which is dependent on the rate of change of deep skin
temperature (Martin and Cooper 1978) probably falls under
the category of habituation, which again is a change in the
central nervous system processing of the strong peripheral
stimulus.

It would therefore seem very reasonable to suggest as
an adjunct to all of the biochemical and other research on
cold adaptation, to look very carefully in a variety of
species for inherited and acquired alteration in the
information picked up by sensory receptors in the skin, and
to make a much more detailed study than is at present
available of the central processing of this information in
those animals which live in the cold and in animals which
become cold-acclimatized. The application of modern sophis-
ticated neurophysiological techniques to the solution of
these problems would seem a profitable venture and one
which is scarcely broached at present, but which one hopes,
as the result of this literary effort will gain popularity
in the near future.

REFERENCES

Cooper, K.E., et al. 1949. A comparison of methods for
     guaging the blood flow through the hand. *Clin. Sci.*
     8:217-234.
Cooper, K.E., et al. 1976. Respiratory and other responses
     in subjects immersed in cold water. *J. Appl. Physiol.*
     40:903-910.
Hellon, R.F. 1973. Central nervous responses to cooling and
     warming of the rat scrotal skin. *Arch. Sci. Physiol.*
     27:371-383.
Hensel, H. 1970. Temperature receptors in the skin. Pages
     442-453 in J.D. Hardy, P. Gagge, and J.A. Stolwijk,
     eds. Physiological and behavioral temperature regu-
     lation. C.C. Thomas, Springfield, Ill.
     ______. 1973. Electrophysiology of thermoreceptors in homeo-
     therms. Page 605 in H. Precht, et al. eds., Tempera-
     ture and life. Springer-Verlag, New York.
Hockaday, J.D.R. et al. 1962. Temperature regulation in
     chronic hypothermia. *Lancet 11*:428-432.
Iggo, A. 1964. Temperature discrimination in the skin.
     Nature 204:481-483.

Kawakami, M. et al. 1972. Participation of the limbic-hypothalamic structures in cold adaptation. In S. Itoh, K. Ogata, and H. Yoshimura, eds. Advances in climatic physiology. Springer-Verlag, New York.

Martin, S. and K.E. Cooper. Unpublished. The effect of skin temperature on hyperventilation during cold water immersion.

Shepherd, J.T. 1963. Physiology of the circulation in human limbs in health and disease. Page 98 in Health and Disease. W.B. Saunders, Philadelphia, PA.

DISCUSSION II.
MECHANISMS OF THERMAL TOLERANCE
IN NON-HIBERNATING ENDOTHERMS

Mohamed K. Yousef

Department of Biological Sciences
Desert Biology Research Center
University of Nevada
Las Vegas, Nevada

Mechanisms of thermal tolerance to cold have been investigated primarily in laboratory animals. The question now, is whether or not these mechanisms are "universal," thus extrapolation of these findings to other animals can be made without debate. Limited studies on acclimation and/or acclimatization of wild small animals suggested that extrapolation of available data must be made with caution. Physiological strategies and dicotomy of thermal tolerance to cold can be better understood with additional studies using wild animals and simulating field conditions. In other words, continuous exposure (24 hr) to cold may result in physiological responses that are peculiar to this type of cold stress. Still open to investigation is the role of the kidney, the gastrointestinal tract, and the immune response in physiological adaptations to cold. Additionally, further work on physiological and biochemical responses of large arctic animals to heat is urgently needed, since these animals may suffer from overheating during migration in winter or during the summer season.

Fundamental knowledge of physiological and biochemical adjustments of endotherms to cold and to arctic climates has been reviewed many times (Heroux 1961; Folk 1976). Most investigations have dealt primarily with the understanding of energy exchange and related mechanisms involved in short- and long-term exposure to cold. The objective of

this discussion is not to review the current literature on
mechanisms of thermal tolerance of nonhibernators, but to
focus attention on some discrepencies and inadequacies of
available data and elude to their probable cause(s).

STUDIES USING WILD AND LABORATORY ANIMALS

     Mechanisms of thermal tolerance to cold have been
studied primarily using laboratory animals, specifically
white rats and mice. Many of the reported physiological
responses have been confirmed by various laboratories and
one is tempted to accept them as universal responses or
indices of cold acclimation. However, in line of the few
studies using wild animals, a shadow of doubt is cast over
the universality of these responses. A summary of data on
physiological responses of wild and laboratory rodents to
cold is shown in Table 1. Hypertrophy of organs and loss
of body weight associated with cold acclimation were repeatedly reported and confirmed to the point that it had
lead some investigators (Hsieh 1963) to suggest that the
increased heat production during cold acclimation is a
result of larger livers and smaller body weight. Such an
hypothesis cannot be valid for wild animals since their
body weight increases and they maintain constant organ
weight and size during cold acclimation (Heroux 1961;
Yousef and Dill 1970). Another discrepency between the
responses of laboratory and wild rodents is the changes in
body composition during acclimation to cold. Laboratory
rodents decrease their body fat and protein content (Barnett and Mount 1967). The loss in body fat had led to the
suggestion that fat is the major tissue for energy reserve
utilized during exposure to cold (Hart and Heroux 1956).
However, wild rodents increased their body fat content
(Heroux 1961; Yousef and Dill 1970). The increased fat
depots may represent an adaptive mechanism since a layer of
body fat increases insulation thus results in heat conservation. Also, the stored fat can provide a source of
energy whenever food resources become minimal.
     Similarities in physiological responses of wild and
laboratory rodents are also shown in Table 1. All studied
small mammals response to cold by increasing their heat
production to prevent hypothermia. However, it seems that
the pathways of regulatory thermogenesis are different.
Additional studies utilizing wild animals are needed to
fully understand the mechanisms of thermal tolerance to
cold.

TABLE I. Physiological Responses During Cold Acclimation in Wild and Laboratory Animals (rodents)

		Wild	Laboratory
1.	Body weight	— or ↑	↓
2.	Food intake	↑	↑
3.	Body composition		
	a. fat	↑	↓
	b. protein	—	↓
	c. protein degradation rate/day	—	↓
4.	Organ weights		
	a. liver	—	↑
	b. kidney	—	↑
	c. heart	—	↑
5.	Organ structure (histology)		
	a. liver	lesions	lesions
	b. kidney	lesions	lesions
	c. heart	lesions	lesions
6.	Blood		
	a. $Hc+$	↑	— or ↑
	b. Hb	↑	— or ↑
	c. Ps	↑	— or ↑
7.	Metabolism		
	(VO_2)	↑	↑

↑ = increase;   ↓ = decrease;   — = no change

LABORATORY AND FIELD STUDIES

Experimental design of laboratory studies on cold acclimation has been centered on continuous exposure of animals to a given environmental temperature for either a short or a long period of time. The question here is whether or not the physiological responses under these conditions are the same as those of cyclic or intermittant (field or seasonal) exposure to cold for only a period of time during the 24 hours, a situation which is similar to natural field conditions. There have been few studies dealing with the mechanisms of thermal tolerance of natural or seasonal changes in endotherms. Some of the basic physiological differences in responses to continuous (cold-chamber studies) and cyclic (outdoor or field studies) exposure to cold are summarized in Table 2. For example, the degenerative changes and frostbite seen in cold-acclimated rats are absent in rats kept outdoors in the winter (Heroux 1961). Continuous exposure of a wild rodent, the Merriam's kangaroo rat *(Dipodomys merriami)*, to cold caused various anatomical changes which may be considered pathological. Similar changes were absent in the winter-acclimatized animals (Babero, Yousef, and Wawerna 1971). The reported histo-pathological alterations in cold-chamber kangaroo rats did not appear to be associated with hypertrophy of organs, rather, they appeared to be related to the continuous exposure to cold. The degenerative changes shown in wild and laboratory rodents may offer a better explanation of the shortening life span of the white rats continuously exposed to cold as shown by Johnson et al. (1963). Winter acclimatized rodents exhibit a high degree of cold resistance and metabolic capacity associated with increased insulation. However, in cold-chamber rodents the increased heat production is associated with either decreased or no change in insulation. In large and small mammals, there is no significant change in thyroid activity during the winter season (Heroux 1960; Yousef and Luick 1971). Physiological strategies and dicotomy of cold acclimation and acclimatization can be better understood with additional studies utilizing simmulated field conditions.

TABLE 2. *Comparison Between Physiological Response During Cold Acclimation and Cold Acclimatization.*

		Acclimation (Cold-chamber)	Acclimatization (Winter season)
1.	Metabolism (VO_2)	↑	↑
2.	Organ structure (Liver, Kidney, Heart, Lung)	Lesions	No lesions
3.	Hormonal changes		
	a. Adrenal Cortex	↑ early period — later	↑
	b. Thyroid	↑	— or ↓
4.	Insulation	↓ or —	↑
5.	Internal organ size	↑	—
6.	Body growth	↓	↑ or —
7.	Body fat content	↓	↓

↑ = *increase;*   ↓ = *decrease;*   — = *no change*

STUDIES ON LARGE ARCTIC MAMMALS

Thermal Tolerance To Cold

Maintenance of warmth in large arctic mammals is accomplished by a combination of two mechanisms: effective insulation, and peripheral heterothermy (Irving, 1966). This means that peripheral tissues are functional at temperatures considerably cooler than the interior of their bodies. Over a considerable range of environmental temperatures, large arctic mammals maintain homeothermy with no greater expenditure of metabolic heat due to physical and physiological insulative adaptations. For example, the overall insulation of the white fox (*Alopex lagopus*) at $-50°F$ is about eight times as great as at $25°F$.

Thermal Tolerance to Heat

The efficiency of insulation for heat conservation poses a physiological problem during heat stress. Bartholomew and Wilke (1956) found that fur seal *(Callorhinus ursinus)* bulls may die from over heating when driven at 10°C temperature. It is obvious then that large arctic mammals may experience exposure to heat stress as a result of their insulation even at a moderate environmental temperature, hard exercise due to migration, or escaping a predator at moderate or even cold temperature, and/or by exposure to summer high temperature. However, little is known of their physiological responses to heat stress. Rosenmann and Morrison (1967) concluded that reindeer *(Rangifer tarandus)* have good capacity for heat resistance when water is available, but very poor resistance to water deprivation with or without heat stress. Recently, Yousef and Luick (1975), found that thermal polypnoea represents an adequate mechanism for evaporative cooling in reindeer. Indices of heat tolerance for reindeer were calculated from data on rectal temperature, and respiratory frequency and compared with non-arctic ungulates as shown in Table 3.

TABLE 3. Heat Tolerance of Domestic, Desert, and Arctic Ungulates

	Rhoad's Coefficient, % (a)	Benezra's Index (b)
Domestic		
(Ayrshire)	65	4 - 11
Desert		
(Waterbuck)	55	6
Arctic		
(Reindeer)	69	11

(a) *Rhoad's Coefficient, % = 100 - ((18 (Tre - 38.3))*
(b) *Benezra's Index = Tre/38.3 + R_f/23*
Tre = Rectal Temperature, C
R_f = Respiratory Frequency/minute

It appears from the limited data that reindeer are as heat tolerant as some domestic cattle and wild African ungulates.

To fully understand thermal tolerance of large or small arctic mammals, there is urgent need for studies on their physiological and biochemical responses to heat stress.

MECHANISMS TO BE EXPLORED

Some areas of physiological and biochemical mechanisms of thermal tolerance have been relatively investigated and scrutinized by various workers. However, many other areas still require intensive studies in laboratory and wild animals. A few examples of these areas are as follows.

Role of Kidney

In response to seasonal dietary and nutritional fluctuations, ingestion and excretion of animals acclimated and/or acclimatized to cold offer some challenging avenues of investigation. The significance of shift in diet quality and quantity imposes continuous stress on renal handling of various elements. Also, diuresis is one of the universal responses exhibited by mammals during exposure to cold. The mechanisms underlying disturbances of water and electrolyte metabolism as related to cold diuresis, are not well studied. Limited data are available on kidney clearance rate of various substances; kidney blood flow and filtration rate; changes in osmotic pressure of the various parts of the nephrones; and hormonal control of body fluids and mineral metabolism with specific reference to ADH and aldosterone. The need for these data should offer at least a starting point for thoughts on understanding the specific role of the kidney in adaptations to cold environments.

Role of Gastro-intestinal Tract

Although the energy and nutrient requirements as related to whole body metabolism have been given some attention, little is known about the function of the transport systems in the intestinal mucosa. Intestinal absorption techniques, *in vivo* and *in vitro* are available and few studies on active transport mechanisms as related to cold acclimation of laboratory rodents (rats and hamsters) have revealed interesting information. On the other hand, no data are available on the functional aspects of the intestinal layer of cells which participate in digestion and absorption of food in wild mammals.

Immune Responses

Mammals adapted to cold, display a wide range of physiological and biochemical changes. Also, these animals are insulted by a wide variety of bacterial and viral agents which challenge the host animal to display an alteration in its immune response. Little data are available on the quantitative changes in production of circulating antibodies, and to the degree of resistance to either bacterial or viral infection. Also, the degree of susceptibility of cold-adapted animals needs to be investigated. These data are necessary to understand the host-parasite relationship which is necessary for revealing the mechanisms of survival of arctic and desert species.

These are only three examples of areas which require intensive investigations. Data obtained from these studies will help to elucidate the mechanisms of thermal tolerance in mammals.

REFERENCES

Babero, B.B., M.K. Yousef, and J.C. Wawerna. 1971. Histopathological changes in cold-exposed kangaroo rats, *Dipodomys merriami*. *Comp. Biochem. Physiol.* *39A*:327-366.

Barnett, S.A., and L.E. Mount. 1967. Resistance to cold in mammals. Pages 411-477 in A.S. Rose, ed., Thermobiology. Academic Press, New York.

Bartholomew, G.A., and F. Wilke. 1956. Body temperature of fur seal. *J. Mammal. 37*:327-337.

Carlson, L.D. 1972. Some comments on low temperatures in the desert and at altitude. In M.K. Yousef, et al., eds., Physiological adaptations: desert and mountain. Academic Press, New York.

Chaffee, R.R.J., and J.C. Roberts. 1971. Temperature acclimation in birds and mammals. *Ann. Rev. Physiol. 33*:155-202.

Folk, G.E., Jr. 1969. Physiological research in northern
     Alaska. *Arctic* 22:315-326.

______. 1976. Effects of cold: mammals. Pages 305-309 in
     H.D. Johnson, ed. Progress in biometerology. Vol. I,
     Part I. Swets and Zeitlinger B.V., Amsterdam, The Ne-
     therlands.

Hart, J.S. 1963. Physiological responses to cold in non-
     hibernating homeotherms. Pages 373-406 in J.D. Hardy,
     ed., Temperature - its measurement and control in
     science and industry. Reinhold, New York.

Hart, J.S., and O. Heroux. 1956. Utilization of body re-
     serves during exposure of mice to low temperatures.
     Can. J. Biochem. Physiol. 34:414-421.

Heroux, O. 1960. Adjustments of the adrenal cortex and
     thyroid during cold acclimation. *Fed. Proc.*
     19:82-85.

______. 1961. Climatic and temperature -- induced changes in
     mammals. *Rev. Can. Biol.* 20:55-68.

Hsieh, A.C.L. 1963. The basal metabolic rate of cold-
     adapted rats. *J. Physiol.* 169:851-861.

Irving, L. 1964. Terrestrial animals in cold: birds and
     mammals. Pages 361-377 in D.B. Dill, et al. eds.,
     Adaptation to the environment. Handbook of physiology,
     section 4. *Am. Physiol. Soc.* Washington, D.C.

______. 1966. Adaptations to cold. *Sci. Am.* 214:94-101.

Jansky, L. 1976. Effects of cold: small mammals. Pages 239-
     258 in H.D. Johnson, ed. Progress in Biometerology.
     Vol. I, Part I. Swets and Zertlinger, V.B., Amsterdam.

Johnson, H.D., L.D. Kintner, and H.H. Kibler. 1963. Effects
     of 48°F (8.9°C) and 83°F (28.4°C) on longevity and
     pathology of male rats. *J. Gerontol.* 18:29-36.

Masoro, E.J. 1976. Cellular metabolism, enzymes and other
     cellular changes: effects of cold. Pages 19-26 in H.D.
     Johnson, ed., Progress in Biometerology. Vol. I, Part
     I. Swets and Zeitlinger, B.V. Amsterdam, The Nether-
     lands.

Morrison, P. 1964. Adaptation of small mammals to the
     arctic. *Fed. Proc.* 23:1202-1206.

Rosenmann, M., and P. Morrison. 1967. Some effects of water
     deprivation in reindeer. *Physiol. Zool.* 40:134-142.

Smith, R.E., and D.J. Hoijer. 1962. Metabolism and cellular
     function in cold acclimation. *Physiol. Rev.* 42:60-
     142.

Yousef, M.K., and D.B. Dill. 1970. Physiological adjust-
     ments to low temperature in the kangaroo rat, *Di-
     podomys morriami*. *Physiol. Zool.* 43:132-138.

Yousef, M.K., and J.R. Luick. 1971. Estimation of thyroxine
    secretion rate in reindeer, *Rangifer tarandus*: effects
    of sex, age, and season. *Comp. Biochem. Physiol.*
    40A:789-795.
______. 1975. Responses of reindeer, *Rangifer tarandus*, to
    heat stress. Pages 360-367 in Luick, et al. eds.,
    Proceedings of the First International Reindeer and
    Caribou Symposium. Biol. Papers of Univ. Alaska.
    Fairbanks, Alaska.

IV. COLD-INDUCED ENZYMATIC ADJUSTMENTS IN ECTOTHERMS AND HOMEOTHERMS

B. A. Horwitz
D. R. Hettinger

Department of Animal Physiology
University of California
Davis, California

Several enzymatic mechanisms are utilized by animals living in cold environments. In ectotherms, these mechanisms may, to varying degrees, counteract the depressant effects of low thermal energy on reaction rates and/or result in altered flows through metabolic pathways. Some of the mechanisms seen in ectotherms are observable in heterothermic tissues of homeotherms. However, in most homeothermic tissues, the reported enzymatic adjustments seem to be related to the animal's need for increased rates of heat production rather than a need to operate at low intracellular temperatures. Examples of these various types of enzymatic adjustments are considered in this chapter.

INTRODUCTION

The phenomenon of homeothermy is traditionally associated with organisms that are capable of maintaining their deep body temperature relatively constant despite widely varying ambient temperatures. Exposure of homeotherms to temperatures below their thermal neutral zone evokes increases in heat production that are essential in enabling the animals to sustain their elevated core temperatures. (Among the exceptions to this generalization are the mammals and birds that enter hibernation.) Creatures that rely primarily on the internal generation of heat (via cellular metabolic activity) for body temperature maintenance are often referred to as endotherms, a term that

encompasses the homeothermic birds and mammals as well as
various nonhomeothermic species such as the tuna and the
sphinx moth. In contrast to endotherms are those animals
that derive their body heat primarily from the environment.
Such organisms, called ectotherms, have core temperatures
that closely follow the temperature of their surroundings
and that will thus decrease as ambient temperature falls.
In view of the differing responses in body temperature seen
in cold-exposed ectotherms and homeotherms, the enzymatic
mechanisms associated with each will be considered separately. It is the intent of this chapter to illustrate, by
example, the types of enzymatic adjustments utilized by
ectotherms and homeotherms (arctic and nonarctic) in their
responses to cold rather than to provide a comprehensive
discussion of all the organisms and systems that have been
examined. (For a more extensive treatment of work done to
1973, the reader is referred to the excellent review by
Hazel and Prosser (1974) on temperature compensation in
ectotherms.)

COLD-INDUCED RESPONSES IN ECTOTHERMS

Cold-exposure of ectotherms might be expected to be
accompanied by decreasing metabolic rates simply as a
consequence of the physical effects of temperature on
enzymes and enzymatic reactions. That is, as body temperature falls, the average kinetic energy of the reacting
molecules will decrease and with it, the probability of the
chemical reaction to proceed. In addition, much of the
higher order structure of enzymes depends on temperature-
sensitive interactions between the solvent water and non-
polar amino acid residues in the enzyme molecules. Such
interactions, termed hydrophobic, diminish with falling
temperature, thereby increasing the likelihood of altered
enzyme conformation and altered catalytic activity. Therefore, one might predict that in ectotherms, these temperature-induced effects on the molecular components of enzyme
reactions would be reflected in reaction rates that vary
widely with ambient temperature and are minimal during cold
exposure.
In reality, however, a number of ectotherms are able
to hold their metabolic rates relatively constant over a
wide range of temperatures (Bullock 1955; Hazel and Prosser
1974; Hochachka and Somero 1973; Prosser 1973). Moreover,
ectotherms living at low temperatures often exhibit higher

metabolic rates than would be expected based upon levels of
thermal energy in the system. For example, the rates of
oxygen consumption of fish acclimated to summer temperatures tend to be less than those of winter or arctic fish
(Hochachka and Somero 1973). Thus, operating within these
cold-exposed animals are mechanisms capable of compensating
for at least some of the physical effects of low temperature.

Among the compensatory mechanisms available to ectotherms facing low or fluctuating ambient temperatures are
those that directly alter enzyme activity. Such alterations may reflect: (a) changes in the amounts of existing
enzyme forms; (b) utilization of variant forms (e.g.,
isozymes) of a given enzyme; and/or (c) modification of the
enzymes' microenvironments (e.g., via changes in levels of
cofactors, modulators, pH, or lipid moieties associated
with the protein portions of the enzyme) (Somero 1972).
These mechanisms and their temporal characteristics are
discussed below.

Altered Levels of Existing Enzyme Forms

Increasing the number of functioning enzyme molecules
would, at first glance, seem to be a reasonable mechanism
to alleviate the depressant effects of low temperature.
Consistent with such a possibility are the reports that
increased rates of protein synthesis do accompany the cold
acclimation of various ectotherms. For example, the rates
of incorporation of radioactive amino acids into the liver
proteins of toadfish (Haschemeyer 1968a), rainbow trout
(Dean and Berlin 1969), and goldfish (Das and Prosser 1967)
are elevated after cold acclimation. (The activity of the
liver enzyme catalyzing the addition of amino acids to the
growing peptide chain (aminoacyltransferase I) also appears
to be increased in the cold-acclimated toadfish (Haschemeyer 1968b; Plant et al. 1977) and may account for the
decreased time necessary to assemble the polypeptide
chains in these cells.) In addition, there are a number of
enzymes (including many associated with pathways mediating
substrate oxidation and ATP synthesis) that do exhibit
enhanced activities in preparations derived from ectotherms
living in the cold (Hazel and Prosser 1974). Presumably
these measured enzyme activities represent rates obtained
in the presence of saturating amounts of substrate and as
such, approximate the maximum velocity (V_{max}) of the reaction. Therefore, the increased reaction rates (reflecting

elevated V_{max} values) could be interpreted as indicating
the presence of more enzyme molecules. However, the biological preparations on which these enzyme activities have
been measured have generally not been sufficiently analyzed
to preclude the possibility that the elevated activities
actually reflect the presence of altered enzyme forms or
cytoplasmic modulating factors rather than simply more
enzyme molecules. Thus enhanced reaction rates, in themselves, do not necessarily indicate increased enzyme levels.

In a few systems, however, quantitative changes in
enzyme levels have been demonstrated. For example, Wilson
et al. (1973) has reported a 66 percent increase in the
cytochrome oxidase content of a skeletal muscle from cold-
acclimated goldfish ($5^{o}C$) relative to those maintained at
$25^{o}C$. Similarly, the cytochrome c content in the skeletal
muscle of the green sunfish increases with decreasing
acclimation temperature, a change that appears to reflect
a greater temperature sensitivity of the system responsible
for degrading than for synthesizing cytochrome c (Sidell
1977). That is, in these studies cold exposure was accompanied by a greater decrease in cytochrome c degradation
than in cytochrome c synthesis, leading to a net increase
in the amount of the protein present. This relationship was reversible and as the temperature rose, the rate
of degradation increased more rapidly than did the rate of
synthesis, the net effect being a decrease in the amount of
cytochrome c present (Sidell 1977).

Consistent with changes in rates of protein synthesis
are the temperature-induced ultrastructural alterations
reported to occur in the nuclear envelope of Euglena cells
(Lott et al. 1977). The cells, grown at $30^{o}C$, had a significantly smaller nuclear volume and nuclear surface area
as well as fewer nuclear pores per unit of surface area
than did Euglena grown at $15^{o}C$. Since nuclear pores serve
as a major site of nucleocytoplasmic exchange of macromole-
cules, the greater number of pores present at colder temperatures may enhance the exchange rate of material between
the nucleus and cytosol. Such an increase may be important
in providing material support for the protoplasmic growth
favored at low growth temperatures. This possibility--
namely, the occurrence of an increased intercompartmental
flux and its correlation with temperature-induced changes
in protein synthesis--needs further evaluation in Euglena
as well as in other systems.

Nonetheless, it appears that one mechanism whereby
compensatory changes could occur during ambient temperature
changes operates via alterations in enzyme levels. However, direct evidence bearing on such changes with respect

to specific enzymes is still sparse. Moreover, the effectiveness of this means of temperature-induced compensation is limited by various physical and temporal factors. For example, restrictions exist on the amount of intracellular space available for housing the increased number of enzyme molecules. Additionally, the time required for enzyme levels to change probably precludes this mechanism from contributing significantly to the responses observed in organisms exposed to short-term temperature fluctuations (as would be seen in intertidal organisms). A third limitation on the usefulness of this form of temperature compensation reflects the fact that since the catalytic ability of most enzymes is temperature dependent, a given enzyme molecule will function more effectively at some temperatures than at others. Thus, increasing the enzyme concentration in response to changing temperature may simply result in a greater number of poorly functioning molecules. More advantageous to the organism would be the utilization of enzyme molecules better suited to cope with the temperature of acclimation.

Utilization of Variant Enzyme Forms

In contrast to the paucity of direct evidence related to changes in enzyme levels, there is considerable experimental support for the idea that enzyme variants are used in compensating for the depressant effects of decreased thermal energy in the system. Such variants may be enzyme forms with different amino acid compositions (isozymes) or, in some cases, simply molecules that have undergone changes in conformation ("conformational variants") but not in amino acid makeup. Both types of molecular variants occur in cold-exposed ectotherms and are associated with differing functional properties. In particular, these variant enzyme forms exhibit altered affinities for their substrates as evidenced by experimental determination of their apparent K_m values. (The value of $K_{m(app)}$ is inversely related to the affinity of the enzyme for its substrate. That is, lowered values of $K_{m(app)}$ reflect increased enzyme-substrate affinities while higher $K_{m(app)}$ values denote decreased enzyme-substrate affinities.) Situations involving both types of molecular variation are considered below.

Isozymic Forms. Using electrophoretic analysis, enzyme molecules of differing amino acid compositions have been detected in a variety of preparations from ectotherms

living at different temperatures. In some cases, the
isozymic forms present in the tissue actually may change
with acclimation temperature, while in other cases, the
different isozymes may always be present, but the variant
that predominantly functions may be dependent on the temperature.

For example, in rainbow trout, the particular isozyme
of acetylcholinesterase (AChE) that exists in the brain
depends on the temperature at which the fish have been
living. Baldwin and Hochachka (1970) found that trout
acclimated to cold (2°C) and warm (17°C) temperatures
revealed two distinct electrophoretic bands--one corresponding to the form found in the 2°C trout and the other to
that from the 17°C trout. In fish acclimated to an intermediate temperature (12°C), both molecular forms (i.e., two
electrophoretic bands) were observed. That these struc-
turally different AChE molecules also differed functionally
was indicated by an examination of their kinetic properties. Specifically, it was noted that the minimum values
of $K_{m(app)}$ for the two isozymic forms occurred at temperatures near those corresponding to the acclimation temperature for the fish. That is, the $K_{m(app)}$ of the enzyme
from the 17°C trout appeared lowest (and therefore the
affinity of the enzyme for its substrate was highest) near
17°C. On either side of 17°C, the enzyme-substrate affinity fell (i.e., the curve of apparent K_m vs. temperature
was U-shaped). Similarly, the lowest values of $K_{m(app)}$
for the 2°C preparation were seen near 2°-5°C with significantly higher values occurring at warmer temperatures.
Such variations in $K_{m(app)}$ assume functional significance
since at physiological concentrations of substrate, enzymatic reaction rates will generally increase with increasing enzyme-substrate affinity. Thus there appears to be a
mechanism in trout whereby temperature can elicit the form
of AChE that seems best suited--at least in terms of affinity for substrate--to operate at that temperature.
(This phenomenon, where the minimum value of $K_{m(app)}$ for a
particular enzyme occurs near the temperature of acclimation, has been observed in a number of enzyme systems and
provides one means whereby the rate of the reaction cata-
lyzed by a specific isozyme can be regulated as temperature
increases. For example, as the enzyme variant is exposed
to temperatures higher than the acclimation temperature,
the decreased enzyme-substrate affinity tends to offset the
stimulatory effects of higher levels of thermal energy.)

Other examples of enzymes for which different acclimation temperatures evoke changes in isozyme distribution
include: (a) those like trout AChE where certain isozymic

forms exist at some, but not at all temperatures (e.g.,
liver glucose-6-phosphate dehydrogenase from the creek
chub, Kent and Hart 1977; esterase from sea urchin tube
feet, Marcus 1977; and liver malate dehydrogenase from the
garter snake, Hoskins and Aleksiuk 1973; and (b) systems
where the same isozymic forms are present, but in differing
amounts, at the various acclimation temperatures (e.g.,
liver NADP-isocitrate dehydrogenase, Moon and Hochachka
1971; liver glucose-6-phosphate dehydrogenase in mullet,
Hochachka and Clayton-Hochachka 1973; and skeletal muscle
lactate dehydrogenase from goldfish, (Bolaffi and Booke
1974). In both of these types of enzyme systems, there
appears to occur a temperature-induced shift in the relative levels of specific isozyme forms.

   In contrast are the enzyme systems that do not exhibit
temperature-induced modifications in isozyme patterns. In
some instances, the failure to detect different isozyme
distributions in individuals acclimated to different temperatures may reflect masking by the isozyme polymorphism
inherent in the population (Wilson et al. 1975). That is,
even for individuals living at the same temperature, considerable variation in the isozyme patterns of some enzymes
may occur (Shaklee et al. 1977; Somero 1975; and Wilson et
al. 1975). (In fact, the finding of extensive enzyme
polymorphism in populations of some fish has led to the
suggestion that the changes reported to be temperature-
induced in some studies may actually be genetic variations
(Shaklee et al. 1977). This possibility emphasizes the
importance of using large sample sizes and of having information regarding the phenotypic variability of the enzyme
under study.) However, some of the enzyme systems that do
not exhibit temperature-compensatory changes in isozyme
patterns do exhibit changes in the variant that functions
as indicated by the altered kinetic characteristics of the
preparation at different temperatures (Aleksiuk 1971;
Somero 1973; and Somero and Hochachka 1969). For example,
Somero (1973) noted no alterations in the relative amounts
of the skeletal muscle lactate dehydrogenase isozymes from
8°C and 28°C fish *(Gillichthys mirabilis)*. Nonetheless, at
assay temperatures above 25°C, the affinity of the enzyme
preparation for pyruvate decreased markedly while at lower
temperatures, the affinity for pyruvate was increased.
These results were interpreted as indicating that although
the same isozymes were present at warm and cold temperatures, different forms were functioning. Thus, regardless
of whether temperature alters the isozymic forms that exist
or simply those that operate, the net effect is the same--
namely, the occurrence of functional warm and cold variants.

These cold and warm variants could differ slightly in amino acid makeup of their catalytic site, although it appears more likely that the compositional changes occur at other regions of the molecule. Support for this view comes from studies of the responses of variants of several allosteric enzymes. The warm and cold forms of these enzymes (Table 1) exhibit different enzyme-"modulator" interactions which in turn will alter the rate of catalysis (most commonly by altering the affinity of the enzyme for its substrate). Such changes seem to involve both positive and negative modulators and are discernable from the altered values of K_a or K_i for the modulators. {Modulators, or effectors, are molecules that bind to specific sites (non-catalytic) on regulatory enzymes and as a result of their binding, alter the affinity with which substrate binds to the catalytic site of the enzyme. In general, positive modulators enhance enzyme-substrate binding affinities while negative modulators decrease these affinities. Just as the $K_{m(app)}$ is inversely related to the affinity of the enzyme for the substrate, the value of K_a is inversely related to the affinity of the enzyme for its positive allosteric modulator while K_i is inversely related to the affinity of the enzyme for its negative allosteric modulator. For example, the affinity of mullet liver glucose-6-phosphate dehydrogenase for substrate is decreased in the presence of NADPH which is a negative allosteric modulator} of the enzyme. Hochachka and Clayton-Hochachka (1973), have shown that NADPH has a greater depressant effect on the enzyme-substrate affinity of the warm isozymic variant of the enzyme than of the cold variant. Therefore, as temperature decreases, the inhibitory effect of NADPH on the enzyme is alleviated and the reaction is facilitated.

Conformational Variants. Some enzyme variants appear to differ only in molecular conformation and not amino acid composition. That is, the enzyme form existing at warm temperatures may assume a different configuration at cold temperatures, with the $K_{m(app)}$ for each form being lowest near the respective induction temperature. These enzyme conformational variants can be distinguished from variants differing in amino acid structure by electrophoretic analysis. Thus, an enzyme preparation containing variants differing only in conformation will yield only one electrophoretic band whereas a preparation containing variants comprised of different amino acids will yield bands corresponding to each of the isozymes. Furthermore, in sys-

TABLE 1. Effects of Temperature on Allosteric Interactions of Isozymic Forms

Enzyme	Species	Organ	Modulator	Characteristics of Variants at Cold Temperatures	References
Glucose-6-phosphate dehydrogenase	Mullet	Liver	NADPH	Decreased affinity for NADPH. Since NADPH is a negative modulator and decreases the affinity of enzyme for substrate, at low temperatures this inhibitory effect will be diminished.	Hochachka and Clayton-Hochachka (1973)
	Arctic tanner crab	Gill	Na^+ and NH_4^+	Increased stimulation by these ions.	Behrisch (1972)
NADP-dependent isocitrate dehydrogenase	Rainbow trout	Liver	ADP	Increased affinity of the positive modulator, ADP, at low concentration.	Moon and Hochachka (1971)
Phospho-6-gluconate dehydrogenase	Arctic tanner crab	Gill	Na^+ and NH_4^+	Increased stimulation by these ions.	Behrisch (1972)

NADPH = Reduced nicotinamide adenine dinucleotide phosphate (NADP);
ADP = adenosine 5'-diphosphate.

TABLE II. Effect of Low Assay Temperature on Allosteric Interactions of Enzyme Conformational Variants

Enzyme	Species	Organ	Modulator	Type	Effect of Cold	References
Phospho-fructokinase	Alaskan king crab	muscle	ATP	Negative	Increased affinity for ATP.	Freed (1971b)
	Alaskan king crab	muscle	Citrate	Negative	Increased affinity for citrate.	Freed (1971b)
	Goldfish	muscle	Citrate	Negative	Increased affinity for citrate.	Freed (1971a)
	Goldfish	muscle	ATP	Negative	Increased affinity for ATP.	Freed (1971a)
Fructose- 1,6-diphosphatase	Rainbow trout	muscle	5′-AMP	Negative	Increased affinity for 5′-AMP.	Behrisch and Hochachka (1969a)
	Arctic tanner crab	muscle	5′-AMP	Negative	Increased affinity for 5′-AMP above and below 8°C.	Behrisch (1971)
	Arctic tanner crab	muscle	Phosphoenol-pyruvate (PEP)	Positive	At 0°C, PEP causes a very sharp increase in the affinity of the enzyme for substrate such that 5′-AMP inhibition is partly overcome.	Behrisch (1971)
	Arctic tanner crab	muscle	PEP	Positive	In addition to decreasing the K_m of the enzyme for substrate, at 6°C PEP abolishes the bimodal substrate saturation curve (i.e., at 6°C, PEP changes the nature of the enzyme-substrate interaction).	Behrisch and Johnson (1974b)

Enzyme	Organism	Tissue	Modulator		Effect	Reference
	Arctic tanner crab	muscle	NH_4^+		NH_4^+ stimulates at $0°$ and $15°C$, but has no effect at $8°C$.	Behrisch (1971)
	Arctic tanner crab	muscle	K+		K+ stimulates the enzyme at $0°C$, but is inhibitory at $8°$ and $15°C$.	Behrisch (1971)
Pyruvate kinase	Rainbow trout	muscle	Ca^{++}		Low Ca^{++} works with Mg^{++} to activate the enzyme. In the cold, the activation by these ions is increased. High Ca^{++}, on the other hand, is inhibitory to the enzyme.	Somero and Hochachka (1968)
NADP-dependent malate dehydrogenase	Rainbow trout	liver	NADPH	Negative	Decreased inhibition.	Baldwin and Reed (1976)
Lactate dehydrogenase	Yellowfin sole	muscle	Pyruvate	Negative	Increased affinity for pyruvate.	Behrisch (1972)
	Frog	muscle	Pyruvate	Negative	Increased affinity for pyruvate.	Enig et al. (1976)
	Gillichthys (mudsucker)	muscle	Pyruvate	Negative	Increased affinity for pyruvate.	Somero (1973)
Glyceraldehyde 3-phosphate dehydrogenase	Arctic tanner crab	muscle	K+		K+ stimulates the enzyme at $-2°C$. At $+3°C$, the enzyme is unaffected by the ion. At $+10°C$, K+ is inhibitory.	Behrisch (1972)
Citrate synthase	Rainbow trout	liver	ATP	Negative	Increased affinity for ATP.	Hochachka and Lewis (1970)

ATP = adenosine 5'-triphosphate; 5'-AMP = adenosine 5'-monophosphate; NADPH as in Table I.

tems with conformational variants, intermediate forms can
often be detected. An example of a system that appears to
respond in this way, the enzyme pyruvate kinase from the
muscle of the Alaskan king crab, is considered below.

Although initial kinetic observations indicated the
existence of a cold and a warm form of this enzyme in the
Alaskan king crab, electrophoretic and electrofocusing
analyses revealed only one band or peak of activity (Hoskins and Aleksiuk 1973). The enzyme form existing near the
lower end of the thermal range of the crab's habitat (4-
5°C) exhibited minimum $K_m(app)$ values for its substrate at
5°C while the form operating at 12°C (near the upper end of
the thermal range) had its lowest $K_m(app)$ near 10-12°C.
However, despite these two kinetically-distinct forms, only
one protein species could be distinguished, thereby supporting the idea that the two variants of the pyruvate
kinase are interconverted via a temperature-dependent
process. As the temperature is decreased, the warm form
would be converted to the cold form and the associated
increase in enzyme-substrate affinity would compensate for
the effects of decreased thermal energy in the system. The
overall result is a reaction rate that is relatively
temperature-independent over the range at which the crab
lives (Somero 1969).

Analogous changes in conformation and enzyme-substrate
affinity have been observed in several other systems and
could involve conformational changes that directly alter
the catalytic and/or the allosteric sites of regulatory
enzymes. As discussed above, the net result of either type
of change is the same--i.e., modification of the interaction of substrate or cofactor with enzyme and therefore
modification of the rate of the catalytic event. However,
the mechanisms involved differ. Of the systems that have
been studied, altered interactions at allosteric sites
appear to be quite common (Table 2).

Changing temperature may also induce changes in
enzyme conformation that directly affect the affinity of
the enzyme for its substrate or required cofactors (i.e.,
changes directly affecting the catalytic site rather than
indirectly via alterations at the allosteric site).
Examples of such changes, however, are not well documented.

Altered Intracellular Environments

A third mechanism whereby compensation for low temperatures might occur is via enzyme activity changes resulting from modification of intracellular factors rather than from modification of the enzyme molecules directly. Alterations in the intracellular environment could involve changes in intracellular pH, in ion distribution, in the levels of positive and negative allosteric modulators that normally modulate the activity of the enzyme, and/or in the case of membrane-bound enzymes, changes in the adjacent lipid environment.

Temperature-dependent changes in plasma pH have been reported for such vertebrates as the carp, bullfrog, toad, turtle, etc. In these organisms, pH increases as temperature decreases, paralleling falling levels of plasma pCO_2 (Howell et al. 1970; Reeves 1969; Robin 1962). (These changes are a manifestation of a system that regulates the OH^-/H^+ ratio rather than the concentration of H^+ unlike the case in mammals where pH is regulated.) That such changes in pH could significantly influence enzyme activity is illustrated by several examples of enzymes (Table 3) whose activities are enhanced at the elevated pH levels occurring with decreasing temperatures. Sometimes this increased activity is associated with altered affinity of the enzyme for its allosteric modulators, increased affinity of the enzyme for its required cofactors or substrate, and/or a shift to an enzyme variant with a higher pH optimum. In some cases, the pH-induced effect on the rate of enzyme action may actually be sufficient to completely compensate for the depressant effects of temperature, and the enzyme activity remains relatively constant over a wide temperature range (Behrisch and Johnson 1974b; Hochachka and Lewis 1970; Hoskins and Aleksiuk 1973). Although the molecular mechanisms underlying the pH-sensitivity of various enzymes are still uncertain, Shindler and Tipton (1977) have recently proposed a model for regulatory enzymes that involves ligand-induced changes in the ionization constants of groups influencing the enzyme-substrate binding site.

Changes in ion levels may also affect enzyme activities. Such effects could be direct, as would occur if the ion served as a cofactor or modulator for the enzyme, or indirect, as might be seen if intracellular osmolalities were altered. In the latter case, elevated ion levels could lead to increased osmolalities which in turn could affect the solvent capacity of the cell. (Along these

TABLE III.   *The Effects of Changing pH on Enzyme Activity*

Enzyme	Species	Effects of increased pH	References
Fructose-1,6-diphosphatase	Arctic tanner crab	Decreased affinity for the negative effector, 5'-AMP (leading to increased inhibition by 5'-AMP).	Behrisch and Johnson (1974a)
	Arctic tanner crab	Increased affinity for Mg++, a required cofactor for the enzyme.	Behrisch and Johnson (1974b)
	Tropical lungfish	Increased affinity for Mg++ and Mn++ (required cofactors for enzyme activity) along with a large increase in V_{max}.	Behrisch and Hochachka (1969a)
	Salmon	Increased affinity for Mn++ (a required cofactor), decreased affinity for AMP (a negative modulator) and greatly increased V_{max} for the reaction.	Behrisch (1969)
Phosphofructokinase	Goldfish	Sharp increase in enzyme activity.	Freed (1971a)
Lactate dehydrogenase	Garter snake	Increased pH optimum of enzyme (allowing maintenance of activity as pH increases).	Aleksiuk (1971)
	Yellowfin sole	Alteration of substrate saturation curve from hyperbolic to sigmoidal.	Behrisch (1972)
Citrate synthase	Rainbow trout	Decreased affinity for the negative effector, ATP (leading to relief of ATP inhibition).	Hochachka and Lewis (1970)
Acetylocholinesterase	Rainbow trout	Increased reaction rate.	Baldwin and Hochachka (1970)
Malate dehydrogenase	Garter snake	Increased pH optimum allowing maintenance of activity as pH increased.	Hoskins and Aleksiuk (1973)

lines, Low et al. 1978, has suggested that proteins from
cells with high salt concentrations have high water binding affinities to facilitate adequate hydration.) Several
studies have in fact reported that plasma concentrations
of Na^+, K^+, Ca^{++}, $Cl-$, and Mg^{++} are altered in some cold-
acclimated ectotherms (Fletcher and Campbell 1978; Hazel
and Prosser 1974; Rao 1962). However, the prevalence of
such changes, and whether they reflect altered intracellular ion levels are not yet clear (Hazel and Prosser
1974; Hickman et al. 1964; Toews and Hickman 1969).
Therefore, the extent to which temperature-induced changes
in intracellular ion concentrations may contribute to the
compensatory responses of the ectotherm needs further
evaluation.

Similarly, although altered concentrations of allosteric modulators would affect enzyme activity, the
possibility that temperature-induced changes in modulator
levels occur has not been systematically examined.

In contrast, numerous studies have focused in recent
years on the characteristics of membrane components in
cold-exposed organisms. Receiving particular attention
have been the membrane phospholipids and their saturation
levels. It has been demonstrated that the degree of
phospholipid saturation tends to be lower in animals
living at cold temperatures (Chapelle et al. 1977; Chapelle 1978; Fukushima et al. 1977; Irving and Watson 1976;
Nozawa and Kasai 1978; Prosser 1973; Roots 1968) and that
increased lipid unsaturation enhances membrane fluidity
and reduces the temperature at which the membrane undergoes a phase shift to a less fluid condition (Blazyk and
Steim 1972; Raison et al. 1970). Since it is generally
thought that the activities of membrane-associated enzymes
may vary with the membrane phase, with optimal activity
occurring under fluid conditions (Charnock and Simonson
1977; Kimelberg and Papahadjopoulos 1974; Singer and
Nicholson 1972; Tanaka and Teruya 1973), such changes in
phospholipid saturation may facilitate catalysis at low
temperatures.

Consistent with the view that temperature-induced
alterations in phospholipid fluidity may significantly
affect enzyme function are the observations that membrane-
bound enzymes from ectotherms living at warm temperatures
(as well as from homeotherms) often display Arrhenius
plots that cannot be expressed as single linear functions
(Charnock and Bashford 1975; Kimelberg and Papahadjopoulos
1972; Lyons and Raison 1970; Pye et al. 1976; Raison et
al. 1970; Vandenheede et al. 1973). That is, occurring in

the plots are discontinuities that delineate temperature ranges over which values of the activation energy (E_a) of the reaction differ (i.e., the E_a value is calculated from the slope of an Arrhenius plot and the discontinuities represent points where the slopes change). In such systems, values of E_a are greater at the lower end of the temperature range than at the higher end, suggesting a decrease in the catalytic effectiveness of the enzyme at low temperature (a more precise interpretation of E_a values is discussed later in this chapter). The assay temperatures at which these discontinuities occur appear to depend upon the temperature at which the enzyme normally functions. For example, Smith (1967) reported that the Na^+/K^+-ATPase from the intestinal mucosa of goldfish acclimated to 30°C exhibited a break in its Arrhenius plot at 21°C while for the enzyme isolated from goldfish acclimated to 8°C, the discontinuity occurred at 12°C.

These discontinuities and increased values of E_a may reflect a shift of the membrane lipids from a fluid to a gel-like phase (Hazel 1972a; Irving and Watson 1976; Lyons and Raison 1970; Pye et al. 1976; Raison et al. 1970; Vandenheede et al. 1973). In accordance with this interpretation, the finding that membrane-associated enzymes from ectotherms that function in the cold often display no such discontinuities (Irving and Watson 1976; Lyons and Raison 1970: Vandenheede et al. 1973) has been attributed to an increased degree of unsaturation in these membranes, with a concomitant decrease in the temperature at which a phase change will occur (Smith 1967). On the other hand, that factors other than lipid phase changes may be involved in the discontinuities observed in some Arrhenius plots of membrane-bound enzymes is the conclusion reached from recent work with homeothermic mitochondrial enzymes (Blazyk and Steim 1972; Cannon et al. 1975; Lenaz et al. 1972), as well as with bacterial plasma membrane preparations (Silvius et al. 1978). For example, in the latter study temperature-dependent changes in substrate-binding affinity that appeared to be independent of membrane-lipid composition could account for the breaks in Arrhenius plots observed when the Na^+ dependent Mg^{++}-ATPase activity was assayed at a fixed substrate (ATP) concentration. Such findings emphasize the need to consider alternatives to changes in membrane fluidity when interpreting breaks in Arrhenius plots of membrane-bound enzymatic reactions.

However, that cold-induced changes in lipids can indeed alter enzyme behavior has been clearly demonstrated by Hazel (1972a,b) using succinate dehydrogenase (SDH), an enzyme bound to the inner mitochondrial membrane. Upon comparison of muscle SDH preparations from goldfish acclimated to 5oC and to 25oC, he found the 5oC enzyme to have higher activity. Moreover, removal of the lipid moieties from the preparations led to a greater decrease in specific activity, E_a value, binding affinity for succinate, and inhibitory effectiveness of oxaloacetate (a negative allosteric modulator) of the 5oC enzyme than of the 25oC enzyme. Further support for the idea that the lipids were at least partly responsible for the differences seen in the cruder preparations came from Hazel's "reconstitution" experiments (1972b). Adding back the lipid from the 5oC preparation to either the 5oC or 25°C lipid-free fraction was more effective in reactivating SDH than was addition of the lipid from the 25°C fish. That is, the magnitude of the final SDH activity depended on the source of the lipid rather than the source of the protein. Furthermore, the finding that the lipid from the 5ºC preparation was more unsaturated than was the 25oC lipid led to the suggestion that the modulating effect of the lipid fraction depended on the degree of phospholipid saturation.

These data from Hazel provide direct evidence for the influential role of lipids on enzyme activity. They also complement reports that the specific activities of membrane-associated enzymes from fish living at low temperatures tend to be higher than those from warm-acclimated fish (Caldwell 1969; Freed 1965; Irving and Watson 1976), the increased activities being accompanied by greater unsaturation of the lipids present in the preparations (Caldwell and Vernberg 1970; Irving and Watson 1976).

A second enzymatic system whose temperature characteristics are markedly affected by membrane phospholipid is the Na^+/K^+-ATPase. The requirement of this enzyme for associated phospholipid has been demonstrated in a number of studies (e.g., Charnock et al. 1973; Kimelberg and Papahadjopoulos 1972, 1974; Tanaka 1969; Tanaka and Sakamoto 1969; Tanaka and Teruya 1973); and evidence has been presented to support the view that the activity of this enzyme is enhanced under conditions of increased membrane fluidity (Charnock and Simonson 1977; Kimelberg and Papahadjopoulos 1974; Tanaka and Teruya 1973). Moreover, in an experiment similar to that of Hazel (1972b), Tanaka and Teruya

(1973) have shown that the temperature-sensitivity of
the Na^+/K^+-ATPase (from bullfrog kidney and bovine cerebral
cortex) depends on the nature of the phospholipid and not
the protein source.

Altered activities of membrane-bound enzymes are not
the only possible consequence of changes in membrane
phospholipid saturation. Such changes may also affect the
rates at which various substances are translocated across
membranes. For example, the increased degree of unsaturation and the maintenance of membrane-fluidity at low ambient temperatures may ensure that molecules are able to
move between appropriate intra-and extracellular compartments at reasonable rates. Although the potential effects
of lipid changes on the passive transport properties of
membranes from cold-exposed ectotherms have been considered
by several investigators (Chapelle 1978; Hazel and Prosser
1974; Pye et al. 1976), there are few experimental measurements of membrane permeabilities and transport characteristics in these animals.

Notwithstanding the need for additional information
relevant to membrane properties and other areas as well,
the examples discussed above illustrate several of the
mechanisms that operate in various ectotherms acclimated to
low temperatures, namely: changes in the amounts of enzymes, in the form of the enzyme utilized, and/or in the
microenvironment of the enzyme. One result of such changes
is an increase in the activity of the affected enzyme
molecules and compensation (partial or complete) for the
depressant effects of low temperature on reaction rates.
In addition, differential effects of temperature on various
enzyme molecules may lead to altered flows through metabolic
pathways. For example, as indicated from the changes in
Table 2 and the comments in Table 4, changes occurring in
several systems could lead to a shunting of glucose from
glycolysis to the hexose monophosphate shunt, to a flow of
pyruvate away from lactate production toward mitochondrial
oxidation, and to an increased dependence on the aerobic
oxidation of free fatty acids as fuel. Thus, cold-induced
alterations of enzymes, their effectors and/or their cofactors, may result in enhanced reaction rates and/or
shifts to alternate metabolic pathways in response to the
changing internal environment of the ectotherm.

Temporal Aspects of Temperature Compensation in Ectotherms

Although the basic mechanisms that have been considered above are inducible (to varying degrees) in individuals transferred from warm to cold environments, a variable amount of time is required for the mechanisms to
become effective. On the one hand, weeks may be needed to
reach the altered steady state associated with the new
ambient temperature if compensation occurs via changes in
the amounts of existing enzymes or by the synthesis of new
isozymic forms. In contrast, temperature-induced intercon-
versions of enzyme forms provide a means for "instantaneous" acclimation without encumbering the cell with the
bulk required by the presence of a variety of isozymic
forms or increased numbers of enzyme molecules. This
latter mechanism would thus be useful to organisms exposed
to short-term temperature fluctuations (e.g., intertidal
ectotherms) while the former (i.e., alterations in the net
synthesis of particular enzyme molecules) would be of
greater value to organisms exposed to prolonged periods of
low ambient temperature (as might occur seasonally).

In addition to these mechanisms that can be invoked
instantaneously or after a period of acclimation, there
appear to be compensatory changes in enzyme function that
have evolved over generations. Such adaptations (see
below) may even enhance reaction rates to the point where
similar rates of metabolic processes occur in ectotherms
from quite different thermal regions (e.g., organisms from
different latitudes) when measured at their normal functioning temperature (Hochachka and Somero 1973). These
adaptive changes in enzyme function may reflect: (a) alterations of intracellular factors influencing enzyme activity
(e.g., the degree of membrane phospholipid saturation tends
to be lower in species living at low temperatures than in
those living in warmer climates, Irving and Watson 1976);
and/or (b) the presence of enzyme molecules with optimal
ligand affinities and/or enhanced enzyme efficiencies.
With respect to enzyme-ligand affinities, there have
been reported several systems where the apparent K_m for a
particular enzyme, when assayed over a range of temperatures, is lowest at a temperature near that of the habitat. For example, in the case of both pyruvate kinase
(Somero and Hochachka 1968) and lactic dehydrogenase
molecules (Somero and Hochachka 1969) from skeletal muscle
of the Antarctic teleost, *Trematomus*, and Alaskan king
crab, respectively, the minimum value of $K_{m(app)}$ occurs
between 0°-5°C while that for the enzyme molecules from
rainbow trout raised at 10-15°C is between 10°C and 15°C.

TABLE IV.   Temperature Effects on Metabolic Flow

Enzyme	Source	Effect of Cold	References
Phosphofructo-kinase (PKF)	Alaskan king crab (muscle)	Increased ATP and citrate inhibition of PFK promoting decreased glycolysis.	Freed (1971b)
	Goldfish (muscle)	High temperature decreases PFK-ATP affinity which decreases ATP inhibition; high temperature decreases citrate inhibition. This suggests that at high temperature, glycolysis increases.	Freed (1971a)
Fructose- 1,6-diphosphatase (FDPase)	Arctic crab (muscle)	Increased affinity for the negative modulator, AMP, causing significant inhibition of FDPase and subsequent decrease in gluconeogenesis. (This effect may be partly offset by increasing pH or cofactor concentration in which case the enzyme would be relatively temperature independent.)	Behrisch and Johnson (1974b)
Glucose-6-phosphate dehydrogenase	Mullet (liver)	Increased affinity for glucose-6-phosphate & NADP; decreased NADPH inhibition of enzyme facilitating increased operation of hexose monophosphate pathway.	Hochachka and Clayton Hochachka (1973)
Lactate dehydrogenase (LDH)	Frog (muscle)	Increased sensitivity to pyruvate inhibition and increased affinity for pyruvate. This may channel pyruvate into the TCA cycle rather than to lactate.	Enig et al. (1976)
	Mudsucker (muscle)	Shift of LDH from "M_4" to "H_4" forms. (H_4 is more sensitive to pyruvate inhibition and has higher affinity for pyruvate, thereby promoting channeling of pyruvate to TCA cycle rather than to lactate.)	Somero (1973)
	Goldfish (muscle)	Decreased LDH form that is at least sensitive to pyruvate inhibition.	Bolaffi and Booke (1974)
Glyceraldehyde phosphate isomerase (GPI), aldolase, glyceraldehyde-3-phosphate dehydrogenase (GAPDH), pyruvate kinase (PK), LDH	Green sunfish (muscle)	Decreased activity of glycolytic enzymes as a group.	Shaklee et al. (1977)

TABLE IV. (cont)

Enzyme	Source	Effect of Cold	References
Succinate dehydrogenase, cytochrome oxidase, cytochrome C	Green sunfish (muscle)	Increased activity of oxidative enzymes as a group.	Shaklee et al. (1977)
GAPDH, GPI, aldolase, PK, LDH, malate dehydrogenase, succinate dehydrogenase	Green sunfish	Increased activity of glycolytic enzymes as a group (perhaps reflecting the dependency of neural tissue on glucose metabolism).	Shaklee et al. (1977)
Glycolytic enzymes	Trematomus (muscle)	Iodoacetate produces twice as much inhibition of metabolism in warm-adapted fish as in cold, and cyanide inhibits more in the cold. This suggests that in the cold there is a decreased dependence on glycolysis and an Increased dependence on oxidative metabolism via the electron transport system.	Somero et al. (1968)

However, if one compares the magnitude of the $K_{m(app)}$ for homologous enzymes from various species, there does not seem to be a correlation between the actual value of the $K_{m(app)}$ and the habitat temperature. In other words, although the apparent K_m for enzyme X from a species living at 5°C may be lower when measured at 5°C than at 15°C, it may be similar in magnitude to the apparent K_m measured at 15°C for enzyme X from a species living at 15°C. This similarity of apparent K_m values supports the view that an enzyme-ligand affinity appropriate for proper enzymatic function and regulation has been rigorously conserved through evolution (Somero and Low 1976). In fact, it appears that the enzyme-substrate affinities for a number of enzymes {pyruvate kinase, phosphofructokinase, acetylcholinesterase, fructose diphosphatase, lactic dehydrogenase (Hochachka et al. 1976; Somero and Low 1976)} from ectotherms living at low temperatures are no greater than those from homologous enzymes from warm-adapted species when such affinities are measured at the habitat temperature.

On the other hand, evolutionary adaptation may have led to the selection of enzyme forms that, in ectotherms living at cold temperatures, are more efficient catalytically than homologous enzymes from warm-adapted organisms. Although this hypothesis has received considerable attention, only recently have data been acquired that can be quantitatively related to catalytic efficiencies. That is, the catalytic effectiveness of an enzyme is directly correlated with its ability to lower the free energy of activation of the reaction, $\Delta G^{\ddagger}$. Since $\Delta G^{\ddagger}$ represents the energy barrier of the reaction, the hypothesis stated above regarding enzyme efficiencies predicts that enzymes operating in cold-adapted ectothermic species would have lower $\Delta G^{\ddagger}$ values* than would those from animals living in

---

Use of activation energies (E_a) rather than $\Delta G^{\ddagger}$ values to compare catalytic effectiveness requires the assumption that the entropies of activation ($\Delta S^{\ddagger}$) are similar in the systems being compared. (Although E_a is directly related to the activation enthalpy ($\Delta H^{\ddagger} = E_a - RT$), the free energy of activation, $\Delta G^{\ddagger}$, is equal to $\Delta H^{\ddagger} - T\Delta S^{\ddagger}$). In view of recent data indicating that alterations in activation entropy as well as in activation enthalpy contribute to the compensatory changes in $\Delta G^{\ddagger}$ in cold-adapted ectotherms (Low et al. 1973; Somero and Low 1976), it is clear that E_a values may not accurately measure the enzyme's ability to reduce the energy barrier of the reaction (Low et al. 1973).

warm habitats (Johnston and Goldspink 1975; Low et al. 1973; Somero and Low 1976). Consistent with this hypothesis is the fact that for the few enzyme systems that have been examined (pyruvate kinase, lactic dehydrogenase, glyceraldehyde-3-phosphate dehydrogenase, glycogen phosphorylase b, and Mg^{++}-activated myofibrillar ATPase), those from low temperature species have lower $\Delta G^{\ddagger}$ values (and higher substrate turnover numbers) than do their homologues operating at warmer temperatures (Johnston and Goldspink 1975; Somero and Low 1976). Thus, the currently available data, although sparse in quantity, are uniform in their support of the idea that the catalytic efficiencies of enzymes correlate inversely with the temperatures at which they function (Borgmann and Moon 1975).

Since $\Delta G^{\ddagger} = \Delta H^{\ddagger} - T\Delta S^{\ddagger}$, it is not surprising that the decreases in catalytic efficiencies observed in the systems discussed above reflect changes in both activation enthalphy ($\Delta H^{\ddagger}$) and activation entropy ($\Delta S^{\ddagger}$). The lower $\Delta H^{\ddagger}$ values seen in the cold-adapted enzymes (Borgmann and Moon 1975; Johnston and Goldspink 1975; Somero and Low 1976) are in accord with the decreased E_a values reported for other enzyme systems operating in the cold (Aleksiuk 1971; Feeney and Osuga 1976; Hochachka and Clayton-Hochachka 1973; Somero and Hochachka 1968; Somero et al. 1968) although decreases in E_a values for cold-adapted ectothermic enzymes do not occur universally (Feeney and Osuga 1976; Hochachka and Lewis 1970; Irving and Watson 1976). Partially offsetting the effect on $\Delta G^{\ddagger}$ of the decreased $\Delta H^{\ddagger}$ are concomitant decreases in $\Delta S^{\ddagger}$. In fact, for the four enzyme systems analyzed, the decreases in $\Delta H^{\ddagger}$ and $\Delta S^{\ddagger}$ of the enzyme homologues {e.g., pyruvate kinases from cells operating at $-2^{\circ}C$ (*Trematomus borchgrevinki*, an Antarctic fish), at $8-17^{\circ}C$ (*Scorpaena gutatta*, a cold water teleost), at $18-30^{\circ}C$ (*Mugil cephalus*, a mullet fish, at $25-32^{\circ}C$ (*Bufo marinus*, toad) and at $37-39^{\circ}C$ (rabbit and chicken)} co-varied with a $\Delta H^{\ddagger}/\Delta S^{\ddagger}$ ratio between $300-340^{\circ}K$. That is, a change of one entropy unit was associated with an enthalpy change of $300-340^{\circ}K$ (Low and Somero 1974; Somero and Low 1976).

The relative constancy of this ratio led to the suggestion that the occurrence, to varying degrees, of one class of reaction--the formation of weak bonds--could account for the observed decreases in $\Delta G^{\ddagger}$. {In fact, Low and Somero (1974) have calculated that an enzyme operating at $5^{\circ}C$ would lower its $\Delta G^{\ddagger}$ by about 650 cal/mol if only one more weak bond were formed during catalysis. This difference in ΔG due to the additional bond approximates the differences experimentally observed between homologous enzymes functioning normally at 5°C and 37°C, respectively.} These weak bonds may involve interactions between the amino acid sidechains (or other residues) of the enzyme with each other and/or with the surrounding water molecules. Recent analyses of the activation volumes of lactic dehydrogenase (Hochachka et al. 1976; Somero and Low 1977) and pyruvate kinase (Low and Somero 1975) support the suggestion that differences in ΔG values between enzyme homologues may reflect alterations in the degree of hydration.

Further support for the existence of adaptive modifications in ectothermic enzymes functioning at low cellular temperatures derives from comparisons between such enzymes and their homeothermic counterparts. From such studies, ectothermic enzymes have been noted to exhibit the following differences {relative to homeothermic enzymes (Bormann and Moon 1975; Johnston and Goldspink 1975; Komatsu and Feeney 1970; Low and Somero 1974; Somero and Low 1976)}: (a) lower ΔG values; (b) greater substrate turnover numbers; (c) lower activation energies (E_a) and enthalpies (ΔH); and (d) more flexible structures (perhaps because of fewer weak bonds). These adaptations (for which the molecular bases are currently under investigation) have resulted in ectothermal enzymes that tend to be more efficient (lower ΔG), less heat stable and less temperature sensitive (lower E_a) than the corresponding enzymes from homeothermic tissues, characteristics that contribute to effective catalysis at the low cellular temperatures at which they operate (Borgmann and Moon 1975; Komatsu and Feeney 1970).

COLD-INDUCED RESPONSES IN HOMEOTHERMS

Since homeotherms generally maintain their deep body temperature relatively constant, most of their cells are normally exposed to temperatures between 37°C and 40°C despite varying ambient conditions. In some cases, however, mammalian and avian cells do experience low or fluctuating intracellular temperatures. For example, enzymes from hibernating species are called upon to function at cold and at warm temperatures (some of the characteristics of these enzymes are considered by Dr. Roberts in the following paper). In addition, many species undergo after birth, a developmental transition from ecto-to endotherm as

central regulatory and/or peripheral thermogenic effector
and heat conservation mechanisms reach maturation; and even
in adult homeotherms, cells in the extremities may be
subjected to temperatures that vary and that may fall to
near ambient levels. In the ensuing discussion, the enzymatic responses of such adult "heterothermic" cells as
well as their homeothermic counterparts will be considered.

Enzymes from Heterothermic Cells

Although relatively few studies have examined avian or
mammalian (non-hibernating) enzymes from cells that function at low temperatures, the information currently available indicates that temperature-dependent compensatory
mechanisms similar to those observed in ectothermic systems
may operate. For example, Behrisch and Percy (1974) have
reported the presence of two variants of 6-phosphogluconate
dehydrogenase in the Arctic spotted seal *(Phoca vitulina)*.
One variant predominates in the deep body fat while the
other can be isolated in greater amounts from adipocytes in
the flipper. The latter exhibits an affinity for its
substrate and negative allosteric modulator (NADPH) that is
relatively temperature insensitive over the range that the
cells would normally experience. {This relative temperature independence of enzyme-ligand affinity is characteristic of enzymes from intertidal ectotherms exposed to
fluctuating ambient temperatures (Hazel and Prosser 1974).}
In contrast to the flipper variant, that found in the deep
body fat exhibits an affinity for substrate and NADPH that
increases with decreasing temperature. These changes are
such that if the variant found in the deep body fat were
exposed to the temperatures seen by the flipper enzyme,
it would be completely shut off due to the large increase
in affinity for NADPH. Thus, the deep body enzyme appears
unsuited for catalysis at the low intracellular temperatures at which the flipper enzyme functions.

Two other enzyme systems from the same species of seal
have been examined and have also been found to behave more
like their ectothermic than their homeothermic (rat, rabbit) homologues (Somero and Johansen 1970). These enzymes,
pyruvate kinase and lactic dehydrogenase from renal and
flipper arterial smooth muscle, exhibited decreased apparent K_m values with decreasing assay temperature. This
reciprocal relationship between temperature and enzyme-ligand affinity was such as to reduce by about two-thirds the
depressant effects of lowered temperature and to help stabil-
ize rates of catalysis between the 7o-37oC range examined.

Another similarity between heterothermic and ectothermic systems is the increase in phospholipid unsaturation observed in membranes from cells functioning at cold temperatures (Meng et al. 1969; Zar 1977). Such changes promote membrane fluidity at low temperatures and may be important in providing an environment for adequate catalysis by membrane-bound enzymes.

Thus, the few heterothermic enzyme systems that have been examined appear to utilize compensatory mechanisms similar to ectothermic enzymes, although the type of mechanism may differ among enzymes as well as among tissues and species.

Enzymes from Homeothermic Cells

Aside from the responses of heterothermic cells to low temperatures, the cold-induced adjustments occurring in most mammalian and avian tissues are related (directly or indirectly) to the organism's need to conserve and/or to generate heat. Whereas, heat conservation may be accomplished by non-metabolic adjustments (e.g., by behavioral changes, piloerection, use of counter-current blood flow systems, altered blood flow patterns), increased rates of heat generation require accelerated rates of substrate oxidation. Hence, likely sites of temperature-induced enzymatic adjustments would be tissues contributing to the increased generation of heat, primarily brown fat and muscle in mammals and muscle in birds. As was the case with ectotherms, enzymatic adaptations in these thermogenic tissues could involve increased amounts of critical enzymes, use of enzyme variants and/or modifications of enzyme activity via changes in the intracellular environment. Although little information is available regarding these possibilities in birds, mammalian tissues and brown fat in particular have received some attention.

Brown Adipose Tissue. In mammals possessing brown fat, cold exposure results in catecholamine-stimulation of heat generating pathways. This stimulation is initiated by the binding at the plasma membrane of norepinephrine released from sympathetic nerves innervating the brown adipocytes. Following such interaction, the activities of at least two membrane-bound enzymes are altered: namely, adenyl cyclase (catalyzing the synthesis of cAMP), and Na^+/K^+-ATPase (the enzyme associated with the active transport of Na^+ and K^+ across the plasma membrane). Subsequently, the rates of triglyceride lipolysis are enhanced, providing free fatty acids for mitochondrial oxidation. From this oxidation comes the chemical energy needed to support such energy-requiring processes as the translocation of Na^+ and K^+ across the plasma membrane, as well as the heat that is generated by the tissue. The extent to which the heat generation reflects altered mitochondrial coupling and/or increased ATP turnover is currently being examined (for a more extensive discussion, see Himms-Hagen 1978; Horwitz 1978).

The cellular events described above occur immediately upon transfer of the animal from a warm to a cold environment. After prolonged cold exposure, culminating in acclimation to the new habitat temperature, several changes are apparent. Prominent among these are brown fat hyperplasia (Cameron and Smith 1964) and an increased number of mitochondria per cell (Roberts and Smith 1967). In the rat, these changes in total brown fat mass reflect varying magnitudes of increase at the different brown fat sites (i.e., interscapular, cervical, thoracic, renal) with the final total amounting to about a 50 percent increase in tissue mass relative to that of the intact rat (Smith and Roberts 1964).

That the increased brown fat mass can be partly accounted for by greater numbers of cells is indicated by the incorporation of ^{3}H-thymidine into the DNA of the brown adipocytes (Cameron and Smith 1964). Concomitant with the cold-induced increase in DNA synthesis is an elevation of tissue nitrogen (Smith and Roberts 1964), and of the amount of mitochondrial nitrogen per gram of brown fat (Roberts and Smith 1967). One consequence of these changes (i.e., greater numbers of cells and more mitochondria per cell) is an increase in the levels of mitochondrial enzymes such as those associated with substrate oxidation (Himms-Hagen 1978; Smith and Horwitz 1969). Moreover, the greater numbers of cells would be expected to result in elevated amounts (total) of cytosol and plasma membrane enzymes as well. For example, Muirhead and Himms-Hagen (1971) have demonstrated that even though there is no increase in specific activity of brown fat adenyl cyclase after cold acclimation, the total activity is enhanced. In addition, increases in several mitochondrial proteins have been noted in the cold-acclimated rat. One of these proteins, a molecule of about 32,000 daltons, has been proposed to serve as a binding site for purine nucleotides and to be involved in the regulation of mitochondrial coupling (see Himms-Hagen 1978). Although this latter suggestion is

still under investigation, there seems to be little doubt
that the other observed changes, namely the elevated number
of cells and mitochondria per cell, result in augumentation
of the oxidative and therefore thermogenic capacity of the
brown fat in mammals at cold temperatures.

In contrast to the evidence documenting such quantitative increases in the enzymes associated with substrate
oxidation and energy dissipation, little information is
available regarding the cold-induced appearance of different
enzyme forms or changes in the intracellular environment of
brown fat or of any other homeothermic tissue. Shifts in
lactic dehydrogenase to the M subunit type (favoring shunting of glucose to lactic acid rather than pyruvate) have
been reported to occur in the brown fat of cold-acclimated
hamsters (Allen and Chaffee 1964), but it is possible that
such shifts are associated with preparation for hibernation,
rather than maintenance of homeothermy at low ambient temperatures. Similarly, although changes in the intracellular
environment of brown adipocytes have been seen after cold
acclimation {e.g., decreased mitochondrial Ca^{++} content;
altered phospholipid saturation (Himms-Hagen 1978)} the
adaptive significance of the changes has not yet been demonstrated.

Muscle. Changes analagous to those seen in brown fat
have not been observed in skeletal muscle. In cold-acclimated rats, the mitochondria are increased in number but are
smaller, resulting in no significant enhancement of mitochondrial mass (Himms-Hagen 1978). On the other hand,
increases have been reported (Himms-Hagen 1978) in the
activities of various enzymes such as: the Na^+/K^+-ATPase,
the membrane Mg^{++}-ATPase, and the adenine nucleotide trans-
locase of the inner mitochondrial membrane (the latter
enzyme being associated with the transfer of ATP and ADP
between the mitochondrial matrix and the cytosol). Additionally, the active transport of Ca^{++} across the mitochondrial membrane of cold-acclimated rats appears to be
enhanced (Greenway and Himms-Hagen 1978). Notwithstanding
these increases in activity, there is insufficient information to determine whether they reflect increases in enzyme
levels, operation of different enzyme variants, and/or
altered membrane properties.

SUMMARY

Enzymes from homeothermic tissues appear to differ from their ectothermic homologues with respect to characteristics that reflect the different thermal conditions under which they function. Nevertheless, cold exposure of homeotherms as well as of ectotherms leads to compensatory changes in enzyme activity. In homeotherms, these changes are generally related to the animals' need to generate more heat and (at least in brown fat) appear to primarily reflect increased numbers of protein molecules associated with cellular thermogenic pathways (e.g., enzymes involved in substrate turnover and energy dissipation). In contrast, the mechanisms employed by ectotherms appear to include modification of enzyme levels, forms, and intracellular surroundings such as to facilitate effective catalysis at lowered intracellular temperatures.

ACKNOWLEDGMENTS

The authors would like to thank Ms. Mary Lou Rodriguez for her valuable assistance and patience in the preparation of this manuscript. D. R. Hettinger is the recipient of a National Research Service Award 5 T32 GM07416-02.

REFERENCES

Aleksiuk, M. 1971. An isozymic basis for instantaneous cold compensation in reptiles: lactate dehydrogenase kinetics in *Thamnophis sirtalis*. *Comp. Biochem. Physiol. 40B*:671

Allen, J.R., and R.J. Chaffee. 1964. The alteration of lactate dehydrogenase isozyme patterns of brown fat in the golden hamster by cold exposure. *The Physiologist. 7*:80.

Baldwin, J., and P.W. Hochachka. 1970. Functional significance of isoenzymes in thermal acclimation: acetylcholinesterase from trout brain. *Biochem. J. 116*:883.

Baldwin, J., and K.C. Reed. 1976. Effect of temperature on the properties of cytoplasmic NADP malate dehydrogenases from liver of warm and cold acclimatized rainbow trout. *Comp. Biochem. Physiol. 54B*:531.

Behrisch, H.W. 1969. Temperature and the regulation of
     enzyme activity in poikilotherms. Fructose diphos-
     phatase from migrating salmon. *Biochem. J.*
     115:687.

———. 1971. Temperature and the regulation of enzyme ac-
     tivity in poikilotherms. Regulatory properties of
     FDPase from muscle of Alaskan king crab. *Biochem. J.*
     121:399.

———. 1972. Molecular mechanisms of adaptation to low tem-
     perature in marine poikilotherms. Some regulatory pro-
     perties of dehydrogenases from two arctic species. *Mar.*
     Biol. 13:267.

Behrisch, H.W., and P.W. Hochachka. 1969a. Temperature and
     the regulation of enzyme activity in poikilotherms.
     Properties of rainbow trout fructose diphosphatase.
     Biochem. J. 111:287.

———. 1969b. Temperature and the regulation of enzyme
     activity in poikilotherms. Properties of lungfish fruc-
     tose-1-6-diphosphatase. *Biochem. J. 112*:601.

Behrisch, H.W., and C.E. Johnson. 1974a. Regulation of
     enzyme activity at low temperature: Ionic influences on
     fructose-1-6-diphosphatase from muscle of the Arctic
     tanner crab, *Chionocetes bairdi.* *Comp. Biochem.*
     Physiol. 47B:417.

———. 1974b. Unusual regulatory characteristics of fruc-
     tose-1-6-diphosphatase from muscle. Studies on the
     kinetic behavior of the enzyme from muscle of the Arctic
     tanner crab, *Chionocetes bairdi.* *Comp. Biochem.*
     Physiol. 47B:427.

Behrisch, H.W., and J.A. Percy. 1974. Temperature and the
     regulation of enzyme activity in homeo- and hetero-
     thermic tissues of Arctic marine mammals: some regu-
     latory properties of 6-phosphogluconate dehydrogenase
     from adipose tissue of the spotted seal (*Phoca vi-*
     tulina). *Comp. Biochem. Physiol. 47B*:437.

Blazyk, J.F., and J.M. Steim. 1972. Phase transitions in
     mammalian membranes. *Biochem. Biophys. Acta.*
     266:737.

Bolaffi, J.L. and H.E. Booke. 1974. Temperature effects on
     lactate dehydrogenase isozyme distribution in skeletal
     muscle of *Fundulus heteroclitus (Pisces: Cyprinidon-*
     tiformes). *Comp. Biochem. Physiol. 48B*:557.

Borgmann, U., and T.W. Moon. 1975. A comparison of lactate
     dehydrogenases from an ectothermic and an endothermic
     animal. *Can. J. Biochem. 53*:998.

Bullock, T.H. 1955. Compensation for temperature in the metabolism and activity of poikilotherms. *Biol. Rev.* *30*:311.

Caldwell, R.S. 1969. Thermal compensation of respiratory enzymes in tissues of goldfish (*Carassius auratus L.*). *Comp. Biochem. Physiol.* *31*:79.

Caldwell, R.S., and J.F. Vernberg. 1970. The influence of acclimation temperature on the lipid composition of fish gill mitochondria. *Comp. Biochem. Physiol.* *34*:179.

Cameron, I.L., and R.E. Smith. 1964. Cytological responses of brown fat tissue in cold-exposed rats. *J. Cell Biol.* *23*:89.

Cannon, B., et al. 1975. The fluidity and organization of mitochondrial membrane lipids of the brown adipose tissue of cold-adapted rats and hamsters as determined by nitroxide spin probes. *Arch. Biochem. Biophys.* *167*:505.

Charnock, J.S., and C.L. Bashford. 1975. A fluorescent probe study of the lipid mobility of membranes containing sodium- and potassium-dependent adenosine triphosphatase. *Mol. Pharm.* *11*:766.

Charnock, J.S., and L.P. Simonson. 1977. Differential lipid control of $(Na^+ + K^+)$-ATPase in homeotherms and poikilotherms. *Comp. Biochem. Physiol.* *58B*:381.

Das, A.B., and C.L. Prosser. 1967. Biochemical changes in tissues of goldfish acclimated to high and low temperatures. I. Protein synthesis. *Comp. Biochem. Physiol.* *21*:449.

Dean, J.M., and J.D. Berlin. 1969. Alterations in hepatocyte function of thermally acclimated rainbow trout (*Salmo gairdneri*).   *Comp. Biochem. Physiol.* *29*:307.

Enig, M., J. Ramsey, and D. Eby. 1976. Effect of temperature on pyruvate metabolism in the frog: the role of lactate dehydrogenase isoenzymes. *Comp. Biochem. Physiol.* *53B*:145.

Feeney, R.E., and D.T. Osuga. 1976. Comparative biochemistry of Antarctic proteins. *Comp. Biochem. Physiol.* *54A*:281.

Fletcher, G.L., and C.M. Campbell. 1978. The effects of hypophysectomy on seasonal changes in plasma freezing-point depression, protein 'antifreeze', and Na^+ and Cl^- concentrations of winter flounder (*Pseudopleuronectes americanus*).   *Can. J. Zool.* *56*:278.

Freed, J.M. 1965. Changes in activity of cytochrome oxidase during adaptation of goldfish to different temperatures. *Comp. Biochem. Physiol.* *14*:651.

Freed, J.M. 1971a. Properties of muscle phosphofructokinase of cold- and warm-acclimated *Carassius auratus*. *Comp. Biochem. Physiol. 39B*:747.

______. 1971b. Temperature effects on muscle phosphofructokinase of the Alaskan king crab, *Paralithodes camtschatica. Comp. Biochem. Physiol. 39B*:765.

Fukushima, H., et al. 1977. Studies on *Tetrahymena* membranes. Palmitoyl-coenzyme A desaturase, a possible key enzyme for temperature adaptation in *Tetrahymena* microsomes. *Biochim. Biophys. Acta. 288*:442.

Greenway, D.C., and J. Himms-Hagen. 1978. Increased calcium uptake by muscle mitochondria of cold-acclimated rats. *Am. J. Physiol. 234*:C7.

Haschemeyer, A.E.V. 1968a. Comparison of liver protein synthesis in temperature-acclimated toadfish *(Opsanus tau). Biol. Bull. 135*:130.

______. 1968b. Rates of polypeptide chain assembly in liver *in vivo*: relation to the mechanism of temperature acclimation in *Opsanus tau. Proc. Natl. Acad. Sci. 62*:128.

Hazel, J.R. 1972a. The effect of temperature acclimation upon succinate dehydrogenase activity from the epaxial muscle of the common goldfish *(Carassius auratus L.)*. I. Properties of the enzyme and the effect of lipid extraction. *Comp. Biochem. Physiol. 43B*:837.

______. 1972b. The effect of temperature acclimation upon succinate dehydrogenase activity from the epaxial muscle of the common goldfish *(Carassius auratus L.)*. II. Lipid reactivation of the soluble enzyme. *Comp. Biochem. Physiol. 43B*:863.

Hazel, J.R., and C.L. Prosser. 1974. Molecular mechanisms of temperature compensation in poikilotherms. *Physiol. Rev. 54*:620.

Hickman, C.P. Jr., et al. 1964. Effect of cold acclimation on electrolyte distribution in rainbow trout *(Salmo gairdneri). Can. J. Zool. 42*:577.

Himms-Hagan, J. 1978. Biochemical aspects of nonshivering thermogenesis. In: L. Wang and J. Hudson, eds. Strategies in Cold: Natural Torpidity and Thermogenesis. Academic Press, New York. p. 595.

Hochachka, P.W., and B. Clayton-Hochachka. 1973. Glucose-6-phosphate dehydrogenase and thermal acclimation in the mullet fish. *Mar. Biol. 18*:251.

Hochachka, P.W., and J.K. Lewis. 1970. Enzyme variants in thermal acclimation. Trout liver citrate synthases. *J. Biol. Chem 245*:6567.

Hochachka, P.W., and G.N. Somero. 1973. Pages 179-270 in
    Strategies of Biochemical Adaptation. W.B. Saunders
    Company, Philadelphia, PA. 179-270 pp.

Hochachka, P.W., et al. 1976. Enthalpy-entropy compensation
    of oxamate binding by homologous lactate dehydrogenases.
    Nature. *260*:648.

Horwitz, B.A. 1978. Neurohumoral regulation of nonshivering
    thermogenesis in mammals. In: L. Wang and J. Hudson,
    eds. Strategies in Cold: Natural Torpidity and Thermo-
    genesis. Academic Press, New York. p. 619.

Hoskins, M.A.H., and M. Aleksiuk. 1973. Effects of temper-
    ature on the kinetics of MDH from a cold climate
    reptile, *Thamnophis sirtalis parietalis*.   *Comp.
    Biochem. Physiol*. *45B*:343.

Howell, B.J., et al. 1970. Acid-base balance in cold-blooded
    vertebrates as a function of body temperature. *Am. J.
    Physiol*. *218*:600.

Irving, D.O., and K. Watson. 1976. Mitochondrial enzymes of
    tropical fish: a comparison with fish from cold waters.
    Comp. Biochem. Physiol. *54B*:81.

Johnston, I.A., and G. Goldspink. 1975. Thermodynamic
    activation parameters of fish myofibrillar ATPase
    enzyme and evolutionary adaptations to temperature.
    Nature. *257*:622.

Kent, J.D., and R.G. Hart. 1977. The effect of temperature
    and photoperiod on isoenzyme induction in selected
    tissues of the creek chub, *Semotilus atromaculatus*.
    Comp. Biochem. Physiol. *58B*:109.

Kimelberg, H.K., and D. Papahadjopoulos. 1972. Phospholipid
    requirements for (Na^+ + K^+)-ATPase activity: head group
    specificity and fatty acid fluidity. *Biochim. Biophys.
    Acta*. *282*:277.

______. 1974. Effects of phospholipid acyl chain fluidity,
    phase transitions, and cholesterol on (Na^+ + K^+)-
    stimulated adenosine triphosphatase. *J. Biol. Chem*.
    249:1071.

Komatsu, S.K., and R.E. Feeney. 1970. A heat labile fructose
    diphosphate adolase from cold-adapted antarctic fishes.
    Biochim. Biophys. Acta. *206*:305.

Lenaz, G., et al. 1972. Activation energies of different
    mitochondrial enzymes; breaks in Arrhenius plots of
    membrane-bound enzymes at different temperatures.
    Biochim. Biophys. Res. Comm. *49*:536.

Lott, J.N.A., J.J. Wilson, and C.M. Vollmer. 1977. Temper-
    ature-induced changes in the nuclear envelope of
    Euglena and *Scenedesmus*.   *J. Ultrastruc. Res. 61*:1.

Low, P.S., and G.N. Somero. 1974. Temperature adaptation of enzymes: a proposed molecular basis for the different catalytic efficiencies of enzymes from ectotherms and endotherms. *Comp. Biochem. Physiol. 49B*:307.

______. 1975. Protein hydration changes during catalysis: a new mechanism of enzymic rate-enhancement and ion activation/inhibition of catalysis. *Proc. Natl. Acad. Sci. 72*:3305.

Low, P.S., J.L. Bada, and G.N. Somero. 1973. Temperature adaptation of enzymes: roles of the free energy, the enthalpy, and the entropy of activation. *Proc. Natl. Acad. Sci 70*:430.

Low, P.S., et al. 1978. Protein water binding ability correlates with cellular osmolarity. *Experientia. 34*:314.

Lyons, J.M., and J.K. Raison. 1970. A temperature-induced transition in mitochondrial oxidation: contrasts between cold- and warm-blooded animals. *Comp. Biochem. Physiol. 37*:405.

Marcus, N.H. 1977. Temperature induced isozyme variants in individuals of the sea urchin, *Arbacia punctulata*. *Comp. Biochem. Physiol. 58B*:109.

Meng, M.S., G.C. West, and L. Irving. 1969. Fatty acid composition of caribou bone marrow. *Comp. Biochem. Physiol. 30*:187.

Moon, T.W., and P.W. Hochachka. 1971. Temperature and enzyme activity in poikilotherms. Isocitrate dehydrogenases in rainbow trout liver. *Biochem. J. 123*:695.

Muirhead, M., and J. Himms-Hagen. 1971. Changes in the amount and properties of adenyl cyclase in brown adipose tissue during acclimation of rats to cold. *Can. J. Biochem. 49*:802.

Nozawa, J., and R. Kasai. 1978. Mechanism of thermal adaptation of membrane lipids in *Tetrahymena pyriformis NT-1*. Possible evidence for temperature-mediated induction of palmitoyl-CoA desaturase. *Biochim. Biophys. Acta. 529*:54.

Plant, P.W., J.B.K. Nielsen, and A.E.V. Haschemeyer. 1977. Control of protein synthesis in temperature acclimation. I. Characterization of polypeptide elongation factor 1 of toadfish liver. *Physiol. Zool. 50*:11.

Prosser, C.L. 1973. Comparative Animal Physiology. W.B. Saunders 3rd ed. Philadelphia, PA. 362–428 pp.

Pye, V.I., W. Wieser, and M. Zech. 1976. The effect of season and experimental temperature on the rates of oxidative phosphorylation of liver and muscle mitochondria from the tench, *Tinca tinca.   Comp. Biochem. Physiol. 54B*:13.

Raison, J.K., J.M. Lyons, and W.W. Thomson. 1970. The influence of membranes on the temperature-induced changes in the kinetics of some respiratory enzymes of mitochondria. *Arch. Biochem. Biophys. 142*:83.

Rao, K.P. 1962. Physiology of acclimation to low temperature in poikilotherms. *Science. 137*:682.

Reeves, R.B. 1969. Role of body temperature in determining the acid-base state in vertebrates. *Fed. Proc. 28*:1204.

Roberts, J.C., and R.E. Smith. 1967. Time-dependent responses of brown fat in cold-exposed rats. *Am J. Physiol. 212*:519.

Robin, E.D. 1962. Relationship between temperature and plasma pH and carbon dioxide tension in the turtle. *Nature. 195*:249.

Roots, B.I. 1968. Phospholipids of goldfish *(Carassius auratus L.)* brain: influence of environmental temperature. *Comp. Biochem. Physiol. 25*:457.

Shaklee, J.B., et al. 1977. Molecular aspects of temperature acclimation in fish: contributions of changes in enzyme activities and isozyme patterns to metabolic reorganization in the green sunfish. *J. Exptl. Zool. 201*:1.

Shindler, J.S., and K.F. Tipton. 1977. A model for the allosteric regulation of pH-sensitive enzymes. *Biochem. J. 167*:479.

Sidell, B.D. 1977. Turnover of cytochrone c in skeletal muscle of green sunfish *(Lepomis cyanellus, R.)* during thermal acclimation. *J. Exptl. Zool. 199*:223.

Silvius, J.R., B.D. Read, and R.N. McElhaney. 1978. Membrane enzymes: artifacts in Arrhenius plots due to temperature dependence of substrate-binding affinity. *Science. 199*:902.

Singer, S.J., and G.L. Nicholson. 1972. The fluid mosaic model of the structure of cell membranes. *Science. 175*:720.

Smith, M.W. 1967. Influence of temperature acclimatization on the temperature-dependence and ouabain-sensitivity of goldfish intestinal adenosine triphosphatase. *Biochem. J. 105*:65.

Smith, R.E., and B.A. Horwitz. 1969. Brown fat and thermogenesis. *Physiol. Rev. 49*:330.

Smith, R.E., and J.C. Roberts. 1964. Thermogenesis of brown adipose tissue in cold-acclimated rats. *Am. J. Physiol. 206*:143.

Somero, G.N. 1969. Pyruvate kinase variants of Alaskan king
    crab. *Biochem. J. 114*:237.

______. 1972. Molecular mechanisms of temperature compen-
    sation in aquatic poikilotherms. Pages 55-80 in F.E.
    South, et al. eds. Hibernation and Hypothermia, Per-
    spectives and Challenges. Elsevier, New York.

______. 1973. Thermal modulation of pyruvate metabolism in
    the fish, *Gillichthys mirabilis*: the role of lactate
    dehydrogenase. *Comp. Biochem. Physiol. 44B*:205.

______. 1975. The roles of isozymes in adaptation to varying
    temperatures. Pages 221-234 in C.L. Markert, ed. Iso-
    zymes. II. Physiological Function. Academic Press, New
    York.

Somero, G.N., and P.W. Hochachka. 1968. The effect of
    temperature on catalytic and regulatory functions of
    pyruvate kinases of the rainbow trout and the antarctic
    fish, *Trematomus bernacchii*.   *Biochem. J. 110*:395.

______. 1969. Isoenzymes and short-term temperature compen-
    sation in poikilotherms: activation of lactate dehy-
    drogenase isoenzymes by temperature decreases. *Nature.
    223*:194.

Somero, G.N., and K. Johansen. 1970. Temperature effects on
    enzymes from homeothermic and heterothermic tissues of
    the harbor seal *(Phoca vitulina)*.   *Comp. Biochem.
    Physiol. 34*:131.

Somero, G.N., and P.S. Low. 1976. Temperature: a shaping
    force in protein evolution. *Biochem. Soc. Symp. 41*:33.

______. 1977. Enzyme hydration may explain catalytic ef-
    ficiency differences among lactate dehydrogenase
    homologues. *Nature. 266*:276.

Somero, G.N., et al. 1968. Cold adaptation of the antarctic
    fish *Trematomus bernacchii*.   *Comp. Biochem. Physiol.
    26*:223.

Tanaka, R., and A. Teruya. 1973. Lipid dependence of ac-
    tivity-temperature relationship of (Na^+, K^+)-activated
    ATPase. *Biochim. Biophys. Acta. 323*:584.

Toews, D.P., and C.P. Hickman, Jr. 1969. The effect of
    cycling temperature on electrolyte balance in skeletal
    muscle and plasma of rainbow trout, *Salmo gairdneri*.
    Comp. Biochem. Physiol. 29:905.

Vandenheede, J.R., et al. 1973. Heart mitochondrial oxi-
    dations of an antarctic fish. *Comp. Biochem. Physiol.
    46B*:313.

Wilson, F.R., G.S. Whitt, and C.L. Prosser. 1973. Lactate
    dehydrogenase and malate dehydrogenase isozyme patterns
    in tissues of temperature-acclimated goldfish *(Carassius
    auratus L.)*. *Comp. Biochem. Physiol. 46B*:105.

Wilson, F.R., et al. 1975. Isozyme patterns in tissues of
temperature-acclimated fish. Pages 193-206 in C.L.
Markert, ed., Isozymes. II. Physiological Function.
Academic Press, New York.
Zar, J.H. 1977. Fatty acid composition of emperor penguin
(*Aptenodytes forsteri*) lipids. *Comp. Biochem. Physiol.*
56B:109.

DISCUSSION I. SPECIAL ENZYMATIC
MECHANISMS OF COLD ADAPTATION

Jane C. Roberts

Department of Biology
Creighton University
Omaha, Nebraska

A coverage of all aspects of biochemical adaptation to cold is beyond the scope of this discussion. Therefore, comments will be limited to two general areas. The first might be referred to as Strategies of Studying Cold Acclimation and the second is Hibernation. The intent is not to review the literature in either of these areas, but rather to mention some examples of the kinds of work that have been done and to point out a number of the unanswered questions in cold acclimation and hibernation.

STRATEGIES OF STUDYING COLD ACCLIMATION

Changes in Total Enzyme Activity

Numerous studies performed over the past 25–30 years have shown that significant increases occur in the specific activities of certain key glycolytic and oxidative enzymes in tissues of cold-acclimated mammals, especially rodents (for reviews see Smith and Hoijer 1962; Chaffee and Roberts 1971). However, studies of enzyme specific activity alone do not tell us whether these changes represent quantitative

or qualitative changes in the enzymes involved; nor do they
give us an assay of the total change in an animal's thermogenic capacity.

Jansky's (1963) estimates of total cytochrome oxidase
in whole body and organs of warm- and cold-acclimated rats
reveal that total body cytochrome oxidase activity is similar in the two groups of animals. However, warm-acclimated animals apprently do not use the cytochrome oxidase
system to its full capacity, whereas cold-acclimated rats
do. Estimates of cold-induced increases in total brown fat
enzyme activity have been made for α-ketoglutarate oxidase
in rats (Smith and Roberts 1964) and β-hydroxybutyrate
oxidase in deer mice (Roberts et al. 1966). Steffen and
Roberts (1977) have made similar estimates of total enzyme
activity in liver, kidney, and brown fat of cold-acclimated
and control gerbils. When changes in organ weight and
total protein are taken into account the effect of cold
acclimation on total enzyme activity is often greater than
that on specific activity alone. Such data clearly indicate that in determining the overall effect of cold
acclimation at the biochemical level, changes in total mass
of metabolically active tissue, especially tissue protein,
as well as enzyme specific activity, must be considered.

Many of the earlier studies on organ weight changes in
cold-acclimated rodents did not analyze the nature or
mechanism of the observed weight increases. It has now
been shown that cold acclimation stimulates DNA synthesis
in liver, kidney (Holeckova and Baudysova 1975), and brown
fat (Cameron and Smith 1964) of rats. Thus the increased
weight of these organs in the cold is the result of an
increased cellularity as estimated by DNA synthesis.
Skeletal muscle showed no appreciable change in weight or
DNA content. These data led to the conclusion (Holeckova
and Baudysova 1975) that cold acclimation stimulates mitosis in cells capable of division. The nature of the
stimulus is unknown.

The fact that protein synthesis is involved in cold
acclimation and the development of nonshivering thermogenesis, is further demonstrated by studies using protein
synthesis inhibitors. For example, Himms-Hagen (1971)
found that in rats injected with the mitochondrial protein
synthesis inhibitor oxytetracycline, development of non-
shivering thermogenesis (as measured by an increased metabolic response to norepinephrine) is blocked. Similarly,
protein synthesis inhibitors have been used in experiments
dealing with cold-induction of individual enzymes, as for

example, hepatic tryptophan pyrrolase in rats (Sitaramam and Ramasarma 1975). The activity of this enzyme increases rapidly in the cold, reaches a maximum at 8 hours, and remains elevated above the control level for at least 48 hours. Injection of the protein synthesis inhibitors cycloheximide or actinomycin D totally abolished the cold-stimulated induction of this enzyme. Sitaramam and Ramasarma (1975) suggested the increased enzyme activity in the cold was due to protein synthesis *de novo* and was mediated by corticosteroids, possibly by increasing transcription of a specific mRNA.

Thus, evidence is accumulating which indicates that at least some of the increased enzyme activity seen in cold acclimation is the result of quantitative changes in total enzyme present. However, in most cases these studies do not tell us the extent to which the resulting increase in calorigenic potential is tapped by the organism *in vivo* or which of these changes are essential for survival in the cold.

Acclimation versus Acclimatization

A great deal of cold acclimation physiology and biochemistry has been carried out in the cold room using laboratory-bred species, such as the rat. More information is needed on wild species, and particularly on acclimatization under natural outdoor conditions. Hart, Heroux and their co-workers (Hart 1957, 1964; Hart and Heroux 1953) demonstrated several years ago that there are physiological differences between cold acclimation and cold acclimatization in rodents. Roberts et al. (1966) also showed this in the deer mouse where winter trapped animals were compared with summer caught mice and seasonal changes were then compared with the effects of cold acclimation in the laboratory (Figure 1). From their data it appears that seasonal changes associated with winter have markedly different effects from those of laboratory imposed constant cold, differing not only in magnitude, but often in direction as well. A plot of percent change (cold acclimated/control) versus percent change (winter/summer) gives a correlation coefficient of 0.38 which is not significantly different from zero (P>0.1) (Figure 1). This would seem to indicate that cold acclimation in the laboratory may involve different mechanisms from winter cold acclimatization in the wild. Recent studies by Feist and Feist (1977, 1978) on Alaskan red-backed voles suggest that differences between acclimation and acclimatization also prevail among subarctic species.

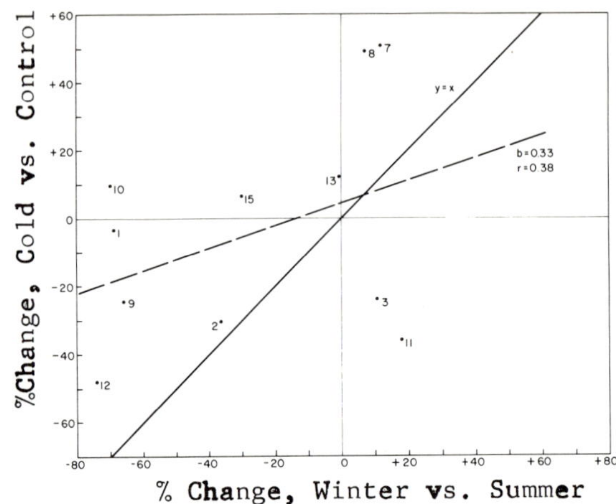

FIGURE 1.   *Percent changes in various metabolic and biochemical parameters of deer mice during cold acclimation
plotted against changes in corresponding parameters measured in wild-caught winter-acclimatized (vs. summer-accli-
matized) animals.   Solid line represents $y = x$.   Adherence
of points to this line would indicate perfect correlation
between laboratory cold acclimation and natural winter acclimatization.   Broken line is least squares regression line
for variables given ($r = 0.38$, $P > 0.1$).   1) qO_2 liver
mitochondria (M_W) with β-hydroxybutyrate (BOH) as substrate;
2) qP liver M_W with BOH; 3) P/O liver M_W with BOH; 7) liver
serine dehydrase; 8) liver threonine dehydrase; 9) liver
DPNH cytochrome c reductase (DPNHccR); 10) liver TPNHccR;
11) DPNH ccR/TPNHccR; 12) Transhydrogenase of liver M_W;
13) qO_2 kidney homogenate with BOH; 15) MR_{32}.   (Data from
Roberts et al. 1966).*

Time-dependence of Cold Acclimation

There is a need to examine time-dependent changes in
cold acclimation, especially changes which occur during
early stages of cold exposure. In this respect, mention
should be made of two recent studies on native arctic
mammals that not only involved wild caught animals, but
also examined cold acclimation beginning with the early
days of cold exposure. In the tundra vole, Ringens et al.
(1977) found that changes in a number of physiological
components of cold acclimation (cold diuresis, food consumption, body weight loss, and urinary pH, K^+ and Na^+)
were completed and had stabilized after only 14 days of
cold exposure. In a study on brown and varying lemmings,
Berberich and Folk (1976) have found that changes in a
number of indices of cold acclimation (liver temperature,

body weight, organ wet and dry weights, and food consumption) are similar in extent to those found in cold-acclimated temperate mammals. However, the rate of acclimation was apparently more rapid in lemmings. In these species compensatory changes in a number of variables were essentially complete within the first week of cold exposure rather than requiring the 6-8 weeks often used in laboratory studies.

These very interesting studies raise a number of questions: Might these animals be considered "preacclimated" to cold because of their normal exposure to a cold environment? Do they have a unique genetic make-up which permits rapid cold acclimation? Do some of the other indices of cold acclimation, i.e., brown fat mass, oxidative enzyme levels, and metabolic response to norepinephrine, also show such rapid changes and short time courses in reaching a new steady state of cold acclimation?

HIBERNATION

Hibernation must certainly be considered an important adaptive strategy for a number of species faced with prolonged periods of cold weather. Tissues of hibernators have the somewhat unusual capacity to change from one operating temperature to another, which may be 30-35°C higher or lower, within a relatively short period of time (Willis 1967). Over this temperature range ATP producing reactions, such as oxidative phosphorylation, continue and the coupling between energy producing and energy requiring, or work performing, processes is maintained. One might ask what biochemical mechanisms are involved in enabling hibernators to tolerate such marked and relatively rapid changes in their body temperature, and what confers upon their tissues a type of cold resistance that permits maintenance of functional integrity at temperatures only a few degrees above freezing? The answers to these and related questions have been sought at the cellular and biochemical level in studies on hibernation-related changes in levels of enzyme activity, and the *in vitro* effects of temperature on enzyme and tissue systems from hibernating and active animals. As an example of such studies, mention of some of the author's work on oxidative enzymes of hamsters and ground squirrels and comparison with the work of others can be made.

Levels of Enzyme Activity

In hamster liver, rates of oxidation of succinate, α-ketoglutarate (αKG) and glutamate are significantly depressed in mitochondria from hibernating compared to either control or cold-acclimated animals (Roberts and Chaffee 1972). This depression or inhibition was apparent at all assay temperatures between 6–37°C. In contrast to these three enzymes no depression was seen in ascorbic acid oxidase, which is also considered to be a measure of cytochrome oxidase activity. These results are in agreement with other reports (Chaffee et al. 1961; Liu et al. 1969; and Frehn 1966) on hamster liver succinoxidase (SO) and cytochrome oxidase. Similarly, in the 13-lined ground squirrel, *Citellus tridecemlineatus*, both kidney and liver SO activity decreased significantly in hibernation (Figure 2) (Garred 1978; Roberts and Steffen unpublished data). These data agree with the findings of Liu et al. (1969) in *C. tridecemlineatus* and Shug et al. (1971) in *C. lateralis* for liver SO.

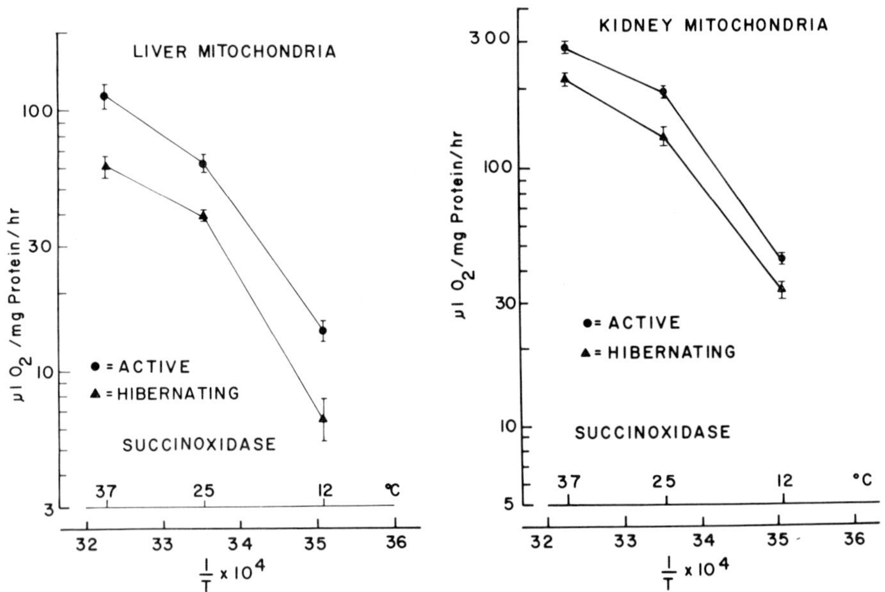

FIGURE 2. *Arrhenius plots of succinoxidase activity of liver and kidney mitochondria from active and hibernating ground squirrels, C. tridecemlineatus. Plotted values = mean ± S.E. (Data on kidney mitochondria from Garred 1978; liver mitochondria from Roberts and Steffen, unpublished).*

At this point one might ask what causes this depression or inhibition of oxidative enzyme activity during hibernation? Why is it found in some tissues, such as liver and kidney, but not in others, such as the heart of hamsters (Roberts and Chaffee 1973; South 1960), and brain and brown fat of hibernators in general (for reviews see Roberts 1971; South and House 1967)? One cannot ignore the fact that the tissues of hibernators not only undergo marked changes in biochemical and physiological activity during hibernation, but some of them play an active role in the subsequent rewarming process. It is possible that a tissue's role in arousal might play an important part in determining the changes that it undergoes in hibernation.

Arousal from Hibernation

One area that has received only limited attention is the changes in enzyme activity which occur during arousal from hibernation (See Burlington 1972; Roberts 1971). As an example of such studies, Roberts (1971; Roberts and Chaffee 1972) has shown that succinoxidase, αKG oxidase and glutamic acid oxidase activities of hamster liver mitochondria all increase rapidly after the initiation of arousal, and reach or exceed the level of the cold-acclimated active animal within 60–90 minutes. The increase in SO activity parallels most closely the changes in metabolism seen in the intact animal during arousal in that it increases rapidly with the onset of arousal, overshoots the normothermic level at about 90 minutes, and then declines (Roberts and Chaffee 1972).

Another of the unanswered questions in hibernation is how does this rapid increase in enzyme activity come about? Is it *de novo* protein synthesis? There is a fair amount of evidence indicating that protein synthesis is markedly depressed during deep hibernation (see Burlington 1972) and that low temperature greatly reduces *in vitro* rates of protein synthesis (Taegtmeyer et al. 1975). The rapidity with which the increase in oxidative enzyme activity occurs during arousal suggests that it does not represent synthesis of new enzymes, at least in the hamster. Does it represent removal of an inhibitory substance from existing enzymes, as originally suggested by Chaffee (1962)? This is a definite possibility although to date, no inhibitory substance has been identified. Does it represent the action of "effectors" as suggested by Burlington (1972)? Or does it represent a qualitative change in

either the lipid or protein portion of these membrane-bound
enzymes, such that the enzymes exist in a state of low
specific activity in hibernation and are rapidly converted
to a state with a high specific activity during arousal?
Again, data required to answer these questions are still
lacking.

In Vitro Effects of Temperature

In addition to hibernation-related changes in levels
of oxidative enzyme activity, considerable attention has
also been given to the *in vitro* effects of temperature on
enzyme and tissue systems from hibernating and active
animals.

Raison and his co-workers (Raison 1973; Raison and
Lyons 1971; Lyons and Raison 1970) have studied the effects
of temperature on liver mitochondrial enzymes from both
homeotherms and poikilotherms. They (Raison et al. 1970)
found sharp discontinuities, or breaks, in Arrhenius plots
of mitochondrial SO activity from liver of rats and active
ground squirrels (*C. lateralis*), while the same enzyme system
from poikilothermic trout and catfish liver had a constant
activation energy (E_a) between 0–40°C. Raison and Lyons (1971)
also examined liver SO from hibernating ground squirrels and
found that, while this enzyme from active squirrels showed
a discontinuity in the Arrhenius plot at between 25–21°C,
mitochondria from hibernating animals showed no such discontinuity. As a result of these and additional studies
using spin labeling techniques, they have concluded that
breaks in Arrhenius plots of homeotherms' enzymes are due
to phase changes in membrane lipids which result in con-
figurational changes in these membrane-bound enzymes. Such
phase changes are not seen in the membrane lipids of poikilotherms, but their appearance in mitochondria from homeotherms was interpreted as, "a barrier that precludes further lowering of the body temperature" and thus, precludes
hibernation (Raison and Lyons 1971). Raison and Lyons
(1971) also propose that in some way hibernating animals
are able to alter the physical properties of the mitochondrial membrane to avoid such lipid phase changes, and
thus can tolerate marked changes in body temperature.

Although our own data on hamster liver SO (Roberts et
al. 1972) also show a relatively straight line for mitochondria from hibernating animals this is not the case for
all hamster liver mitochondrial enzymes (Figure 3), nor is
it the case for SO from tissues of all hibernators (Figure
2).

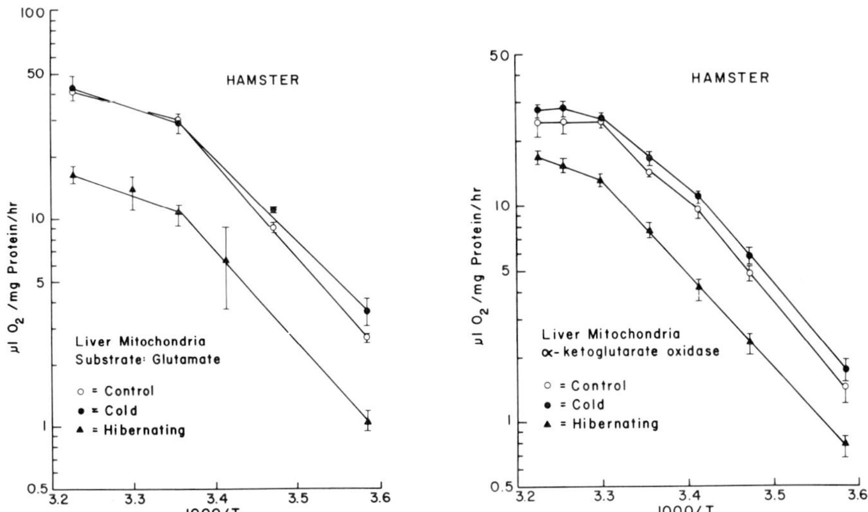

FIGURE 3. Arrhenius plots of glutamic acid oxidase and α-ketoglutarate oxidase activities of liver mitochondria from control, cold-acclimated and hibernating hamsters. Plotted values = mean ± S.E. (Roberts 1971).

For hamster liver mitochondria, Arrhenius plots of two other membrane-bound enzymes, αKG oxidase and glutamic acid oxidase, show breaks even in mitochondria from hibernating animals (Figure 3). Arrhenius plots for αKG oxidase show breaks at 30°C, which in mitochondria from active, but not hibernating, animals seems to correspond to a temperature optimum for the enzyme (Roberts 1971). Arrhenius plots of glutamic acid oxidase activity from liver of both active and hibernating animals show breaks at 25°C such that the E_a below 25°C is more than double the E_a above 25°C (Figure 3) (Roberts 1971). This suggests that the factor(s) that account for a shift toward a nearly uniform E_a for liver SO from hibernating hamsters do not alter the response of glutamic acid oxidase to temperature.

Measurements of SO activity of ground squirrel (*C. tridecemlineatus*) liver and kidney mitochondria also show breaks in Arrhenius plots for enzymes from hibernating animals (Figure 2). Since only three assay temperatures have been examined, the exact temperature at which these breaks occur cannot be specified. However, it is evident that even for SO from hibernating squirrels these are not straight lines.

Thus the theory that hibernators may in some way be able to avoid abrupt changes in the effects of temperature on mitochondrial membrane-bound enzymes must be viewed with some caution. Also, the theory that breaks in Arrhenius

plots of these enzymes are due to phase changes in the
lipid component of the membranes has been questioned
recently by Cannon and Polnaszek (1976). These authors
found no evidence of a lipid phase transition in brown fat
mitochondria from control or cold-acclimated rats or
hamsters over a temperature range from 6–37ºC. They concluded that discontinuities previously observed in Arrhenius plots of brown fat mitochondrial enzyme activity
are, therefore, not due to lipid phase changes. Moreover,
they suggest that the effects of temperature on proteins,
per se, and on protein-lipid interactions should be investigated further.

Isozymes

     Evidence is beginning to accumulate which indicates
that multiple molecular forms of enzymes (isozymes) may be
important in the process of adaptation to environmental
stress. In this respect, variants of some enzymes, specifically lactic dehydrogenase (LDH) and pyruvate kinase
(PyK), have been shown to occur in the tissues of hibernating animals. In the heart of 13-lined ground squirrels,
C. tridecemlineatus, (Burlington and Sampson 1968) and
temporal muscle of bats, *Eptesicus fuscus*, (Brush 1968) an
increase in M-type LDH is seen in hibernation, consistant
with a shift toward more anaerobic metabolism. Brush
(1968) also examined isozyme patterns for several other
enzymes in tissues of the bat and found that, except for
LDH, none showed any qualitative changes in hibernation.
     In Arctic ground squirrels, *C. parryii*, *(undulatus)*
Behrisch (1974) found that the liver of hibernating animals
contains a single isozyme of pyruvate kinase (H-PyK) that
is distinct from the single isozyme of this enzyme (NH-PyK)
found in liver of nonhibernating squirrels. The affinity
of H-PyK for substrates (phosphoenolpyruvate and ATP),
allosteric activator (fructose-1,6-diphosphate) and inhibitor (ATP) are relatively independent of temperature
change. However, the affinity of NH-PyK for both phosphoenolpyruvate and ATP is temperature sensitive (Behrisch
1974). If the two isozymes of pyruvate kinase have adaptive properties, they would be advantageous to an animal
only if it possessed the appropriate isozyme before actually entering either the active or hibernating state. In
this connection Behrisch (1974) was able to assay PyK in
livers of four ground squirrels about 2–3 weeks before

their final spring-summer arousal. He found that these
animals already show considerable NH-PyK isozyme activity,
suggesting they "anticipate" the forthcoming transition to
the active state.

Borgmann and Moon (1976) have also reported several
differences in characteristics of flight muscle and liver
pyruvate kinase from hibernating and active bats, *Myotis
lucifugus*. These include quantitative differences in
isozyme patterns, differences in specific enzyme activity,
discontinuous Arrhenius plots in which the "critical" or
transition temperature is about 10°C lower for the hibernator enzyme, and differences in temperature sensitivity of
the binding of substrates with only the enzyme from normothermic bats showing temperature sensitivity. Their
data, like those of Behrisch (1974), suggest that marked
changes occur in pyruvate kinase during hibernation,
resulting in an enzyme "better" suited to function at
reduced temperatures.

From these few examples it seems that at least some of
the basic enzymatic mechanisms utilized by ectotherms in
adapting to marked temperature changes may also occur in
hibernators, although the extent to which they occur
requires further examination.

REFERENCES

Behrisch, H.W. 1974. Temperature and the regulation of
    enzyme activity in the hibernator. Isozymes of liver
    pyruvate kinase from the hibernating and non-hiber-
    nating Arctic ground squirrel. *Can. J. Biochem.*
    52:894-902.
Berberich, J.J., and G.E. Folk, Jr. 1976. Cold acclimation
    in Arctic lemmings. *Comp. Biochem. Physiol.*
    54A:175-178.
Borgmann, A.I., and T.W. Moon. 1976. Enzymes of the nor
    mothermic and hibernating bat, *Myotis lucifugus*:
    Temperature as a modulator of pyruvate kinase. *J.
    Comp. Physiol. 107*:185-199.
Brush, A.H. 1968. Response of isozymes to torpor in the
    bat, *Eptesicus fuscus.*   *Comp Biochem. Physiol.*
    27:113-120.
Burlington, R.F. 1972. Recent advances in intermediary
    metabolism of hibernating mammals. Pages 3-15 in F.E.
    South, et al. eds. Hibernation-hypothermia, perspec-
    tives and challenges. American Elsevier, New York.

Burlington, R.F., and J.H. Sampson. 1968. Distribution and activity of lactic dehydrogenase isozymes in tissues from a hibernator and a non-hibernator. *Comp. Biochem. Physiol.* 25:185-192.

Cameron, I.L., and R.E. Smith. 1964. Cytological responses of brown fat tissue in cold-exposed rats. *J. Cell Biol.* 23:89-100.

Cannon, B., and C.F. Polnaszek. 1976. The role of mitochondrial membrane lipids in cold adaptation and hibernation. Pages 93-116 in: L. Jansky and X.J. Musacchia, eds. Regulation of depressed metabolism and thermogenesis. C.C. Thomas, Springfield, IL.

Chaffee, R.R.J. 1962. Mitochondrial changes during the process of awakening from hibernation. *Nature.* *196:*789-790.

Chaffee, R.R.J., F.L. Hoch, and C.P. Lyman. 1961. Mitochondrial oxidative enzymes and phosphorylations in cold exposure and hibernation. *Am. J. Physiol.* *201:*29-32.

Chaffee, R.R.J., and J.C. Roberts. 1971. Temperature acclimation in birds and mammals. *Ann. Rev. Physiol.* *33:*155-202.

Feist, D.D., and C.F. Feist. 1977. Catecholamine synthesizing enzymes in adrenals of laboratory acclimated and wild acclimatized subarctic voles. *Fed. Proc.* *36:*419.

______. 1978. Catecholamine-synthesizing enzymes in adrenals of seasonally acclimatized voles. *J. Appl. Physiol. Respir. Environ. Exercise Physiol.* *44:*59-63.

Frehn, J.L. 1966. Mitochondrial respiration in mammalian liver during cold exposure and hibernation. *Trans. Ill. State Acad. Sci.* *59:*342-346.

Garred, J.L. 1978. Activity of oxidative enzymes in the kidney of hibernating and active ground squirrels (*Citellus tridecemlineatus*). M.S. Thesis. Creighton University, Omaha, NEB.

Hart, J.S. 1957. Climatic and temperature induced changes in the energetics of homeotherms. *Rev. Can. Biol.* *16:*133-174.

______. 1964. Insulative and metabolic adaptations to cold in vertebrates. *Symp. Soc. Exptl. Biol.* *18:*31-48.

Hart, J.S., and O. Heroux. 1953. A comparison of some seasonal and temperature-induced changes in *Peromyscus*: cold resistance, metabolism, and pelage insulation. *Can. J. Zool.* *31:*528-534.

Himms-Hagen, J. 1971. Inhibition by oxytetracycline of the development of the enhanced response to noradrenaline in cold-acclimated rats: A new approach to the study of nonshivering thermogenesis. *Can. J. Physiol. Pharmacol.* 49:545-553.

Holeckova, E., and M. Baudysova. 1975. Stimulation of DNA synthesis in rat liver and kidney by cold acclimation. *Physiol. Bohemoslovaca* 24:311-313.

Jansky, L. 1963. Body organ cytochrome oxidase activity in cold- and warm-acclimated rats. *Can. J. Biochem. Physiol.* 41:1847-1854.

Liu, C.-C., J.L. Frehn, and A.D. LaPorta. 1969. Liver and brown fat mitochondrial responses to cold in hibernators and non-hibernators. *J. Appl. Physiol.* 27:83-89.

Lyons, J.M., and J.K. Raison. 1970. A temperature-induced transition in mitochondrial oxidation: contrasts between cold- and warm-blooded animals. *Comp. Biochem. Physiol.* 37:405-411.

Raison, J.K. 1973. The influence of temperature-induced phase changes on the kinetics of respiratory and other membrane-associated enzyme systems. *Bioenergetics.* 4:285-309.

Raison, J.K., and J.M. Lyons. 1971. Hibernation: Alternation of mitochondrial membranes as a requisite for metabolism at low temperature. *Proc. Natl. Acad. Sci.* 68:2092-2094.

Raison, J.K., et al. 1971. Temperature-induced phase changes in mitochondrial membranes detected by spin labeling. *J. Biol. Chem.* 246:4036-4040.

Ringens, P.J., G.E. Folk, and J.J. Berberich. 1977. Cold acclimation in the tundra vole. *Acta Theriologica* 22:67-74.

Roberts, J.C. 1971. Studies on the suppression of mitochondrial respiration during hibernation and its reversal during arousal in the hamster. Ph.D. Thesis. University of California, Santa Barbara.

Roberts, J.C., et al. 1972. Effects of temperature on oxidative phosphorylation of liver mitochondria from hamster, rat and squirrel monkey. *Comp. Biochem. Physiol.* 41B:127-135.

Roberts, J.C., and R.R.J. Chaffee. 1972. Suppression of mitochondrial respiration in hibernation and its reversal in arousal. Pages 101-107 in: R. Em. Smith et al. eds. Proceedings of the International Symposium on Environmental Physiology. Bioenergetics. FASEB, Bethesda, MD.

Roberts, J.C., and R.R.J. Chaffee. 1973. Effects of cold acclimation, hibernation and temperature on succinoxidase activity of heart homogenates from hamster, rat and squirrel monkey. *Comp. Biochem. Physiol.* *44B*:137-144.

Roberts, J.C., R.J. Hock, and R.E. Smith. 1966. Seasonal metabolic responses of deer mice (*Peromyscus*) to temperature and altitude. *Fed. Proc.* *25*:1275-1283.

Shug, A.L. et al. 1971. Changes in respiratory control and cytochromes in liver mitochondria during hibernation. *Biochem. Biophys. Acta* *226*:309-312.

Sitaramam, V., and T. Ramasarma. 1975. Nature of induction of tryptophan pyrrolase in cold exposure. *J. Appl. Physiol.* *38*:245-249.

Smith, R.E., and D.J. Hoijer. 1962. Metabolism and cellular function in cold acclimation. *Physiol. Rev.* *42*:60-142.

Smith, R.E., and J.C. Roberts. 1964. Thermogenesis of brown adipose tissue in cold-acclimated rats. *Am. J. Physiol.* *206*:143-148.

South, F.E. 1960. Hibernation, temperature and rates of oxidative phosphorylation by heart mitochondria. *Am. J. Physiol.* *198*:463-466.

South, F.E. and W.A. House. 1967. Energy metabolism in hibernation. Pages 305-324 in: K.C. Fisher, et al. eds. Mammalian Hibernation III. American Elsevier, New York.

Steffen, J.M., and J.C. Roberts. 1977. Temperature acclimation in the Mongolian gerbil (*Meriones unguiculatus*): Biochemical and organ weight changes. *Comp. Biochem. Physiol.* *58B*:237-242.

Taegtmeyer, H., A.G. Ferguson, and M. Lesch. 1975. Thermoregulation of myocardial protein synthesis. *Am. J. Physiol.* *228*:884-889.

Willis, J.S. 1967. Cold adaptation of activities of tissues of hibernating mammals. Pages 356-381 in: K.C. Fisher, et al. eds. Mammalian Hibernation III. American Elsevier, New York.

V. HORMONAL MECHANISMS

Bruce A. Wunder

Department of Zoology-Entomology
Colorado State University
Fort Collins, Colorado

This review is limited primarily to hormonal mechanisms associated with increases in thermogenic capabilities (especially nonshivering thermogenesis, NST) of homeotherms in meeting cold stress. The major endocrine systems which have been implicated in cold adaptation are: 1) the pituitary, 2) adrenal cortex, 3) adrenal medulla, and 4) the thyroid gland. The actions of these systems on cold acclimation have been severly questioned and are much less clear today than they seemed 10 years ago. It appears that thyroid hormones are necessary to potentiate the actions of catacholamines in stimulating NST, but increased thyroid activity is not necessary. The hormones of the adrenal cortex appear necessary to mobilize energy substrates in the early stages of cold acclimation; and, it appears that increased secretion of epinephrine and norepinephrine and increased sensivitity to their action on brown adipose tissue to stimulate NST are a well-defined response to cold.

Few natural arctic species have been studied, but what data are available suggest that acclimatization responses of animals from the field, in winter, may be different from laboratory acclimation responses to cold alone. One cue which may be responsible for mediating these differences is photoperiod. And one major system associated with photoperiodic responses is the pineal gland and its "hormone" melatonin. Thus, the possible role of photoperiod and melatonin in modulating responses to cold is discussed along with the data suggesting such a role.

INTRODUCTION

    Given the time and space in which to review this topic,
some limits will have to be set on the discussion. Since
the overall theme of the symposium was "Mechanisms of Cold
Adaptation in the Arctic," comments will be restricted to
adptation to cold in the winter (since that is when cold is
most dramatic). Problems of adjusting to summer cold
versus winter in arctic areas will not be discussed. Only
homeotherms will be dealt with, since poikilotherms generally are not active during winter.
    For homeotherms, there are basically three response
patterns which are shown when the animals are faced with
cold: 1) abandon homeothermy for torpor, 2) reduce heat
loss, and 3) increase thermogenic capabilities in order to
meet the stress of cold. A great deal of work has been done
on the hormonal mechanism associated with torpidity in cold
environments which is commonly called hibernation, and many
recent reviews have been written on the topic (Hudson 1973;
Wang and Hudson 1978; South et al. 1972). However, there
has been very little work done with the hormonal mechanisms
associated with reduction in heat loss for either small or
large homeotherms (other than acute vasomotor control);
therefore, most of this review will be concerned with the
hormonal mechanisms associated with increases in thermogenic
capabilities of homeotherms to meet cold stress.
    It should also be pointed out, that there have been
relatively few studies done on natural species from the
wild, and most of what we know concerning the hormonal
mechanisms of adaptation to cold came from studies of
animals exposed to cold in laboratory circumstances. This
is unfortunate, as the primary stressor and primary cue
which animals have in these laboratory circumstances is
cold. That is, we are studying acclimation responses of
animals to cold rather than acclimatization responses of
animals to all of the environmental changes associated with
wintertime in arctic areas (Hart 1967). Certainly, one of
the primary cues other than temperature, which animals may
use and hence which may be involved hormonally in adaptation
to cold, is change in photoperiod. This has been little
studied and will be considered at the end of this review.

HORMONAL MECHANISMS

The major endocrine systems which have been implicated
in adaptation to cold are: 1) the pituitary, which is
responsible for secretion of thyroid stimulating hormone
(THS) and adrenocorticotropic hormone (ACTH), 2) the adrenal
cortex, which secretes adrenalcorticoids (primarily cortisol
and corticosterone) which mobilize free fatty acids from
white adipose tissue and possibly affect the action of
catecholamines. These hormones then are primarily responsible for mobilizing the fuels for increased thermogenesis,
3) the adrenal medulla which secretes epinephrine and norepinephrine (NE), both of which seem to affect non-shivering
thermogenesis (NST), and 4) the thyroid, which secretes
thyroxine and triiodothrionine which are implicated in
changes in basal metabolism.
Ten to 20 years ago the action of these hormones in
adaptation to cold seemed fairly clear, and it was felt that
we had a good understanding of the hormonal mechanisms
associated with cold adaptation. Since that time, however,
our understanding has become more confused. Most of our
interpretations came from studies of rats exposed to cold
(usually about 5°C) in laboratory cold chambers.
The following will outline the response patterns most
often seen, their mechanistic interpretation, and the
problems with these interpretations. When placed at 5°C,
rats first showed an immediate increase in catecholamine
secretion. Levels then declined but were still above pretreatment value up to at least three months following the
initial cold exposure (Leduc 1961). Secondly, plasma
corticosteroids peaked within 30 minutes of cold exposure
but returned to normal within about 30 days (Straw and
Fregly 1967; Boulouard 1963; 1966). There was an increased
thermogenic response to norepinephrine which developed over
a 2 to 5 week period and persisted as long as the animals
were held in cold (Hseih and Carlson 1957; Heroux 1960).
Also the thyroid gland and the adrenal cortex hypertrophied
during a period of 1 to 3 weeks but returned to normal
within 1 to 3 months in the cold (Heroux 1960; Schonbaum
1960). Plasma-bound iodine and thyroid secretion peaked
about 1 to 3 weeks following cold exposure and remained high
so long as the animals were held in cold (Brown-Grant 1965;
Cottle and Carlson 1956; Straw and Fregly 1967). The general interpretation was that the sympathoadrenal system was
essential for the metabolic response to cold. Early in cold
exposure the adrenal cortex (through the actions of cortisol

and corticosterone) was responsible for mobilizing substrates for thermogenesis. Then, as metabolic acclimation
occurred, the increased activity of the thyroid gland would
take over and maintain the high levels of thermogenesis, and
activity of the adrenal cortex was reduced. Further, the
hyperthyroidism was felt to be essential to potentiate the
thermogenic action of catecholamines and an increase of
metabolism in thermoneutrality.

Webster (1974) reviewed some of the problems with, and
questions of this interpretation. One problem was the fact
that most of the studies from which these conclusions were
drawn involved young rats which were placed in cold environments and, hence, were growing during the experiments.
Thus, it might be argued that the effective cold stress was
reduced as the animals grew larger. However, more recent
studies show that this probably is not the case and that
increased catecholamine secretion is indeed a definite
hormonal response to cold (see Smith and Horwitz 1969).
Further, Boulouard (1963, 1966) showed that the plasmacorticoids, which are responsible for increasing free fatty
acid mobilization, were present only when rats were in
negative energy balance; once the animals adjusted their
eating levels to being in the cold and began to gain weight
again, the plasmacorticoid levels decreased. Therefore,
cold *per se* may not be a stimulus for the secretion of
plasmacorticoids but rather negative energy balance is.

The role of the thyroid has become even more confusing.
Heroux (1969) has shown that increased thyroid weight and
thyroxine turnover rate may be due to animals growing on
commercial rat chow, which is high in thyroxine and fiber
content, and not due to exposure to cold *per se*. In experiments with low fiber diet, free of thyroxine, rats in the
cold were found to have a normal thyroxine turnover rate,
no hyperthyroidism, and a normal basal metabolism (Heroux
and Petrovic 1969). Under those conditions the animals
showed a much greater cold tolerance than animals fed lab
chow (Deslauriers et al. 1971). Deslauriers et al. (1971)
then reached the conclusion that increased thyroid activity
is not essential for cold acclimation, or "it could even be
detrimental" since in its absence, rats are more resistant
to cold. LeBlanc and Villemaire (1970) also studied the
role of the thyroid in cold adaptation. Rats were given
daily injections of either thyroxine, norepinephrine, or a
combination of the two. Those groups, plus control and
thyroidectomized animals, were kept at 26ºC. After 40 days
the metabolic responses of the animals to thyroxine, norepinephrine and thyroxine, and norepinephrine were measured.

The results showed that basal metabolism is affected by the
state of thyroid activity but sensitivity does not depend on
the thyroid secretion level (Figure 1). However, metabolic
responses to norepinephrine do depend on normal thyroxine
levels, are reduced in the absence of thyroxine, and are
enhanced in the presence of high thyroxine levels (Figure
2). Thus, whereas normal thyroxine levels are necessary to
potentiate the thermogenic action of catecholamines, hyper-
tyroid activity is not essential for cold acclimation,
and when it occurs, it appears to be diet-dependent rather
than due to cold *per se.*

We can further consider the adrenosympathetic system.
It has already been mentioned that plasmacorticoid levels
increase during acclimation to cold, but it is difficult to
interpret their action in adaptation to cold as pointed out
by Boulouard (1963, 1966). The other two compounds usually
associated with this system and its role in adaptation to
cold are the catecholamines, epinephrine and norepinephrine
(NE). It has traditionally been felt that NE is more important in stimulating or potentiating non-shivering thermogenesis than epinephrine. However, Himms-Hagen (1969) has

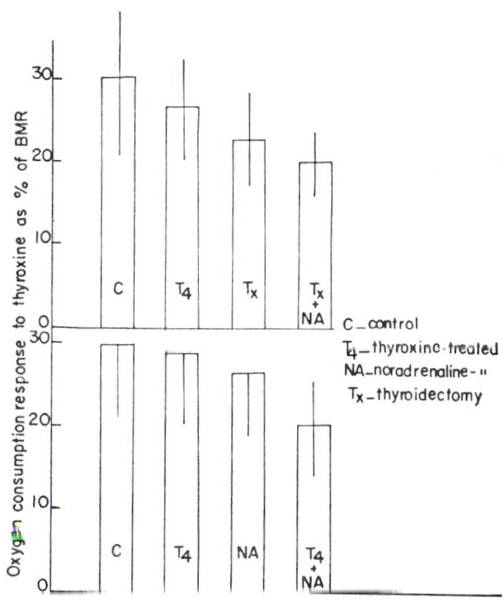

*FIGURE 1. Increase in oxygen consumption as percentage
of basal metabolic rates in response to thyroxine (75 pg/100g)
in different groups of rats: controls, injected daily with
thyroxine (50 pg/rat), injected daily with noradrenaline
(300 pg/kg) and injected daily with both thyroxine and noradrenaline for 40 days in all groups. Similar responses were
measured in thyroidectomized animals and in animals thyroidectomized subsequently injected for 40 days with noradrenaline
(300 pg/kg) (From LeBlanc 1975).*

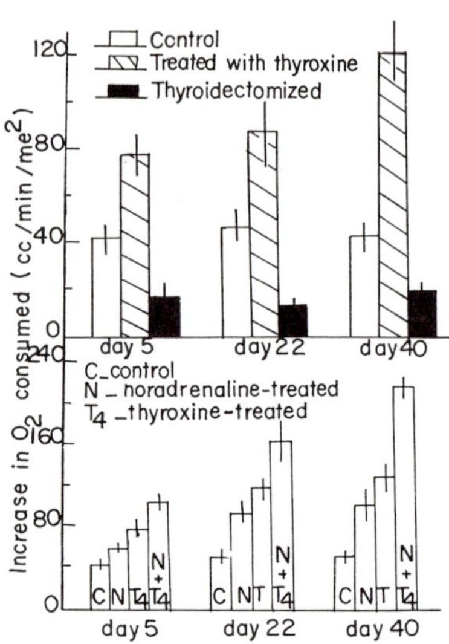

FIGURE 2. *Increase in oxygen consumption above basal values in response to noradrenaline (300 pkg) injected subcutaneously at various times (5, 22, N.D. and 40 days) in different groups of rats: controls, injected daily with thyroxine (50 pg/rat), injected daily with noradrenaline (300 pg/kg), injected daily with thyroxine and noradrenaline and thyroidectomized (from LeBlanc, 1975).*

recently shown that either hormone can affect non-shivering thermogenesis (NST) provided that they are infused intravenously rather than injected intramuscularly (Figure 3). In the later case, only NE stimulates the NST response.

One of the problems in trying to determine the relative roles of these two hormones in adaptation to cold is in assaying their activity. Traditionally they have been examined as either epinephrine or norepinephrine in urinary samples, but these account for less than one percent of the compounds as they are excreted (Himms-Hagen 1975). Much of the excretion is as metabolites; and unfortunately, the breakdown of either epinephrine or norepinephrine frequently goes through the same process and many of the metabolites are the same for the breakdown of both compounds. Therefore, it is not possible to assay these compounds and their metabolites in urine and differentially determine their excretion rates in response to cold (Himms-Hagen 1975). Rather all we can tell is that there is a change in the epinephrine-nonrepinephrine system, but we cannot determine the quantities of each or even both hormones. The only

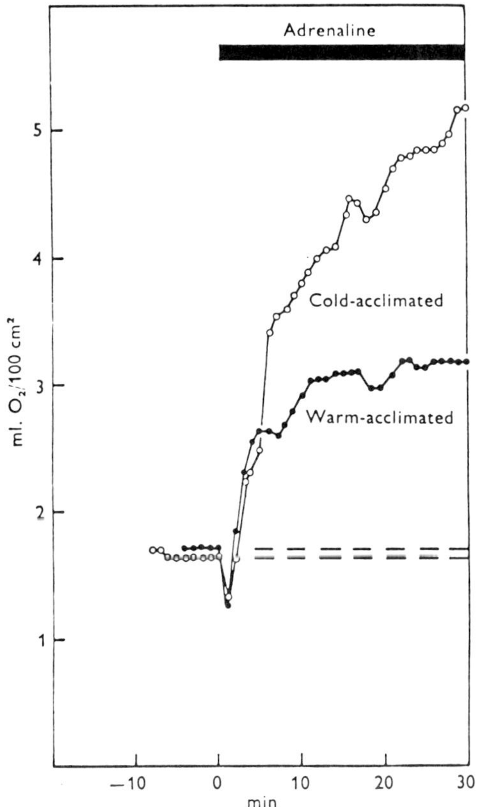

FIGURE 3a. Oxygen uptake of a warm- and a cold-acclimated rat during infusion of noradrenaline. Noradrenaline was infused intravenously, 0.5 pg/100 cm^2/min, from 0 to 30 min. The rats were lightly anesthetized with sodium pentobarbital. The warm-acclimated rat weighed 478 g and the cold-acclimated rat weighed 373 g; they had lived at room temperature (25°-28°C) and in the cold (4°C) respectively for 13 weeks and their weights at the start of the acclimation period were 202 g and 193 g. The values of 80.5 and 22.0 on the graph are obtained from the area under the curve during the 30 min of infusion of noradrenaline and they represent the total increase in oxygen uptake in ml O_2/100 cm^2 in 30 min. The increases shown are typical of rats kept under these conditions (From Himms-Hagen 1970).

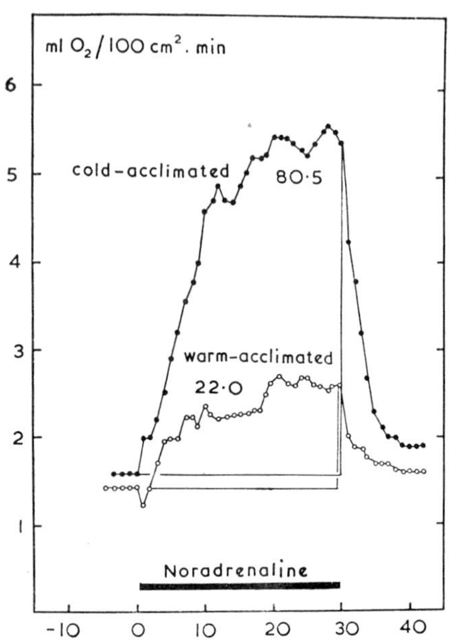

*FIGURE 3b. Effect of intravenous infusion of adrenaline
on oxygen uptake of a warm- and cold-acclimated rat. Adrenaline, 0.5 pg/100 cm^2 body surface, min was infusted via
a tail vein from 0-30 min. The warm-acclimated rat weighed
415 g, the cold-acclimated rat weighted 331 g; they had
lived at room temperature (25-28ºC) or at 4ºC for 9 weeks
and the cold-acclimated rat received the infusion within
one hr of removal from the cold (From Himms-Hagen 1969).*

clear response with regard to the role of the adrenosympathetic system in adaptation to cold is that there is an
increased thermogenic sensitivity to norepinephrine in small
mammals following exposure to cold.
     One last topic bears discussion and that is Himms-
Hagen's (1969, 1970) suggestion that brown adipose tissue
itself may have an endocrine role in potentiating adjustment to cold and NST associated with acclimation to cold.
In these experiments, Himms-Hagen found that when brown adipose tissue was dissected from animals, a high NST was still
found for a day following the surgical procedure and then
was lost following four days. Subsequently, one can re-
implant the brown adipose tissue in the peritoneal cavity
and demonstrate again an increase in NST. These data suggested that BAT had an endocrine role (Figure 4); but recent

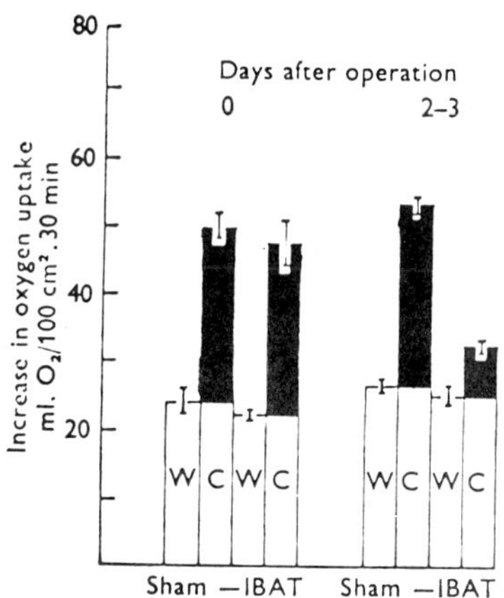
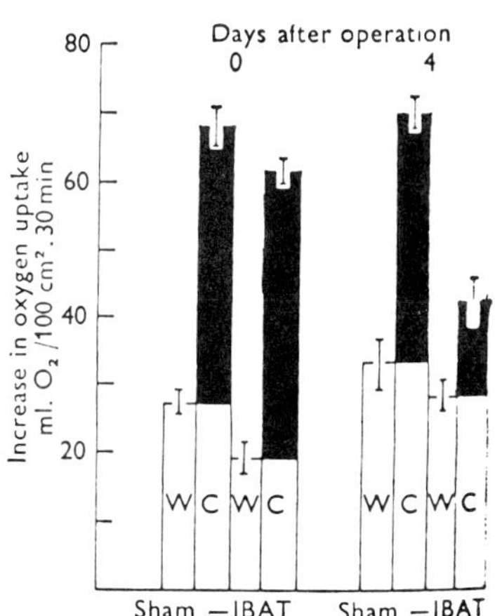

FIGURE 4. *Effect of removal and replacement of the interscapular brown adipose tissue of warm- and cold-acclimated rats on the calorigenic response to noradrenaline. Values shown represent the total increase in oxygen uptake during a 30 min infusion of noradrenaline (see caption to Fig. 1). Each bar is the mean of 4 experiments ± standard error of the mean. Cold-acclimated rats are represented by the bars with the black upper portions and warm-acclimated controls are represented by the open bar to the left. The blackened portion of the bar is the difference between the responses of the cold-acclimated rats and the response of the corresponding control warm-acclimated rats; it is a measure of the acclimation to cold. Cold-acclimated rats had lived in the cold for 10 to 13 weeks; warm-acclimated rats had lived at room temperature during this same period. Rats were removed from cold or warm rooms on day 0, anesthetized with either and operated upon as follows: (i) sham-operated, (ii) interscapular brown adipose tissue removed, (iii) interscapular brown adipose removed, cleaned and weighed, cut into 4 or 5 portions and replaced in the peritoneal cavity; a similar laparotomy was performed also on rats in groups (i) and (ii). Rats remained at room temperature until the infusion of noradrenaline (From Himms-Hagen 1970).*

strong evidence has been obtained which makes superfluous the hypothesis that that BAT of the cold-acclimated rat plays an intermediary role as an endocrine gland whose secretory product modifies the ability of other tissues to respond calorigenically to catecholamines. Foster and Frydman (1978) have demonstrated that BAT can now be regarded as the dominant site of the calorigenesis induced by the administration of NE to the cold-acclimated rat.

In summary then, the endocrine glands and hormones involved in modifying the thermogenic responses shown by mammals in response to cold are still fairly unclear. The only things we seem to know for certain are that the thyroid gland and thyroid hormones are necessary in order to potentiate the action of catecholamines in stimulating NST, but increased thyroid activity is not necessary. Secondly, it is suggested but not necessarily proven, that hormones of the adrenal cortex may be necessary to mobilize substrates such as free fatty acids in the early stages of adaptation to cold; and lastly, it appears that increased secretion of epinephrine and NE and increased sensitivity to their actions on BAT stimulating NST are also a response to cold exposure.

NATURAL SPECIES

Hormonal mechanisms of adaptation to cold have been studied in few natural species and almost none directly from the field. Heroux (1961) found that wild Norway rats *(Rattus norvegicus)* captured in winter did not show enlargement of the thyroid, adrenals, or pituitary. This is similar to the response of rats kept in groups of 20 in cages placed outside in winter, but is in contrast to the response of rats acclimated to cold in the laboratory. During the winter, wild Norway rats developed metabolic adjustments similar to lab cold-acclimated white rats. The rate of shivering in winter-captured wild rats was lower than in summer-captured animals, indicating the presence of non-shivering thermogenesis and the metabolic response to norepinephrine was enhanced in the winter (Heroux 1962). Also the adrenal steroid secretion rate of wild rats captured in winter was greater than during summer. The seasonal adjustment in adrenal activity is similar to that found in white rats which are kept in groups in cages exposed outdoors to natural summer and winter, but it differs from that found in white rats acclimated to cold in the laboratory. In such rats adrenal activity is lower than normal after cold acclimation (Willmer and Heroux 1963). Aleksuik and Frohlinger (1971) found that the weight of the thyroid in muskrat *(Ondatra zibethicus)* decreases during winter, suggesting lower activity. These very different responses from laboratory studies indicate that much more work should be done on native arctic or subarctic species, both acclimated to cold

(for comparison to lab animals) and acclimatized to winter.
Feist and Rosenmann (1975) studied the role of catecholamines in cold acclimation of snowshoe hares, *(Lepus americanus)*. They found an increased sensitivity to norepinephrine and enhanced NST in winter hares.

There are very few data which have been gathered on
changes in NST in homeotherms other than mammals. The usual
conclusion is that birds do not show NST. However, Freeman
(1971) has argued that it is equivocal whether domestic
chicks, during the neonatal period, produce heat by non-
shivering thermogenesis. In a more recent article (Freeman
1976) he reviews the arguments for NST in neonate chicks,
pointing out that the mechanism for NST must be different
from mammals since BAT has not been demonstrated in any
bird. He points out that there are three lines of evidence
supporting the NST argument and that two are now capable of
alternate explanations. One of these, the reduced thermoregulatory ability with propranolol, has been one of the
strongest arguments. However, it has been pointed out that
propranolol reduces heart rate and that may be why the birds
show impaired thermoregulatory ability (Marley and Stephenson 1975). But Freeman (1976) mentions that ". . . the
observations that the hypothermic activity of propranolol
declines with age (Wekstein and Zolman 1969) remains to be
explained and indicates that a description of its activity
solely in terms of reducing heart rate . . . is too simple."

All of Freeman's arguments for NST are based on neonate
fowl, and even he agrees that NST probably does not exist in
adult birds. However, if one looks carefully at the studies
of West (1972a) and Pohl and West (1973) with redpolls *(Acanthis flammea)* studied near Fairbanks, these birds do have
increased metabolic rates in the winter compared to summer.
West (1972b) has also reported higher thermoneutral metabolic rates of willow ptarmigan *(Lagopus lagopus)* in winter
than in summer. These increases are apparent not only at
low temperature but also within the thermoneutral zone.
Although it is possible that the animals may be shivering,
it seems more likely that they may be a candidate for studying some form of NST which is not associated with brown
adipose tissue.

ACCLIMATIZATION

     One of the major differences between acclimatization
and acclimation to cold is that coincident with changes in
temperature as winter approaches are changes in photoperiod.
This allows that there may be other hormonal systems involved in acclimatization responses not involved in acclimation responses. It has already been mentioned that
muskrat show different thyroid responses in winter than
might be predicted based on acclimation responses to cold of
lab animals. Unfortunately, no studies of norepinephrine
were performed on these animals. Also, Heroux (1963) compared white rats held as individuals or in groups placed in
the cold, to white rats held in groups outdoors and wild
rats caught outdoors in the winter. He found that all
groups could withstand -30°C better than summer animals and
had elevated summit metabolism and NST. However, the outside groups had better insulation and were able to withstand
the cold with less incidence of frostbite and other cold
injuries than animals which were exposed to cold in the
laboratory.
     Further, a number of recent studies have shown that
there are very dramatic seasonal differences in thermoneutral metabolism of animals captured in the field at
different times of the year (Lynch 1973; Rosenmann et al.
1975; Wunder et al. 1977). Wunder et al. (1977) has also
found that the acclimation responses of prairie voles
(Microtus ochrogaster) to both heat and cold acclimation are
different when the animals are caught at different times of
the year. Winter animals are capable of showing increases
in metabolism in response to cold acclimation but do not
show changes in metabolism when heat acclimated. Conversely,
summer animals do not show acclimation to cold whereas they
do show a decrease in metabolism when acclimated to heat.
Recently, Heldmaier (pers. comm.) found seasonal changes in
the metabolism of djungarian hamsters *(Phodopus sungorus)*
when these animals are held at constant temperature in the
laboratory but exposed to natural photoperiods, by being
kept in a room with a window to the outside. Hence, the
changes in metabolism which he measures are not a response
to changes in seasonal cold but seem to be cued primarily by
changes in photoperiod. All of these studies suggest that
something other than ambient temperature is driving the
seasonal adjustment of these small mammals. One of the most
obvious factors other than temperature is photoperiod. If
photoperiod is being used as a cue by animals to adjust
seasonally to potentially different thermal environments,

what organs and what hormones may be involved? We already
know from a number of studies with lizards that the parietal
eye and its associated structure, the pineal gland, will
affect the ability of those animals to thermoregulate be-
haviorally.

We also know in mammals that melatonin can in some
instances mimic the effects of photoperiod on development of
brown adipose tissue. Heldmaier and Hoffman (1974) found
that they could modify the amount of brown adipose deposited
in djungarian hamsters by changing photoperiod and that by
implanting melatonin infusion pellets, they could simulate
short-day photoperiods in animals which were simultaneously
seeing long-day photoperiods. Secondly, Lynch and Epstein
(1977) have shown that warm-exposed, white-footed mice *(Peromyscus leucopus)* given melatonin implants, will show all
the characteristics of animals entering a fall or winter
period. They show regression of the reproductive system, a
fall molt, hypertrophy of brown adipose tissue, increases in
basal metabolic rate, and increased sensitivity to NE as
shown by increased NST. Similar responses to melatonin have
been shown in the djungarian hamster by Hoffman (1973).
Thus, it appears that in natural systems it may be important
to look at the interaction of the pineal and melatonin on
the actions of catecholamines and the thyroid in controlling
the adaptations of homeotherms to cold stress in arctic
winters.

REFERENCES

Aleksiuk, M., and A. Frohlinger. 1971. Seasonal metabolic
     organization in the muskrat (*Ondatra zibethicus*). I.
     Changes in growth, thyroid activity, brown adipose
     tissue and organ weights in nature. *Can. J. Zool.*
     *49:*1143-1154.
Boulouard, R. 1963. Effects of cold and starvation on adren-
     ocortical activity of rats. *Fed Proc.* 22:750-753.
          . 1966. Adrenocortical activity during adaptation
          to cold in the rat: role of Porter-Silber chro-
          mogens. *Fed. Proc.* 25:1195-1199.
Brown-Grant, K. 1965. Changes in thyroid activity of rats
     exposed to cold. *J. Physiol.* 131:52-57.
Cottle, M., and L.D. Carlson. 1956. Turnover of thyroid
     hormone in cold exposed rats determined by radioactive
     iodine studies. *Endocrinology.* 59:1-11.

Desalauriers, R., et al. 1971. Caloric uptake and cold
     resistance in cold-acclimated rats fed commercial chow
     or semipurified diet. *Can. J. Physiol. Pharmacol.*
     49:707-712.

Feist, D., and M. Rosenmann. 1975. Seasonal sympatho-adrenal
     and metabolic responses to cold in the Alaskan snowshoe
     hare *(Lepus americanus macfarlani).*   *Comp. Biochem.*
     Physiol. 51A:449-455.

Foster, D.O. and M.L. Frydman. 1978. Non-shivering thermo-
     genesis in the rat.   II.   Measurements of blood flow
     with microspheres point to brown adipose tissue as the
     dominant site of the calorigenesis induced by nor-
     adrenaline. *Can. J. Physiol. Pharmacol.56*:110-122.

Freeman, B.M. 1971. Body temperature and thermoregulation.
     Pages 1115-1151 in D.J. Bell and B.M. Freeman, eds.
     Physiology and biochemistry of domestic fowl. Academic
     Press, New York.

______. 1976. Thermoregulation in the young fowl *(Gallus*
     domesticus).   *Comp. Biochem. Physiol. 54A*:41-144.

Hart, J.S. 1957. Climate and temperature-induced changes in
     the energetics of homeotherms. *Rev. Can. Biol.*
     16: 133-174.

Heldmaier, G. 1977. Personal Communication.

Heldmaier, G., and K. Hoffman. 1974. Melatonin stimulates
     growth of brown adipose tissue. *Nature 247*:224-225.

Heroux, O. 1960. Adjustments of the adrenal cortex and
     thyroid during cold acclimation. *Fed. Proc. 19*:82-85.

______. 1961. Seasonal adjustements in captured wild rats. I.
     Organ weidghts, ear vascularization, and histology of
     epidermis. *Can. J. Biochem. Physiol. 39*:1865-1870.

______. 1962. Seasonal adjustments in captured wild Norway
     rats. II. Survival time, pelt insulation, shivering,
     and metabolic and pressor responses to noradrenaline.
     Can. J. Biochem. Physiol. 40:538-545.

______. 1963. Patterns of morphological, physiological and
     endocrinological adjustments under different environ-
     mental conditions of cold. *Fed. Proc. 22*:789-792.

______. 1969. Diet and cold resistance. *Fed. Proc. 28*:955-959.

Heroux, O., and V.M. Petrovic. 1969. Effect of high and low
     bulk diets on the thyroxine turnover rate in rats with
     acute and chronic exposure to different temperatures.
     Can. J. Physiol. Pharmacol. 47:963-968.

Himms-Hagen, J. 1969. The role of brown adipose tissue in
     the calorigenic effect of adrenaline and noradrenaline
     in cold-acclimated rats. *J. Physiol. Lond.*
     205:393-403.

Himms-Hagen, J. 1970. Regulation of metabolic processes in
    brown adipose tissue in relation to non-shivering
    thermogenesis. Pages 131-151 in G. Weber, ed. Advances
    in enzyme regulation. Pergamon Press., Oxford.
______. 1975. Role of the adrenal medulla in adaptation to
    cold. Pages 637-665 in H. Blaschko, G. Sayers, and A.D.
    Smith, eds. Handbook of physiology, endocrinology, VI.
    Amer. Physiological Society, Washington, D.C.
Hoffman, Klaus. 1973. The influence of photoperiod and mela-
    tonin on testis size, body weight, and pelage colour in
    the Djungarian hamster *(Phodopus sungorus)*. *J. Comp.
    Physiol. 85*:267-282.
Hsieh, A.C.L., and L.D. Carlson. 1957. Role of adrenaline
    and noradrenaline in chemical regulation of heat pro-
    duction. *Amer. J. Physiol. 190*:243-246.
Hudson, J.W. 1973. Torpidity in Mammals. Pages 98-165 in
    G.C. Whittow, ed. Comparative physiology of thermore-
    gulation. Vol. III. Special aspects of thermoregula-
    tion. Academic Press, New York.
LeBlanc, J. 1975. Man in the cold. C. C. Thomas, Spring-
    field, Il.
LeBlanc J., and A. Villemaire. 1970. Thyroxine and noradren-
    aline on noradrenaline sensitivity, cold resistance,
    and brown fat. *Am. J. Physiol 218*:1742-1745.
Leduc, J. 1961. Catecholamine production and release in
    exposure and acclimation to cold. *Acta Physiol. Scandia
    53*:Suppl. 183.
Lynch, G.R. 1973. Seasonal changes in thermogenesis, organ
    weights, and body composition in the white-footed
    mouse, *Peromyscus leucopus.   Oecologica.
    13*:363-376.
Lynch, G.R., and A.L. Epstein. 1977. Melatonin-induced
    changes in gonads, pelage, and thermogenic characters
    in the white-footed mouse, *Peromyscus leucopus.
    Comp. Biochem. Physiol. 53C*:67-68.
Marley, E., and J.D. Stephenson. 1975. Effects of noradrena-
    line infused into the chick hypothalamus on thermore-
    gulation below thermoneutrality. *J. Physiol. Lond.
    245*:289-303.
Pohl, H., and G.C. West. 1973. Daily and seasonal variation
    in metabolic response to cold during rest and forced
    exercise in the common redpoll. *Comp. Biochem. Physiol.
    45A*: 851-867.
Rosenmann, M., P. Morrison, and D. Feist. 1975. Seasonal
    changes in the metabolic capacity of red-backed voles.
    Physiol. Zool. 48:303-310.
Schonbaum, E. 1960. Adrenocortical function in rats exposed
    to low environmental temperatures. *Fed. Proc. 19*:85-88.

Smith, R.E., and B.A. Horwitz. 1969. Brown fat and thermo-
    genesis. *Physiol. Rev. 49*:330-425.
South, F.E., et al. 1972. Hibernation and hypothermia, per-
    spectives and challenges. American Elsevier, New York.
Straw, J.A., and M.J. Fregly. 1967. Evaluation of thyroid
    and adrenal-pituitary function during cold acclima-
    tion. *J. Appl. Physiol 23*:825-830.
Wang, L., and J.W. Hudson, eds. 1978. Strategies in cold:
    natural torpidity and thermogenesis. Academic Press,
    New York. 715 pp.
Webster, A.J.F. 1974. Adaptation to cold. Pages 71-106 in
    D. Robertshaw, ed. Environmental physiology. Univ.
    Park Press, Baltimore, MD.
Wekstein, D.R., and J.F. Zolman. 1969. Ontogeny of heat
    production in chicks. *Fed. Proc. 28*:1023-1028.
West, G.C. 1972a. The effect of acclimation and acclimati-
    zation on the resting metabolic rate of the common
    redpoll. *Comp. Biochem. Physiol. 43A*:293-310.
______. 1972b. Seasonal differences in resting metabolic
    rate of Alaskan ptarmigan. *Comp. Biochem. Physiol.
    42A*: 867-876.
Willmer, J.A. and O. Heroux. 1963. Seasonal adjustments in
    captured wild Norway rats. III. Production of adrenal
    steroids *in vitro.   Can J. Biochem. Physiol.
    41*: 1147-1153.
Wunder, B.A., D.S. Dobkin, and R.D. Gettinger. 1977. Shifts
    of thermogenesis in the prairie vole *(Microtus och-
    rogaster).   Oecologia. 29*:11-26.

DISCUSSION I. THE ROLE OF HORMONES IN THE RESPONSES OF HOMEOTHERMS TO COLD[1]

Melvin J. Fregly

Department of Physiology
University of Florida
College of Medicine
Gainesville, Florida

Acclimation to cold air involves changes in the rates of secretion of many endocrine glands in mammals. These changes appear to be directed toward facilitation of body temperature regulation. While the rates of secretion of hormones by individual endocrine organs of mammals exposed to cold are important, an understanding of their interaction one with another is equally important. Present knowledge of hormonal interactions occurring during exposure to cold suggest that most hormones may interact positively to facilitate body temperature regulation but a few may interact negatively. The need to characterize these interactions and their importance to both body temperature regulation and cold acclimation is stressed.

HORMONAL RESPONSES TO COLD

There is agreement among investigators in the field of thermoregulation that hormones are important in the maintenance of the core temperature of homeotherms. Experimentally, most emphasis has been placed on the effect of certain hormones on heat production and, to a lesser extent, on heat loss. However, hormones play other, less well studied, roles in the responses of homeotherms to cold. A few of these many roles may include: changes in cellular permeability and transport processes; alteration in central nervous system responsiveness and activity; increases in

[1]Supported by the Office of Naval Research Contract N0014-68-A-0173-0007 with funds provided by the Naval Bureau of Medicine and Surgery.

food intake; regulation of fluid balance; changes in reproductive behavior, implantation, gestation, lactation, etc.
     At the present time, only incomplete information is
available for any single hormone and its role in the
responses. Complete hormonal information for a given hormone and a given homeothermic species requires a knowledge
of the rates of production and secretion, mechanisms of
transport and binding to plasma proteins, receptor responsiveness, as well as the rates of peripheral "utilization"
and hepatic conjugation of the hormone.
     Thus, the rates of secretion of thyroxins (Cottle and
Carlson 1956; Sellers and You 1950; Straw and Fregly
1967), glucocortidoids (Heroux and Hart 1954; Straw and
Fregly 1967; Boulouard 1966; Schönbaum 1960) and catecholamines (Leduc 1961; Lutherer et al. 1969; Sellers et al.
1971) are reported to increase in cold-exposed animals
while the rate of secretion of antidiuretic hormone may
decrease (Itoh et al. 1959; Moore and Seeger 1966). Less
information is available for aldosterone, gonadal hormones,
insulin, glucagon, growth hormone, parathyroid hormone,
melatonin, serotonin and prostaglandins. Because hormonal
secretion rates are well known to follow circadian cycles
in animals exposed to a thermoneutral temperature, measurements made at intervals throughout a 24-hour period will be
required to determine whether exposure to cold changes
either the amplitude of the cycle or its periodicity or
both. Because of the cyclic variation in hormonal secretion
rates, it is theoretically possible to increase the mean
hormonal secretion rate during cold exposure without affecting maximal hormonal secretion rate. Such measurements
are not available for any single hormone in cold-exposed
animals. The diurnal variation in secretion rate emphasizes the necessity for more than one measurement of blood
hormone concentration during a 24-hour cycle.
     Steroid hormones and thyroxine are poorly soluable in
saline and are transported in blood bound to plasma protein. The bound hormone is in equilibirum with the free
hormone. The binding of the different hormones to plasma
protein varies from approximately 20 percent for aldosterone to 99.9 percent for thyroxine. It is important to
recognize that it is the free hormone, and not the bound
hormone, that is important physiologically. We have only
very incomplete information regarding the effects of exposure to either acute or chronic cold on the extent of
binding of hormones to proteins, the binding characteristics
of the proteins themselves, as well as their rates of
production and metabolism.

Hormonal, drug, and other responses are elicited by way of receptors. The possibility that the responsiveness of hormonal receptors may be altered by cold seems real, since such an increase in responsiveness for catecholamine constitutes the definition of nonshivering thermogenesis. In this regard, the question whether α or β adrenergic receptors, or both, play a role in non-shivering thermogenesis requires additional study. Here the imaginative use of pharmacologic agents, adrenergic agonists and antagonists, as well as synthetic steroids and steroid antagonists, offer promise for a better understanding of changes in hormonal receptor responsiveness during cold exposure. The possibility that circadian, and other cycles, affect receptor sensitivity will also require consideration in both warm- and cold-adapted animals.

A significant proportion of every hormone produced by an endocrine gland is conjugated and inactivated by the liver before it can exert its characteristic effect. Other organs, including the kidney and peripheral tissues, also contribute. A reduction or an increase in blood level of a hormone can occur as easily by changes in rates of metabolism as by rates of production of hormones. We need more information regarding the effects of acute and chronic cold exposure on rates of metabolism and conjugation of the various hormones.

Any study of the role of hormones in the survival of the cold-exposed animal must include the interaction of hormones one with another, as well as their individual roles (Fregly 1975). Thus, thyroxine has been shown to interact with glucocorticoid hormones in the maintenance of body temperature of cold-exposed, adrenalectomized rats. Such animals could not be rendered "normal" with respect to their ability to maintain body temperature during exposure to cold even by large doses of either cortisone acetate or whole adrenal cortical extract. However, simultaneous administration of a small dose of thyroxine with the adrenal cortical hormone returned responsiveness of the adrenalectomized rat to that of intact controls (Fregly 1960).

Another well-known hormonal interaction is that of either norepinephrine or epinephrine with thyroxine. Swanson (1956) showed that hypothyroid rats manifest a poor metabolic response to administration of epinephrine that is correctable by the simultaneous administration of thyroxine. Studies carried out in our laboratory have expanded our understanding of this observation by showing that it is the β-adrenergic receptor whose responsiveness is affected by the presence of thyroid hormones (Fregly et al. 1975;

1976; Lutherer et al. 1978). A large body of experimental
evidence suggests a metabolic interrelationship between
thyroid hormones and catecholamines (Lutherer et al. 1969;
Sellers et al. 1971; Bray and Goodman 1965). Many of the
metabolic effects of catecholamines are associated with an
increase in the level of cyclic 3',5' -AMP (adenosine
monophosphate) mediated through activation of the enzyme,
adenylcyclase, which converts adenosine triphosphate to
cyclic 3',5' -AMP. Recent studies of Armstrong et al.
(1974) and Van Inwegen et al. (1975) suggest that increasing blood levels of thyroxine inhibit the enzyme, phos-
phodiesterase, which metabolizes cyclic 3',5' -AMP to 5' -
AMP. This effectively increases the half-life of cyclic
3',5' -AMP. This interaction may be of considerable significance to the cold-exposed animal and may constitute a
mechanism by which non-shivering thermogensis develops.

Others (Parvez and Parvez 1972; Pohorecky and Wurtman
1971) have shown that glucocorticoid hormones play an
important role in regulating adrenal medullary epinephrine
content as well as the content of the adrenal medullary
enzyme, phenylethanolamine-N-methyltransferase, which
catalyzes the methylation of norepinephrine to epinephrine.
Glucocorticoids also appear to maintain a normal pattern of
urinary excretion of catecholamines and of 3-methoxy-4-
hydroxy mandelic acid (VMA). The latter depends mostly on
the amount of norepinephrine present in the sympathetic
nerves and adrenal glands. Parvez and Parvez (1972) also
suggest that the presence of glucocorticoid hormones serves
as a brake on the activity of the enzyme, monoamine oxidase; loss or reduction of glucocorticoid activity increases catecholamine degradation. This interaction between glucocorticoids and catecholamines should be of
considerable importance to the animal exposed to cold but
has not yet been studied in such animals.

There is a large body of experimental evidence to
support the statement that either glucocorticoid hormones
or norepinephrine can inhibit the antidiuretic action of
exogenously administered pitressin (Gaunt et al. 1957; Levi
et al. 1973; Sadowski et al. 1972). This interaction is of
particular interest since cold-exposed animals are well
known to have an increase in blood levels of both glucocorticoid hormones and norepinephrine. These hormones may
inhibit the responses to endogenous antidiuretic hormone
and thereby contribute, at least partially, to the cold-
induced dehydration observed in animals (Fregly 1967, 1968;
Fregly and Waters 1966). This possibility is currently
under study (Fregly 1978).

A further interaction that may occur during exposure to cold relates to the effect of the increased thyroid hormone secretion rate on hepatic conjugation of glucocorticoid hormones. Yates et al. (1958) showed that an increased thyroid hormone secretion rate was accompanied in the laboratory rat by an increased rate of hepatic metabolism of glucocorticoid hormone and in turn, by an increased rate of secretion of glucocorticoids stimulated by way of the feedback mechanism and pituitary ACTH. An increased glucocorticoid secretion rate may affect thyroid hormone secretion rate by inhibiting uptake of iodide by the thyroid gland as well as by increasing urinary excretion of iodide (Berson and Yalow 1952). Glucocorticoids also appear to inhibit peripheral conversion of thyroxine to triiodothyronine (Chopra et al. 1975) as well as production of thyrotrophin releasing hormone at the level of the hypothalamus (Wilber and Utiger 1969; Haigler et al. 1971). An understanding of the significance of these interactions in any single species during cold exposure will require much additional information.

Both Hollander et al. (1967) and Radomski et al. (1968) have reported that an increase in serum fatty acid concentration may displace thyroxine from serum protein binding sites. This interaction would also seem appropriate during exposure to cold when both an increased availability of energy substrate and thyroxine appear to be important in the maintenance of body temperature. The increased circulating levels of free thyroxine could result in the increased hepatic and peripheral matabolism of this hormone observed during cold exposure in laboratory rats (Intoccia and Van Middlesworth 1959; Kassenaar et al. 1959; Heroux and Brauer 1965; Cottle and Veress 1966, Galton 1971).

Some of the hormonal interactions discussed here are shown in the form of a schema in Figure 1. This schema does not represent all of the hormonal interactions occurring during exposure to cold and is largely representative of information available for the laboratory rat. The schema suggests that the acute and chronic responses of the whole animal to cold are not a function of a single endocrine gland or hormone but of a number of endocrine glands and hormones acting in concert. While one may generally assume these hormonal interactions enhance the beneficial effects of each alone, there are interactions that are inhibitory (Figure 1). The initial common pathway, if one exists, or if not, the sequence of events stimulating the various endocrine organs to activity during exposure to cold needs further study as does the role and interaction

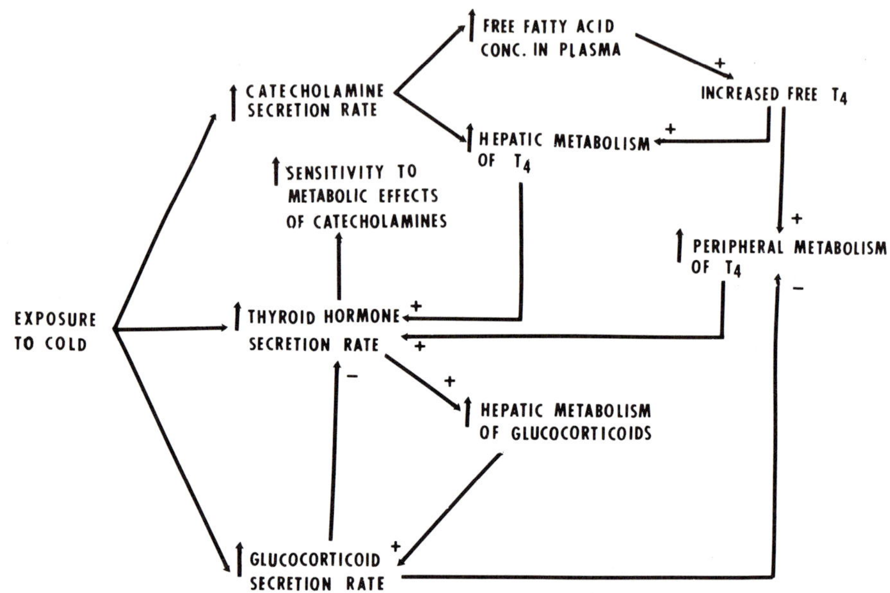

FIGURE 1.   *A schema illustrating some hormonal interactions that may occur in rats exposed to cold. An increased catecholamine secretion rate is accompanied by an increased fatty acid concentration in plasma as well as by increased hepatic metabolism of thyroxine (T₄). The effect of an increased free fatty acid concentration in plasma is to increase free T₄ concentration in plasma. This results in increased hepatic and peripheral metabolism of T₄. The accompanying reduction in plasma T₄ results in an increased thyroid hormone secretion rate by way of the pituitary TSH feedback mechanism. The increased thyroid hormone secretion rate is accompanied by an increased sensitivity to the metabolic effects of catecholamines. Increased thyroid hormone secretion is also accompanied by an increased hepatic metabolism of glucocorticoids (Yates et al. 1958) and, in turn by an increased glucocorticoid secretion rate, initiated by way of pituitary ACTH. An increased glucocorticoid secretion rate may affect thyroid hormone secretion rate by inhibiting uptake of iodide by the thyroid gland as well as by increasing urinary excretion of iodide (Berson and Yalow 1951). Glucocorticoids also appeare to inhibit peripheral conversion ot T₄ to T₃ (Chopra et al. 1975), and to inhibit production of TRF in the hypothalamus. This schema is incomplete and represents only some of the better known hormonal interactions occurring during exposure to cold.*

of other hormones such as aldosterone, growth hormone, glucagon, insulin, serotonin, melatonin, prostaglandins, parathyroid hormones and calcitonin. The final common pathways for hormonal interaction relate to both maintenance of substrates for heat production and prevention of heat loss. Hormonal influences on physiological functions other than those related directly or indirectly to heat production and heat loss require a great deal more attention than they have received heretofore. These include reproduction (implantation, gestation, birth, lactation); maintenance of fluid and electrolyte balance; central nervous system activity and behavior;and cardiovascular activity. It is to all these areas that future research must be directed for a more complete understanding of hormonal interactions in the maintenance of body temperature of homeotherms exposed to cold.

REFERENCES

Armstrong, K.J., et al. 1974. Effects of thyroid hormone deficiency on cyclic adenosine 3',5'-monophosphate and control of lipolysis in fat cells. *J. Biol. Chem.* *249*:4226-4231.

Berson, S.A., and R.S.Yalow. 1952. The effect of cortisone on the iodine accumulating function of the thyroid gland in euthyroid subjects. *J. Clin. Endocrinol. Metab.* *12*:407-422.

Boulouard, R. 1966. Adrenocortical activity during adaptation to cold in the rat: role of Porter-Silber chromogens. *Fed. Proc.* *25*:1195-1199.

Bray, G.A., and H.M. Goodman. 1965. Studies on the early effects of thyroid hormones. *Endocrinol.* *76*:323-328.

Chopra, I.J., et al. 1975. Opposite effects of dexamethasone on serum concentrations of 3,3',5'-triiodothyronine (Reverse T$_3$) and 3,3',5-triiodothyronine (T$_3$). *J. Clin. Endocrinol. Metab.* *41*:911-920.

Cottle, W.H., and L.D. Carlson. 1956. Turnover of thyroid hormones in cold-exposed rats determined by radioactive iodine. *Endocrin.* *59*:1-11.

Cottle, W.H., and A.T. Veress. 1966. Serum binding and biliary clearance of triiodothyronine in cold-acclimated rats. *Can. J. Physiol. Pharmacol.* *44*:571 574.

Fregly, M.J. 1960. Interaction of adrenals and thyroid in maintenance of body temperature of rats exposed to cold. *Am. J. Physiol. 199*:437-444.

______. 1967. Effect of exposure to cold on evaporative loss from rats. *Am. J. Physiol. 213*:1003-1008.

______. 1968. Water and electrolyte exchange in rats exposed to cold. *Can. J. Physiol. Pharmacol. 46*:873-881.

______. 1975. Hormonal interactions in body temperature regulation. Pages 308-325 in G.J. Mogenson and F.R. Calaresu, eds. Neural integration of physiological mechanisms and behavior. Univ. of Toronto Press, Toronto, Can.

Fregly, M.J., and I.W. Waters. 1966. Water intake of rats immediately after exposure to a cold environment. *Can. J. Physiol. Pharmacol. 44*:651-662.

Fregly, et al. 1976. Effect of hypothyroidism on responsiveness to B-adrenergic stimulation. *Can. J. Physiol. Pharmacol. 54*:200-208.

Fregly, M.J., et al. 1975. Reduced B-adrenergic responsiveness in hypothyroid rats. *Am. J. Physiol. 229*:916-924.

Galton, V.A. 1971. Environmental effects. In S.C. Werner and S.H. Ingbar, eds. The thyroid, a fundamental and clinical text. 3rd ed. Harper and Row, New York.

Gaunt, R., C.W. Lloyd, and J.J. Chart. 1957. The adrenal-neurohypothyseal interrelationship. Pages 233-250 in H. Heller, ed., The neurohypophysis. London, Butterworth.

Haigler, E.D., Jr. et al. 1971. Direct evaluation of pituitary thyrotropin reserve utilizing synthetic thyrotropin releasing hormone. *J. Clin. Endocrinol. Metab. 33*:573-581.

Heroux, O., and R. Brauer. 1965. Critical studies on determination of thyroid secretion rate in cold-adapted animals. *J. Appl. Physiol. 20*:597-606.

Heroux, O., and J.S. Hart. 1954. Adrenal cortical hormone requirement of warm- and cold-acclimated rats after adrenalectomy. *Am. J. Physiol. 178*:453-456.

Hollander, C.S., et al. 1967. Free fatty acids. A possible regulator of free thyroid hormone levels in man. *J. Clin. Endocrinol. 27*:1219-1223.

Intoccia, A., and L. Van Middlesworth. 1959. Thyroxine execretion increase by cold exposure. *Endocrinol. 64*: 462-464.

Itoh, S., Y. Toyomasu, and T. Konno. 1959. Water diuresis in cold environment. *Japan. J. Physiol. 9*:438-443.

Kassenaar, A., L.D.F. Lameyer, and A. Querido. 1959. Studies on the peripheral disappearance of thyroid hormone. VI. The effect of environmental temperature on the distribution of I^{131}-I in thyroidectomized, L-thyroxine maintained rats after the injection of I^{131} labelled L-thyroxine. *Acta Endocrinol. 32*:575-578.

Leduc, J. 1961. Effect of acclimation to cold on the production and release of catecholamines. *Acta Physiol. Scand. Suppl. 53*:1961.

Levi, J, S.G. Massry, and C.R. Kleeman. 1973. The requirement of cortisol for the inhibitory effect of norepinephrine on the antidiuretic action of vasopressin. *Proc. Soc. Exptl. Biol. Med. 142*: 687-690.

Lutherer, L.O., M.J. Fregly, and A.H. Anton. 1969. An interrelationship between theophylline and catecholamines in the hypothyroid rat acutely exposed. *Fed. Proc. 28*:1238-1242.

Lutherer, L.O. et al. 1978. Reversal by theophylline of some changes characteristically accompanying hypothyroidism in rats. *Pharmacol. 17*:21-31.

Moore, W.W., and W.E. Segar. 1966. The effects of change in position and ambient temperature on blood ADH in the human. *Fed. Proc. 25*:253.

Parvez, H., and S. Parvez. 1972. Catecholamine excretion after hypophysectomy and with hydrocortisone administration. *Am. J. Physiol. 223*:1281-1285.

Pohorecky, L.A., and R.J. Wurtman. 1971. Adrenocortical control of epinephrine synthesis. *Pharmac. Rev. 23*: 1281-1285.

Radomski, M.W., A. Britton, and E. Schönbaum. 1968. A free fatty acid-free thyroxine relationship in serum. *Can. J. Physiol. Pharmacol. 46*:119-120.

Sadowski, J., K. Nazar, and E. Szczepanska-Sadowska. 1972. Reduced urine concentration in dogs exposed to cold, relation to plasma ADH and 17-OHCS. *Am. J. Physiol. 222*:607-610.

Schönbaum, E. 1960. Adrenocortical function in rats exposed to low environmental temperatures. *Fed. Proc. 19: Suppl. 5*:85-88.

Sellers, E.A., and S.S. You. 1950. Role of the thyroid in metabolic responses to a cold environment. *Am. J. Physiol. 163*:81-91.

Sellers, E.A., et al. 1971. Thyroid status in relation to catecholamines in cold- and warm-environments. *Can. J. Physiol. Pharmacol. 49*:268-275.

Straw, J.A., and M.J. Fregly. 1967. Evaluation of thyroid and adrenal-pituitary function during cold acclimation. *J. Appl. Physiol. 23*:825-830.

Swanson, H.E. 1956. Interrelations between thyroxin and adrenalin in the regulation of oxygen consumption in the albino rat. *Endocrinol.* *59*:217-225.

Van Inwegen, R.G., et al. 1975. Cyclic nucleotide phosphodiesterases and thyroid hormone. *J. Biol. Chem.* *250*:2452-2456.

Wilber, J.F. and R.D. Utiger. 1969. The effect of glucocorticoids on thyrotropin secretion. *J. Clin. Invest.* *48*:2096-2103.

Yates, F.E., J. Urquhart, and A.L. Herbst. 1958. Effect of thyroid hormones on ring A reduction of cortisone by liver. *Am. J. Physiol.* *195*:373-380.

DISCUSSION II. IMPORTANCE OF STUDIES ON RESISTANCE TO LOW TEMPERATURE

O. Heroux

Division of Biological Sciences
National Research Council of Canada
Ottawa, Canada

Resistance to low temperature of male Sprague Dawley rats (300 g) fed a laboratory chow (P) or a semi-purified diet (H) for 14 days was evaluated by the degree of hypothermia developed under restraint at +1ºC or -4ºC. Cold resistance: 1) followed a circochan as well as a circannual cycle, 2) was much greater with the H diet than with the P diet, 3) was increased by oral administration of tryptophan, and 4) was correlated with geomagnetic activity. The possibility is raised that resistance to cold might depend on the state of activity of the central serotonin nervous system at the time the stress is imposed. These results are presented to support the call for more research on the identification of the system limiting the duration of cold resistance in normal warm-acclimated animals.

Dr. Dave Foster at the Division of Biological Sciences laboratory has shown very convincingly that brown adipose tissue (BAT) accounts for a very large fraction of non-shivering thermogenesis (NST) produced by infusion of noradrenaline (NA) in cold-acclimated rats, which suggests that in those animals NST is rooted mainly in BAT. If this is true in other animals, it would explain why it is insignificant in man, who has only traces of brown fat in the adult stage.

Therefore, in order to be practical, and if this research is to be applicable to man eventually, it appears that in addition to pursuing studies which try to explain

how heat is produced by NST and how NA mediates it, the
following must also be done:

1. Try to understand why brown fat, which is so
   important in newborns, disappears in the adult
   stage in many species including man; and
2. To develop NST in man, a way must be found to
   increase brown fat without having to expose man
   to prolonged cold.

There is, besides cold adaptation, another aspect of
cold physiology which is somewhat neglected and which would
deserve much greater attention (i.e., cold resistance
through shivering, in man and warm-acclimated animals).

In the past three years, it has been found that
normal non-cold-acclimated rats can resist severe cold or
avoid severe hypothermia for a much longer time if they are
fed a semi-purified diet rather than a commercial chow
(Heroux et al. 1972). Since a semi-purified diet has been
shown by other authors to provide small animals with a
greater degree of resistance than commercial chow to other
stresses such as DDT (Ortego 1966), radiation (Hugon and
Bounous 1972), or hypophysectomy (Shaw and Greep 1949),
which have nothing to do with thermogenesis, at least
directly, the conclusion was reached that the protection
given by the semi-purified diet was not through its influence on energy substrates but through its influence on the
control of homeostasis, or on the central nervous system
(Heroux et al. 1972).

It was then found that with one diet or the other, the
degree of cold resistance depended on the time of the day
the animals were cold exposed (Figures 1 and 2).

Since a similar circadian cycle in resistance to other
stresses such as: heat (Vener, Puggard, and Sloan 1976),
white noise (Halberg, Bittner, and Cully 1955), x-rays
(Halberg 1960), and endotoxin (Halberg 1960) has been
observed, it is concluded or hypothesized that resistance
to stress in general varies with the state of activity of
the central nervous system. Cold resistance is at its
minimum at 2:00 p.m. when rats usually sleep and when the
turnover rate of 5HT is at its minimum in the hypothalamus;
cold resistance is at its maximum at 11:00 p.m. when the
rat is usually active and when 5HT is at its lowest in the
hypothalamus or at the time when the turnover rate of 5HT
is at its fastest.

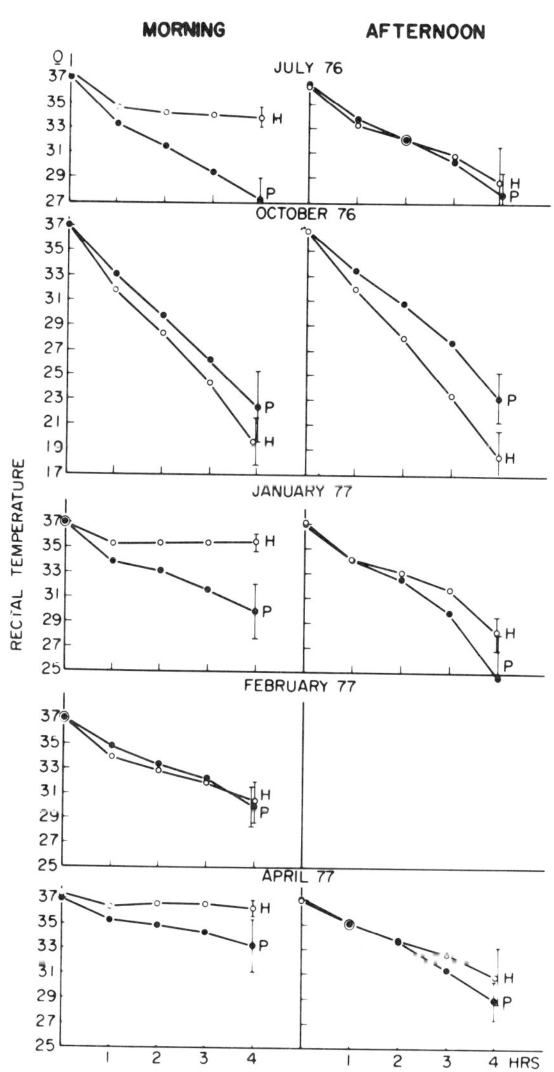

FIGURE 1. Cold resistance, estimated by drop in rectal temperature during 4 hours of restraint at +1°; both in the morning (between 09:00 and 13:00 hours) and in the afternoon (between 14:00 and 18:00 hours) in July, October, January, February, and April in rats fed Purina (P) or H die (H). Bar = standard error of the mean.

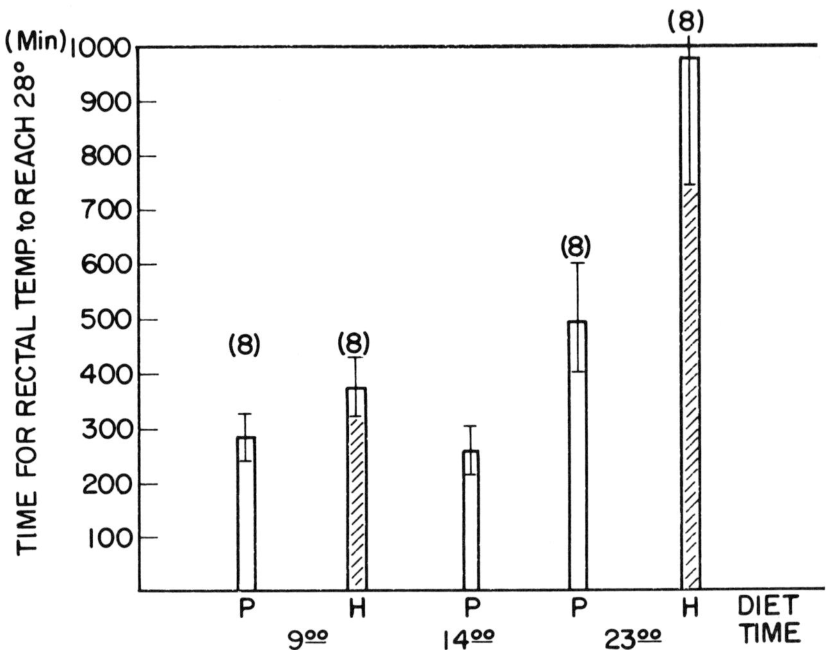

FIGURE 2. *Cold resistance, estimated by the time required for rectal temperature to reach 28° in rats fed P or H diets, when exposed to +1° under restraint at 09:00, 14:00, or 23:00 hours. Bar = standard error of the mean.*

Assuming that resistance to cold in warm-acclimated rats depended on the state of activity of the serotoninergic system at the time cold was imposed, if the turnover rate of 5HT could be stimulated, cold resistance should be increased.

Knowing that an increased turnover rate of 5HT could be achieved by administering ℓ-tryptophan (Movi and Eccleston 1968), the precursor of 5HT, increasing doses of tryptophan, were given to rats by oral intubation at 9:00 a.m. and the maximum dose only at 11:00 p.m. (Figures 3 and 4). On the basis of these two experiments, it appeared that it might be possible to increase cold resistance by oral administration of tryptophan.

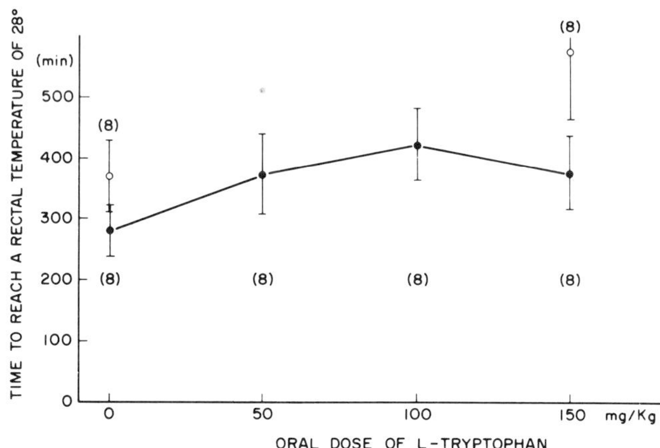

FIGURE 3.   Effect of different oral doses of ℓ-thypto-phan on cold resistance of rats fed Purina or II diet.   Cold resistance was estimated by the time required for rectal tem-perature to reach 28ºC when the animals were exposed to +1º under restraint.   Bar = standard error of the mean, number of animals per group in parenthesis.___. P.   0___0 H diet.

However, these experiments should be repeated and cold resistance should be tested at different times after the administration of tryptophan, not only after one hour, as was the case in these experiments.

Finally, another very interesting observation was made.   In rats that were kept all year around under the same constant temperature, humidity, and light conditions, growth and cold resistance of 300 g rats varied with sea-son, both dropping significantly in spring and fall (Fig-ures 5 and 6).

Since it has been repeatedly shown that the incidence of infarctus (Anderson and Harding Le Riche 1970) and of mental depression (Eastwood and Peacocke 1976) increases in spring and fall, it is very tempting to suggest that perhaps this is so because resistance to stress is reduced at those seasons. It may be possible that a given indivi-dual may become sick in spring or fall because he is then stressed at the most vulnerable times of the circadian and circannual rhythms of his serotoninergic system.   This im-plies that in spring and fall, the metabolism of serotonin would be slower than in summer and winter, resulting in an accumulation of serotonin.   As a matter of fact, increased amounts of serotonin have been found at those two seasons:

in human platelets (Arend et al. 1977), in rat's serum
(Scheving et al. 1972), and in the hypothalamus of hamsters (Kempp et al. 1978).

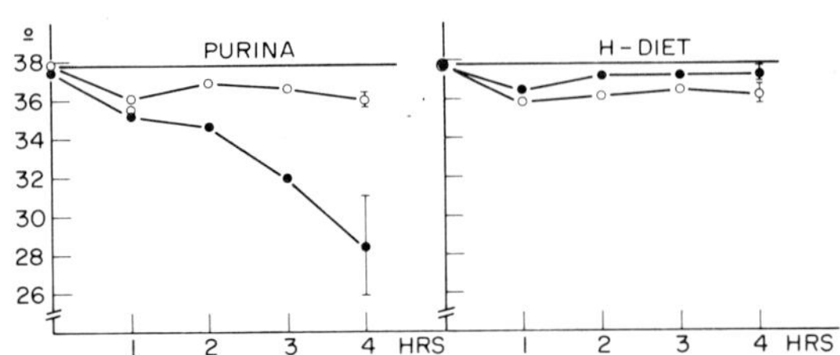

*FIGURE 4.   The effect of the oral administration of
150 mg/kg of ℓ-tryptophan at 22:00 hour on the cold
resistance of rats exposed one hour later (at 23:00 hour)
to -4⁰ under restraint.   Rats had been fed either Purina
(P) or H diet (H) at 28⁰ for 14 days prior to this cold
resistance test.   0___0 rats treated with tryptophan,
.___. controls treated with the vehicle used to dissolve
tryptophan (0. in HCI adjusted to pH7.3).   Bar = standard
error of the mean.*

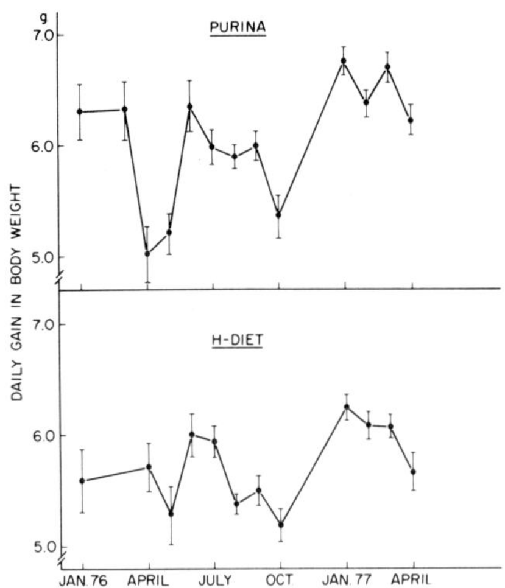

*FIGURE 5.   Average daily gain in body weight at differ-
ent month of the year during the 14 days that 200 g rats
were fed Purina or H diet.   Bar = standard error of the
mean.*

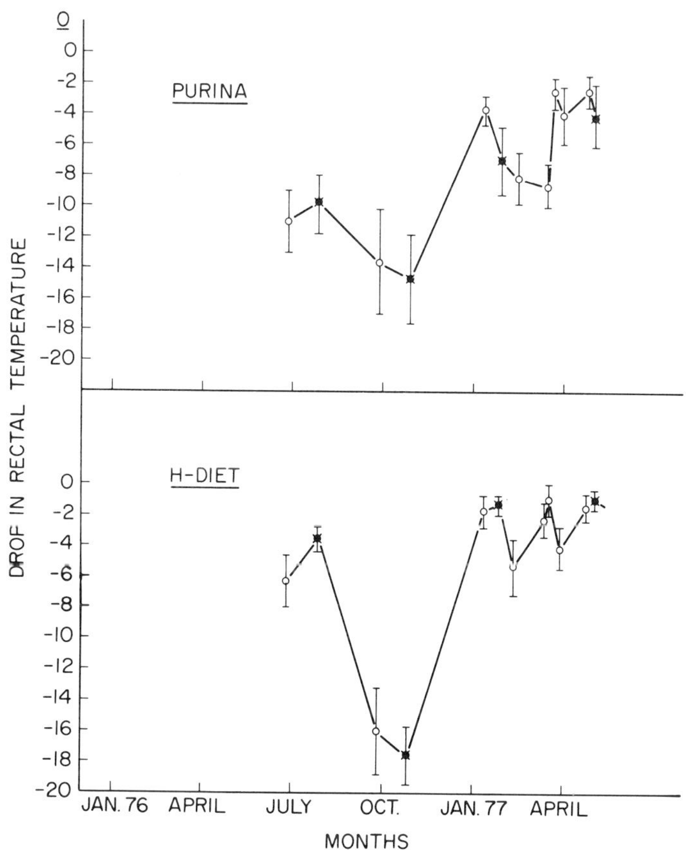

FIGURE 6. Seasonal variations in cold resistance at 09:00 hour (estimated by drop in rectal temperature during a four hour exposure to +1° under restraint of rats fed Purina or H diet. A few circles: rats fed fresh diet, closed squares: rats fed thawed out portions of diets and mates obtained in April and kept frozen at -40° for one year between test. Bar — standard error of the mean.

Because geomagnetic activity varies with the seasons (Figure 7), showing significant increases in spring and fall, the question arose whether any correlation could be seen between this phenomenon and the seasonal variations observed in the growth and cold resistance of our animals.

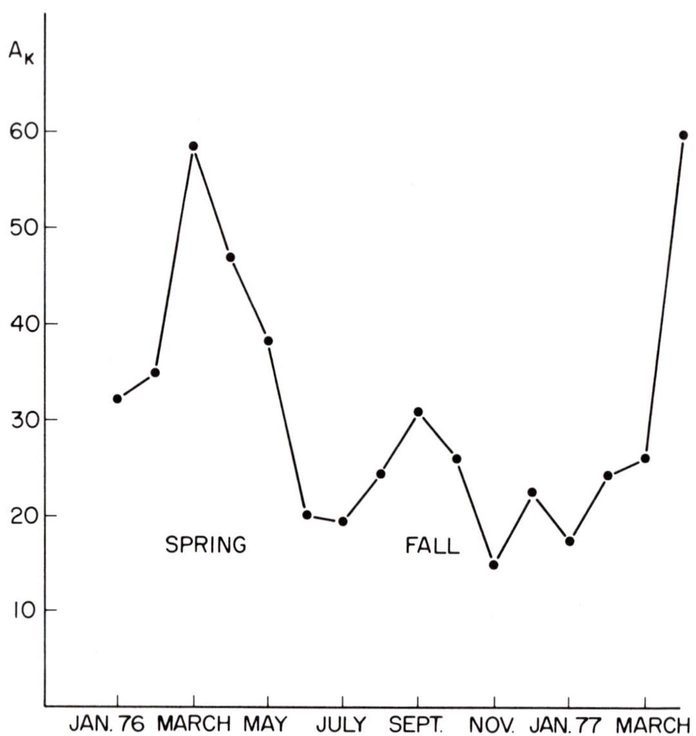

FIGURE 7. Seasonal variations in geomagnetic activity as measured in Ottawa.

It turned out, as shown in Figures 8 and 9, that as geomagnetic activity increased, both growth and cold resistance reached a maximum around 25γ. Above that activity, the situation was reversed; with increasing geomagnetic activity, growth and cold resistance decreased. This could suggest that there is an optimal geomagnetic activity for growth and cold resistance.

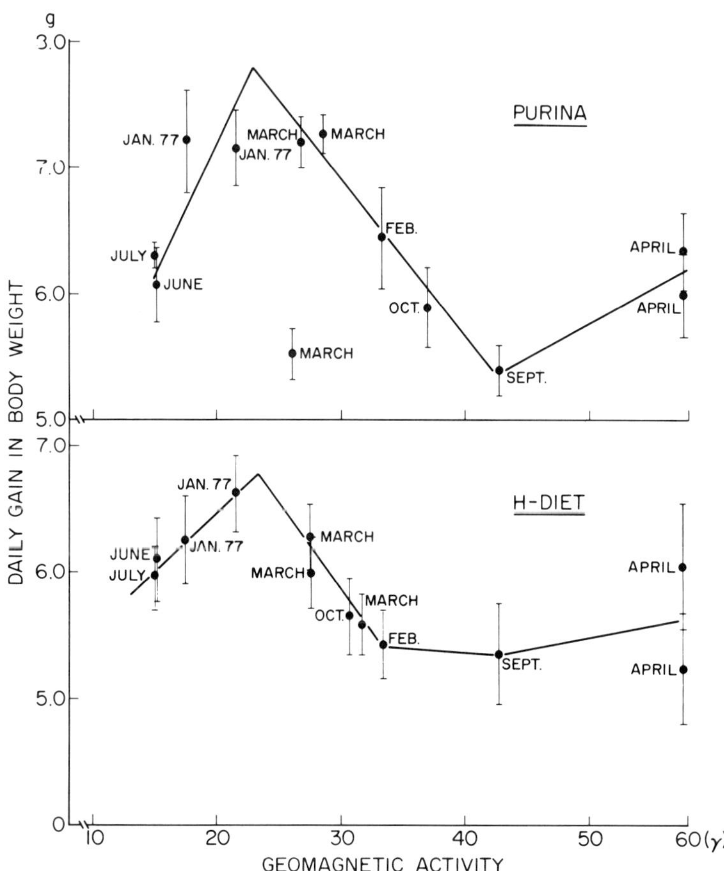

FIGURE 8. Correlations between average daily weight gain observed at different times of the year during the 14 days that the two diets were fed to the rats before being cold tested and the geomagnetic activity during the same periods of time. Bar = standard error of the mean.

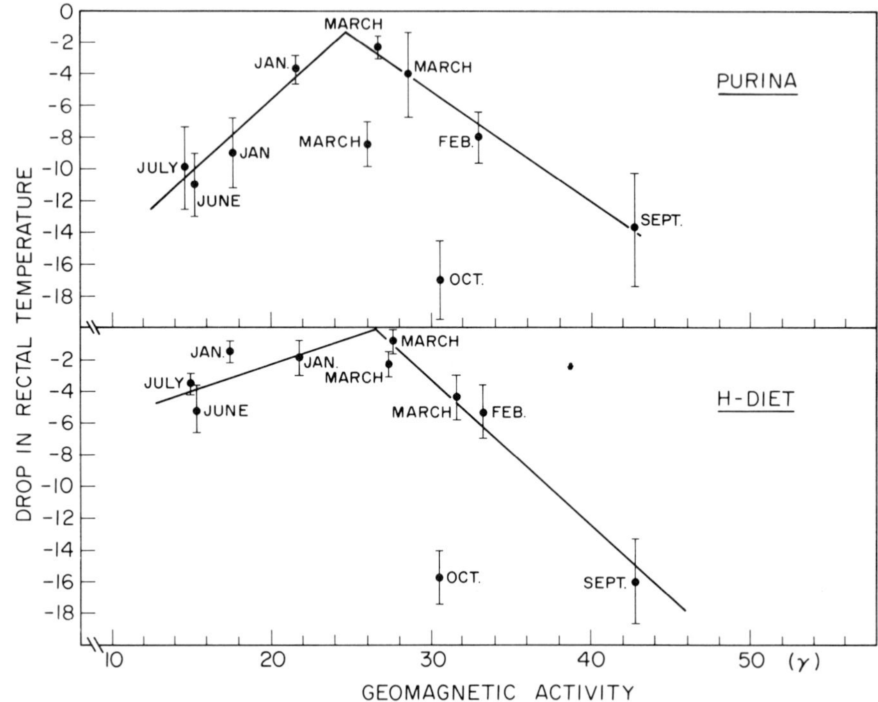

FIGURE 9.   Correlation between cold resistance and the geomagnetic activity prevailing during the 14 days before the cold test.   Bar = standard error of the mean.   Cold resistance was estimated by the drop in rectal temperature after 4 hours of restraint at +1º.

RECOMMENDATIONS

In conclusion, based on these observations, it would be very important to put a greater effort in this line of research. It is time that a greater number of physiologists pay attention to chronopathology and to chronotherapy.

It is a must to: find out what controls resistance to stress in non-cold adapted organisms, and especially resistance to cold; identify the indigenous factors limiting the duration of resistance; identify atmospheric factors other than temperature, humidity, and light that might influence resistance; and identify the nutritional factors that alter resistance.

This knowledge would certainly be useful to man in his fight against disease. Why is one individual more resistant to stress than another one? Why is the incidence of certain diseases increasing in spring and fall?

REFERENCES

Anderson, T.W., and W. Harding Le Riche. 1970. Seasonal variation in ischemic heart disease mortality. *Lancet. Nov. 28*:1140.

Arend, J.A., et al. 1977. Circiadian, diurnal and circannual rhythms of serum melatonin (MT) and platelet serotonin (5HT) in man. *Chronobiologia. 4*:96-97.

Eastwood, M.B., and J. Peacocke. 1976. Seasonal patterns of suicide, depression and electroconvulsive therapy. *Brit. J. Psychiatric. 129*:472-475.

Foster, D.O. 1977. Nonshivering thermogenesis in the rat. II. Measurement of blood flow with microspheres point to brown adipose tissue as the dominant site of the colorigenesis induced by noradrenaline. *Can. J. Physiol. Pharmacol. 56*(1):110-122.

Halberg, F. 1960. Temporal coordination of physiologic function. *Cold spring harbor symposium quant. Biol. 25*:289-310.

Halberg, F., J.J. Bittner, and B.J. Cully. 1955. 24-hour periodic susceptibility to audiogenic convulsions in several stocks of mice. *Fed. Proc. 14*:67-68.

Heroux, O., et al. 1972. Diet-induced cold resistance and focal myocytolysis of heat in warm-acclimated rats exposed to -18º. Pages 185-189 in Proceedings of the International Symposium on environmental physiology. FASEB.

Hugon, J.S., and G. Bounous. 1972. Elemental diet in the management of the intestinal lesions produced by radiation in the mouse. *Can. J. Surg. 15*:18-26.

Kempp, E, et al. 1978. Seasonal changes in the levels and the turnover of brain serotonin and noradrenaline in the European hamster kept under constant environment. *Experientia 34(8)*:1032-1033.

Movi, A.T.B., and D. Eccleston. 1968. The effects of precursor loading on the cerebral metabolism of 5-hydroxyindoles. *J. Neuroch. 15*:1073-1108.

Ortego, P. 1966. Light and electron microscopy of dichlor-
    odiphenyl-trichloroethane (DDT) poisoning in the rat
    liver. *Lab. Invest 15(4)*:657-679.
Schewing, L.E., et al. 1972. Circadian variation in rut
    serum 5-hydroxytryptamine and effects of stimuli
    on the rhythm. *Am. J. Physiol. 222(2)*:252-255.
Shaw, J.H., and B.O. Greep. 1949. Relationships of diet to
    the duration of survival, body weight, and composition
    of hypophysectomized rats. *Endocrinol. 44*:520-535.
Vener, K.J., T. Puggard, and B. Sloan. 1976. Susceptibili-
    ty-resistance cycle to heat stress: chinchillas,
    drugs, biogenic amines. Karger, Basel.

VI. A MODELING APPROACH TO UNDERSTANDING PLANT ADAPTATION TO LOW TEMPERATURES

P. C. Miller
W. D. Billings
W. C. Oechel

Systems Ecology Research Group
San Diego State University
San Diego, California

Department of Botany
Duke University
Raleigh, North Carolina

Systems Ecology Research Group
San Diego State University
San Diego, California

Research on plant adaptations to arctic environmental conditions has focused on determining the existence and magnitude of biological compensations of temperature-dependent physiological processes. Systems modeling can provide a framework for integrating information on diverse processes and morphological attributes as they are affected by arctic conditions.

Modeling has proceeded at various levels from ecosystem models to detailed processes based on models of radiation, plant water relations, and photosynthesis and to models of growth and allocation in various life forms.

Some insights into the classical views of tundra plant adaptation have been gained through the use of simulation models. A cost-benefit approach allows quantification of particular plant adaptations, but complete data sets to carry out simulation experiments are often lacking.

INTRODUCTION

Since temperatures generally decline with increasing
latitude, the rate of many temperature-dependent physical
and physiological processes can also be expected to decrease unless biological compensation occurs. Much of the
interest in the Arctic has been to determine existence
and magnitude of these compensations. The search is not
straight-forward. When the annual mean temperature is
below 0°C, permafrost can occur. The surface layers of the
soil are seasonally thawed, but the deeper layers are
permanently frozen. If water is present, it also is frozen. The ice impedes drainage and more water and ice
accumulate. The soils can become water-logged, anaerobic,
slow to thaw in summer, and cold because the temperatures
are closely coupled to the thawing frost. The cold temperatures and anaerobic soils lead to slow rates of decomposition, decreased rates of root activity, increased
accumulation of soil organic matter, and decreased nutrient
availability. Thus, low temperatures, especially as annual
temperatures drop below 0°C, affect diverse processes which
will affect plant growth and adaptation.

Low temperatures affect plants directly and indirectly.
Some of the direct effects of low temperatures on plants
are decreased *total* photosynthesis, decreased growth of
above- and belowground parts, decreased transpiration,
decreased root uptake of water and inorganic nutrients,
possibly decreased translocation, and slower or complete
lack of germination. Some of the indirect effects are on
photosynthesis and growth, via decreased transpiration and
altered water stress; on growth via decreased decomposition, nutrient release, and nutrient uptake; on growth via
shortened lengths of the growing season; on various plant
processes via soil thaw, and the water-logged, organic,
acid, and anerobic soils. The direct and indirect effects
of low temperature and the possible interactions between
diverse plant responses imply that the study of plant
adaptation to low temperatures should be set in a conceptual framework capable of integrating direct and indirect
effects and possible interactions between plant responses.
Plant adaptations should, therefore, be studied in the
context of the ecosystem.

The early recognition of plant adaptation to tundra
often did not consider these possible interactions and was
oriented very broadly towards the tundra ecosystem. Warren
Wilson (1957) attempted to sort out the various factors
affecting plant growth in the arctic. Tundra research has

benefited from periodic reviews of tundra literature (Bliss
1962; Billings and Mooney 1968; Savile 1972; Bliss 1971;
Billings 1974). We will not attempt to rereview this
literature here. In the early work, adaptations were recognized in both plant structure and processes. In the
Arctic, temperatures close to the ground surface are warmer
than those in the air, possibly 15°C or more in sunshine
(Savile 1972; Billings 1974). The short stature of tundra
plants and the location of meristematic regions close to the
ground surface means warmer plant temperatures and faster
growth. *Saxifraga oppositifolia* was observed to flower and
grow although air temperatures were below 0°C (Billings
1974). Photosynthesis in wet meadow grasses and sedges
proceeds at about 40 percent of the maximum rate of 0°C but
probably is near zero at -4°C (Tieszen 1973). Lichens
photosynthesize at much lower temperatures (Lange 1965). At
0°C, photosynthesis in mosses ranges from near compensation
early in the season, to as much as 56 percent in late season
(Oechel 1976). The optimum temperature for photosynthesis
in wet meadow grasses and sedges is about 15°C (Tieszen
1973) and in *Dryas integrifolia* it is near 10°C (Mayo et al.
1973). The temperature optima in mosses vary seasonally and
may be above these values (Oechel 1976). In some species
photosynthesis may be increased by the diffusion of respiratory CO_2 through the hollow stems to the site of photosynthesis (Billings and Godfrey 1967).

Growth rates may be increased by morphological adaptation which increases temperatures and by physiological
adaptation increasing the rate at relatively low temperatures. Krogg (1955) discussed the thermal advantage in
willow catkins of light-colored hairs and a dark-colored
ovary surface. Hocking and Sharpin (1965) discussed the
thermal advantage to the development of the ovary of para-
bolic-shaped flowers which reflect solar radiation onto the
ovary. Tikhomirov et al. (1960) found that the temperature
inside blue flowers was about 2°C higher than in white
flowers. Kleinendorst and Brouwer (1972) distinguished the
effect of soil on growing point metabolism from the effect
of soil temperature on water uptake and decreased growth
because of water stress. Billings (1974) has pointed out
the high respiration rate of tundra plants, which correlate
with high growth rates. And it has been shown that root
growth can occur with soil temperatures at 0°C (Billings et
al. 1973; Shaver and Billings 1977).

Translocation to roots can occur when soil temperatures
are at 0°C (Allessio and Tieszen 1975) and is relatively
unaffected by temperature (Tieszen et al. 1978). Water
uptake may be reduced by cold temperatures (Kuiper 1964) or

may be independent of temperatures (McNaughten 1974; Kramer 1952), and in water-logged soils, this reduction may not be apparent (Stoner and Miller 1975). Nutrient uptake is reduced at low temperatures (Chapin et al. 1979). Shortening photoperiod may also be a regulator in this and other processes in late season. Seed germination takes longer and may be lower at low temperatures; optimal temperatures for germination are 20–30°C for many arctic plants (Billings 1974).

The study of the direct and indirect effects and interactions between adaptation is aided by the cost-benefit analysis approach. The costs and benefits of changing plant form or physiology can be expressed in terms of carbon (Mooney 1972) or nutrients (Chapin et al. 1979). The carbon analysis is more common, even though plants are limited more by nutrients than by carbon or energy. The carbon or nutrient balance is presumably related to the production of offspring.

However, these analyses often proceed as a search for an optimal morphology or physiology, which would be competitively advantageous, or they attempt to determine the mode of survival given the constraints of the physical environment (Miller and Stoner 1979). Systems modeling can provide a framework for integrating information on diverse processes and morphological attributes, particularly if the relationships are amenable to some degree of quantification. The heritage of research on primary production processes in agriculture can be brought to bear on problems in natural communities. The modeling approach involves the development of the model structure and the validation of the model in total. The methodology of such an approach is to establish the relations included in the model with experiments and measurements. Validation experiments should be carried out in the field to test the ability of the model to predict the behavior of the plants under natural conditions. The validation acts as a check that all critical reactions have been included.

REVIEW OF TUNDRA MODELS

Ecosystem Models

The tundra ecosystem, with its apparent simplicity, has led investigators to think in terms of the total system rather than remaining caught up in the details of individual processes (Miller et al. 1975). The most complete of the

early conceptual models was the nutrient recovery hypothesis
of Schultz (1969). Later, word models were developed early
in the U.S. Tundra Biome Program (Brown et al. 1970; Brown
1971).

The first mathematical representation of the tundra was
based on these verbal descriptions (Timin et al. 1973), but
did not include all the compartments which they identified.
Instead, it simulated in detail the populations of lemmings
and their predators, but approximated the seasonal course of
aboveground and belowground portions of vascular plants and
omitted insects and decomposition. The model encoded the
hypothesis that the tundra could be understood as a pair of
Lotka-Volterra relationships: weasels and jaegers as predators of lemmings, and the lemmings as predators of the
graminoid plants (Timi et al. 1973). Simulations with the
model tended to contradict this hypothesis because the
predictions of the model were inconsistent with field data.
Later ecosystem models were developed by the Decomposition
Working Group of the Tundra Biome (Bunnell and Dowding 1974;
Bunnell and Scoullar 1975), and therefore, placed considerable emphasis on the simulation of decomposition. Nutrients were not followed by these models. Thus, they
assume that energy and biomass are sufficient controlling
factors to account for the dynamics of the tundra.

After IBP, tundra decomposition models have been
continued by Barkley et al. (1976). This model includes the
flows of nitrogen, phosphorus, and calcium, as well as of
energy and biomass. Primary productivity is determined by
nutrient and energy availability. The interaction of soil
and plants, constrained by the physical environment, define
the dynamics of the total system. Herbivores exert only a
small effect upon these key components. The model includes
detailed submodels for mosses, vascular plants, surface
energy exchange, and soil temperature. Below-ground composition is affected by nutrient level and substrate accessibility, as well as by temperature, moisture, and aeration. Aboveground decomposition is calculated as before.

Primary Production Models

Models of primary production processes and of interactions between the vegetation and the physical environment
have been developed and used in research with the Tundra
Biome and the U.S. Origin and Structure of Ecosystem programs of the IBP, in addition to AEC and ERDA research. The
models form a complex set of hypotheses about the mechanisms

and behavior of the plant production system based on experimental results. The modeling effects have provided a basis
both for developing fundamental concepts and for applying
research results in a diversity of vegetation types. The
models include (Figure 1):

1.   the effects of the vegetational structure on
     profiles of solar and infrared radiation, wind,
     air and soil temperatures, and vapor density;

2.   the effects of these microenvironmental variables
     and plant surface properties on the exchange of
     heat, water vapor, and carbon dioxide by leaves
     and stems in different positions in the canopy;

3.   the effects of internal morphological and phy-
     siological plant properties on rates of photo-
     synthesis and respiration, and on the uptake and
     loss of water and minerals;

4.   the effects of daily and seasonal patterns of
     allocating carbohydrates and minerals in plant
     growth on primary production, including some of
     the physiological mechanisms and microenviron-
     mental variables affecting these patterns;

5.   the effects of different reproductive patterns of
     plant survival and dispersal in different environ-
     ments; and,

6.   the accumulative effects of 2, 3, 4, and 5 on
     community composition, canopy structures, soil
     thermal and chemical properties, and the competi-
     tion of plant species within an ecosystem for
     light, water, nutrients, and heat.

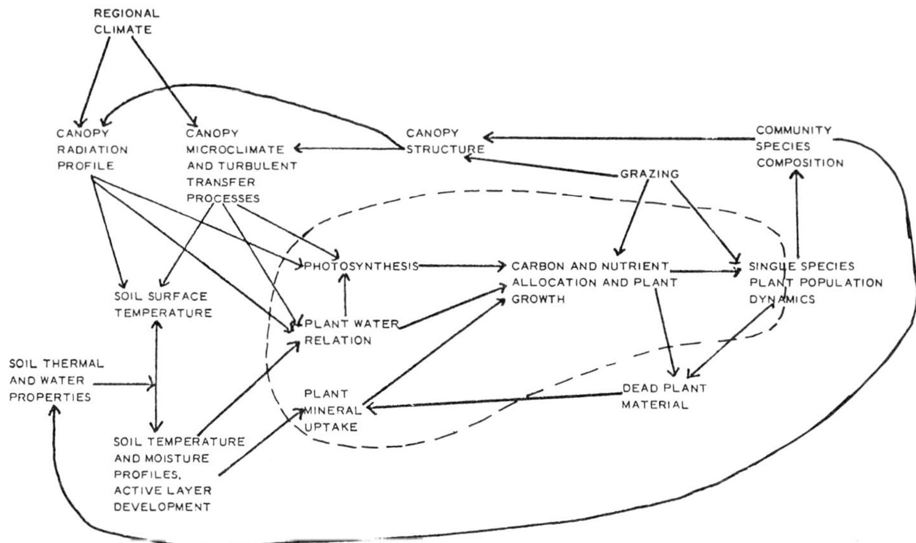

FIGURE 1. *Flow diagram of interactions between sub-models on sub-systems affecting the primary production. (From Tieszen et al. 1978).*

Radiation. The canopy radiation model was developed in the red mangrove ecosystem in south Florida (Miller 1974) and applied to the arctic tundra (Tieszen et al. 1978; Miller et al. 1976; Ng and Miller 1977; Stoner et al. 1978). The radiation models were developed from concepts of light penetration in continuous vegetation canopies which have a broad base of usage in temperate zone crops (Monsi and Saeki 1953; Davidson and Phillip 1958; de Wit 1965; Duncan et al. 1967; Anderson and Denmead 1969; Yim et al. 1969) calculate infrared radiation within the canopy, a component not found in previous models and includes the effects of stems on the radiation distribution within the canopy, a factor which has rarely been considered. The model has not been tested rigorously in the Arctic since radiation profiles are difficult to measure because of the low stature of the canopy. Existing observations and measurements (Oechel and Sveinbjornsson 1974; L.L. Tieszen 1978; Stoner et al. 1978) and the moderately-successful usage of the model in calculating canopy photosynthesis (Miller et al. 1976) and soil temperatures (Ng and Miller 1977) indicate that the model is reasonable. However, a rigorous test should be made in the Arctic because the low sun angles there may accentuate biases in calculating interception and reflection.

Turbulent Transfer. Models of physical processes affecting canopy microclimate have been proposed by Denmead (1964), Waggoner and Reifsnyder (1968), Waggoner et al. (1969), Murphy and Knoerr (1970, 1972), and Stewart and Lemon (1972) and tested with data from red clover, millet, pine, and corn canopies with reasonable success. The model of Miller et al. (1976) calculates profiles of air temperature and humidity between the top to the canopy and a reference height above the soil surface with the algorithms proposed by Waggoner and Reifsnyder (1968) and Waggoner et al. (1969). The model was used to calculate leaf and ground surface temperatures and soil temperatures in mangroves, chaparral, and arctic tundra. Calculations of temperature and humidity profiles and ground surface temperatures require knowledge of the resistances to evaporation at the ground surface. These resistances are poorly known for unsaturated, moist soils (Ng and Miller 1977; Goudriaan and Waggoner 1972; Denmead 1973). Values for the turbulent exchange coefficient and the soil resistances, when varied through the range of variation of present measurements, reversed the effect of the vegetation canopy of the depth of thaw in the Arctic. These values are critical and will require careful measurements of the air temperature and humidity profile, along with simultaneous measurements of net radiation and heat conduction into the ground.

Soil Surface and Subsurface Temperatures. The surface temperature is the link between canopy heat exchange processes and heat conduction in the soil. Most canopy or soil temperature models involve only the canopy or soil, and use the surface temperature as one of the boundary conditions. However, the effect of the canopy on surface temperature is significant in open, natural canopies, because of the reciprocal effect of surface temperature on air and leaf temperature; although it may be less important in agricultural crops with dense canopies. The surface temperature involves a solution of the energy exchange processes at the soil surface. The ground or soil surface is defined as the level at which turbulent exchange processes become negligible and conduction becomes the dominant process of heat exchange. The resistance of the air boundary layer to the exchange of heat and water vapor and the resistance of the soil to the transfer of water vapor from the effective evaporating surface to the actual ground surface are essential when calculating soil surface temperatures, but are poorly known. The surface temperature model has been used in mangroves, chaparral, and arctic tundra (Miller et al. 1974; Ng and Miller 1975; Ng and Miller 1977).

A soil temperature model including thawing of the permafrost was developed by Nakano and Brown (1972) and adapted by Miller (1972), Ng and Miller (1975, 1977) to explore the interactions between canopy structure and depth of thaw. Validation of the soil model was presented by Nakano and Brown. Validation of the combined canopy-soil model is given by Ng and Miller (1977).

Soil Water Movement. The movement of water in the soil has been modeled by Hanks and Bowers (1962), Liakopoulous (1965), Amerman (1971), and de Wit and Goudriaan (1974). Guymon and Luthin (1974) have been developing a model of soil moisture flow in the Arctic. Soil moisture in the wet meadow system is almost always abundant, but in inland tundra, soil moisture flow may become more critical to plant survival. The resistance of the soil to the flow of water to plant roots is important to clarify measurements of the plant root resistance to water uptake. In inland tundra, with generally better-drained soils and greater topographic diversity than in the coastal wet meadow, soil moisture was the first environmental complex segregating species and life forms (Brown et al. 1979).

Plant Temperature. The leaf and stem temperature models are one of the interfaces between the environment and plant processes. The models are adaptations of the energy budget equation for individual leaves (Gates 1962, 1965). The models calculate the physical processes of solar and infrared radiation absorption, infrared loss from leaves, convectional exchange, and transpirational exchange. The leaf temperature is calculated by balancing the incoming and outgoing fluxes of energy. Calculated leaf temperatures have been tested only cursorily against measured temperatures in tundra (Tieszen unpublished data). Air boundary layer resistance in natural conditions is still an area for research, especially when tightly-clustered leaves require calculation on a branch or stem basis, similar to Gates et al. (1965) and Tibbals et al. (1964) energy exchange processes, and leaf temperatures are calculated for sunlit and shaded leaves at different levels in the canopy.

Water Relations of Vascular Plants. A model of plant water relations was used in alpine (Ehleringer and Miller 1975a, 1975b) and arctic tundras (Stoner and Miller 1975;

Miller et al. 1976). The model is essentially that of
Honert (1948), modified by Rawlins (1963), and described in
nonmathematical terms by Jarvis and Jarvis (1963). It
allows for nonequal rates of water uptake and loss and the
development of plant water stress. The model requires
measurements of relations between leaf resistance to water
loss and the water content of the leaf, measurements of the
root resistance to water uptake and its environmental or
plant controls. The relation of leaf resistance to leaf
water content or water potential has been measured in six
arctic tundra species (Stoner and Miller 1975); five alpine
tundra species (Ehleringer and Miller 1975a,b); and boreal
forest tree species (Jarvis and Jarvis 1963). The relation
can be partially inferred for conifers from Luposhinsky
(1969); heath species from Bannister (1971); and alpine
plants from Klikoff (1965). The relation between leaf water
content and leaf water potential has been measured on some
tundra and boreal forest species. Matrix osmotic and
turgor components have been measured in only a few tundra
species (Courtin and Mayo 1974; Kedrowski 1976). The root
resistances have been measured crudely in arctic species
(Stoner and Miller 1975). Slatyer (1967) and Kramer (1969)
discuss the relation of root resistance to the rate of water
uptake. Fiscus and Kramer (1975) have updated the analytical models of root resistance. The relation of root resistance to temperature, root morphology, and rate of water
uptake, should be clarified. Stoner and Miller (1975)
showed no relation of root resistance to soil temperature in
the wet meadow species. More precise measurements are
needed to verify this observation and determine whether the
independence of resistance to temperature is an adaptation
of cold temperature plants, or due to root morphology in
waterlogged soils. The measurement on plant water relations
indicate that plants restricted to wetter habitats show high
leaf resistances at higher water potentials than plants of
dryer habitats (Poole 1974; Ehleringer and Miller 1975a,b),
that plants of wetter habitats show less sensitivity of leaf
water potential (Poole 1974; Stoner and Miller 1975) to leaf
water content.

Vascular Canopy Photosynthesis. A canopy photosynthesis model for the wet meadow tundra (Miller and Tieszen
1972; Miller et al. 1976) is based on the energy exchange
processes of a single leaf located at various heights in the
canopy. Previous canopy photosynthesis models have also been
based on the photosynthesis relations of a single leaf
within the canopy. Monsi and Saeki (1953) assumed uniformly
overcast skies and a constant rate of light penetration into

the canopy, regardless of the time of day or depth in the canopy, and related leaf area index, leaf inclination, incoming light intensity, and primary production. Davidson and Phillip (1958), with similar assumptions, analyzed the interrelations between leaf area index, grazing schedules, and primary production. De Wit (1965) included direct and diffuse light, light scattering in the canopy, variable solar altitudes, variable leaf inclinations, and variable resistances of the air above the canopy to the transport of carbon dioxide into the canopy, and calculated the annual course of potential production at different latitudes. Monteith (1965) pointed out the error in calculating photosynthesis for the total canopy using a single exponential equation for average solar intensity, as used previously, and proposed a binomial expansion to express the decrease in direct and diffuse solar radiation within the canopy. Anderson (1966) developed the analysis of solar radiation in the canopy further to include direct and diffuse solar radiation and nonuniform sky radiation. Duncan et al. (1967) proposed a model which included carbon dioxide and the sources of carbon dioxide sink into the canopy, which show the relative contribution of different levels in the canopy to the production of the crop, and the significance of the production of carbon dioxide by the soil.

Growth and Allocation. Models of growth of individual plants and of allocation of carbon and nutrients in the plant for different life forms (single-shooted graminoid, deciduous shrub, evergreen shrub, and graminoid tussock) are in various stages of development for the tundra. These models are based on data from several investigators in the IBP, RATE, and ERDA programs. The models are driven by the seasonal patterns of growth in various structures. Growth is affected by temperature and the plant sugar, nitrogen, phosphorus, and calcium status. The carbon and inorganic nutrient costs associated with the growth of new tissue are calculated from the biochemical composition of young mature tissue and the growth respiration costs of creating different biochemical constituents (Penning de Vries 1972, 1973, 1974; Penning de Vries et al. 1974).

Three models of growth and allocation for the single-shooted graminoid with different levels of sophistication were developed during the U.S. Tundra Biome research (Miller 1972; Miller et al. 1975, 1976).

The present model for a single-shooted graminoid (*Dupontia fisheri*) (Miller et al. 1976) follows a five-member tiller system; three aboveground tillers, one newly-initiated growing rhizome, and one old dying rhizome. Each of the three aboveground tillers includes leaves, stem base, rhizome, and roots. A tiller during its life produces 12 leaves before the shoot meristem differentiates into an inflorescence. In its first and second years aboveground, a tiller will initiate four to five leaves each growing season. In its third year the tiller will normally produce one to two leaves and an inflorescence. In the year following flowering, the shoot dies and the rhizome and roots begin senescence.

The deciduous shrub model (Stoner et al. 1978), based on *Salix pulchra*, follows an individual comprised of an above- and belowground main stem with attached roots; 2-year older stems attached to the 2-year stems; 1-year stems attached to the 2-year stems; and current stems, leaves, and inflorescences attached to the 1-year-old stems. A main reserve of carbohydrates and nutrients (N, P, and Ca) is shared by all structures except current stems, leaves, and inflorescences. Current stems, leaves, and inflorescences maintain their own reserves by translocation and photosynthesis.

The evergreen shrub model (Stoner et al. 1978) follows an individual *Ledum palustre* composed of an above- and belowground main stem and roots; 2-year and older stems with 2-year and older leaves attached; 1-year leaves and stems attached to 2-year stems; and current leaves, stems, and inflorescences attached to 1-year-old stems. A main reserve pool of carbohydrates and nutrients is partitioned between all structures except current stems, inflorescences, and all classes of leaves. The structures not associated with the main reserves maintain their own reserve pools.

The *Eriophorum* tussock model consists of roots, old senescent rhizomes, new growing rhizomes, mature rhizomes, leaves, and inflorescences. These structures contain all the biomass of all individual members of the tussock. A member of the tussock is either a new individual composed of only a growing rhizome; a mature individual with roots, rhizomes, leaves, and inflorescences; or a senescent dying individual made up of only a rhizome and attached roots. The growth of inflorescences, leaves, rhizomes, and live roots is computed on an individual basis and is multiplied by the number of individuals in each category.

At present, in the above models, there is no senescence of inflorescences or seed set and dispersal. As such, inflorescences are only sinks for available carbohydrates and nutrients.

Plant Population. Quantification and modeling of population processes is a more recent concern than is an understanding of growth, even though animal population ecology was a very early quantitative field in ecology. One of the problems in the study of plant populations has been to ascertain meaningful units to count (Harper and White 1974), since plants show considerable flexibility in their ability to drop parts while the genotype survives. Callaghan and Collins (1976) stress the need for a population orientation in tundra research, apart from the recent heavy emphasis on production processes. A plant population model was developed for the wet meadow graminoids (Lawrence 1974; Lawrence et al. 1978). The ground surface is divided into a grid. Each cell of the grid can have different environmental conditions. Individual tillers in each cell compete for light, heat, water, and inorganic nutrients. The tiller system develops reproductive rhizomes and inflorescences which are dispersed over the surface. Individual tillers in the system interact to support each other in the face of grazing pressures. The population model indicates that rapid growth early in the growing season increases production in the presence of lemming grazing, and that sexual reproduction is not essential in the short-term maintenance of the population. The model will be used to clarify the role of the spatial arrangement in the outcome of competition between differing plant populations. The plant population models were developed for *Dupontia*, a grass. The models must be developed for other life forms in the arctic to complete a picture of vegetation dynamics and the mechanisms underlying the segregation of species and life forms along gradients.

Moss Production. Mosses are a dominant element in arctic vegetation, contributing to the tundra's carbon, water, heat, and nutrient budgets. Below the green moss is a layer of non-green moss material in early stages of senescence and decay. Under this is a peat layer of largely decomposed moss material, where the roots of the vascular plants occur. These layers of live and dead moss material affect the soil's thermal and water regimes (Nakano and Brown 1972; Ng and Miller 1977). The model of Miller et al. (1976) which simulates moss photosynthesis, biomass accumulation, and water relations has been modified to include nitrogen, phosphorus, and calcium.

The exchange of CO_2, water, and nutrients between the moss and the environment occurs through moss surfaces. Water and nutrients move upwards to the green moss from the non-green and peat layers along the exterior of the moss.

Water is absorbed actively. Water is lost from the moss
surface by evaporation and from the moss tissue by transpiration. Photosynthesis is limited by light, temperature,
and tissue water content. Potential growth is controlled
by available carbohydrates. Carbohydrates above 10 percent
of the total moss biomass are used for growth. If nutrients are limiting, growth is reduced to use what is
available. Growth costs are calculated from biochemical
compositions (Penning de Vries 1972). Simulated seasonal
CO_2 uptake is maximum when the optimum temperature for
photosynthesis is 5°C. However, field measurements optimal
temperature range are between 12 and 19°C. Simulations in
connection with field observations indicate that for arctic
moss species, temperature acclimation of photosynthesis
does not tend to maximize CO_2 accumulation (Oechel 1976;
Oechel et al. 1975). This situation does not hold in the
sub-arctic where acclimation in mosses tends to be more
complete (Hicklenton and Oechel 1976).

INSIGHTS INTO THE TUNDRA ECOSYSTEM FROM MODELING

Some insights into the classical views of tundra adaptation have been gained through simulation models. More
can be expected.

Morphological Effects on Energy Exchange and Leaf Tempera-
tures.

In the wet meadow tundra, the intercepted fraction of
solar irradiance increases through the growing season as
the live leaf biomass increases, although the incoming
solar irradiance decreases from June 21 onward. Of the
intercepted solar irradiance, about 20 percent is reflected
back, decreasing by about 2 percent as the darker green
live leaves show above the lighter dead leaves. The fraction of solar irradiance absorbed by the ground under the
canopy decreases. Of the solar energy absorbed by the
canopy about 25 percent supports the net infrared radiation
loss, about 65 percent is lost by convection, and about 10
percent is lost by transpiration (Figure 2).

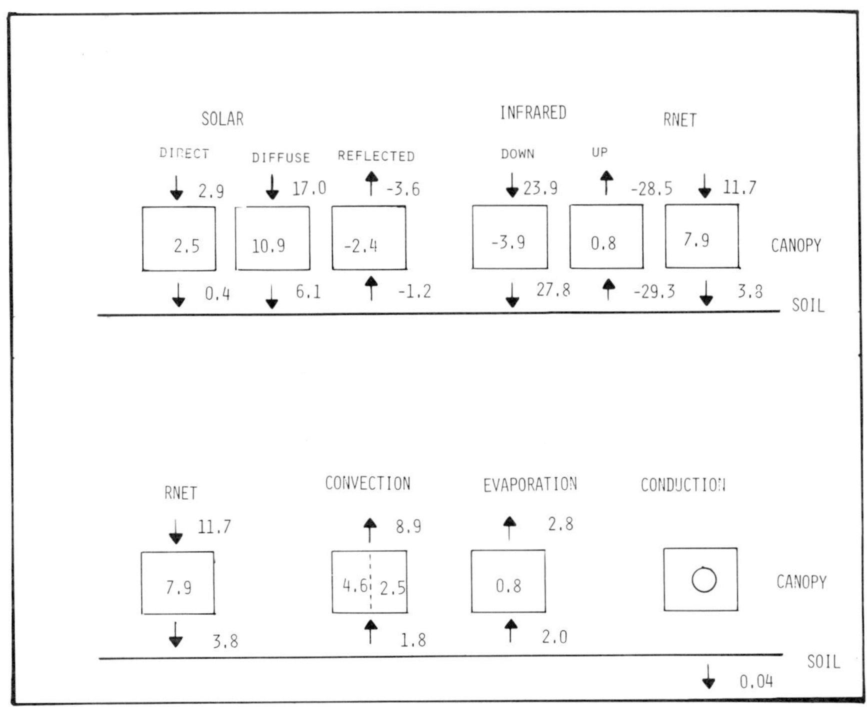

FIGURE 2. Partitioning of incoming radiation into reflected solar and infrared radiation, convection, evapotranspiration, and conduction in the wet meadow ecosystem. Units are MJ m^{-2} day^{-1}

Most of the evapotranspirational loss from the wet meadow is evaporation from the wet moss surface beneath the vascular plant canopy. Transpiration from the vascular plant is only about 10-15 percent of the total evapotranspiration estimated by Weller and Holmgren (1974). These results of the simultions are supported by measurements of Koranda et al. (1978). Most of the convectional loss from the wet meadow is from the standing dead material, which because it is relatively dry, does not evaporate and would be expected to have slightly higher temperatures than live leaves. The convectional exchange of heat from leaves to air, warms the air. Thus, the air within the canopy should be warmer than the air above the canopy. This warming is greatest at the bottom of the canopy and air temperatures are about 3^oC higher than air with no canopy (Figure 3).

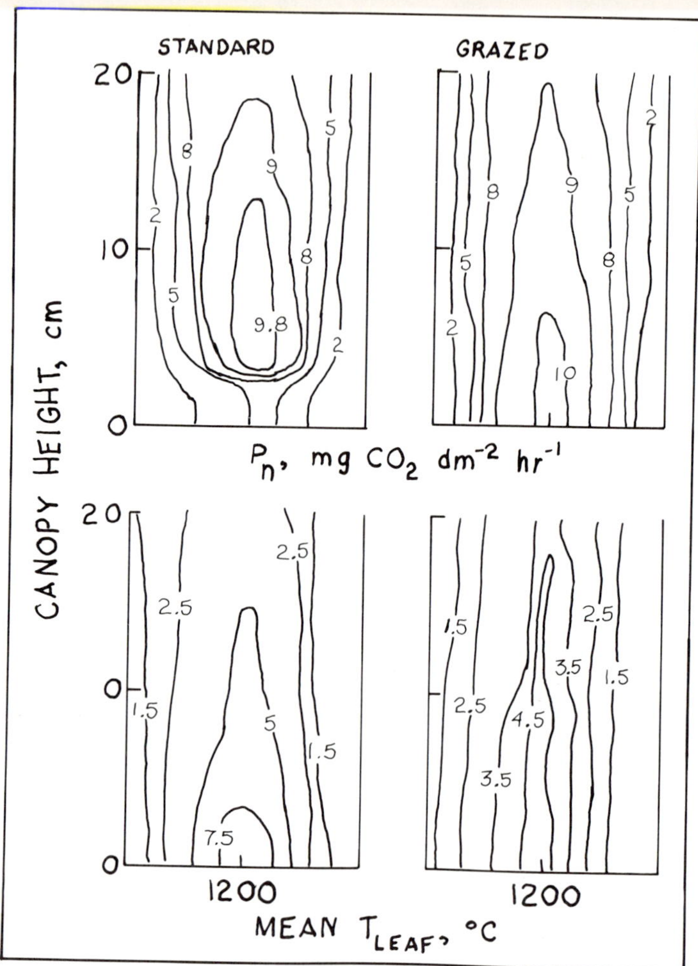

FIGURE 3. Isopleths of air and leaf temperatures with height and time of day.

Physiological Properties Related to Photosynthesis

Photosynthesis is affected by both light and temperature. Light decreases with depth in the canopy and with
time of day away from solar noon. Even at the highest
light intensities absorbed solar irradiance is scarcely
above light saturation levels of vascular plants (Figure
4). Thus, from the temperature limitation, photosynthesis
should be highest at the bottom of the canopy. The net
result is that maximum photosynthesis occurs near the top
and middle (Figure 4c).

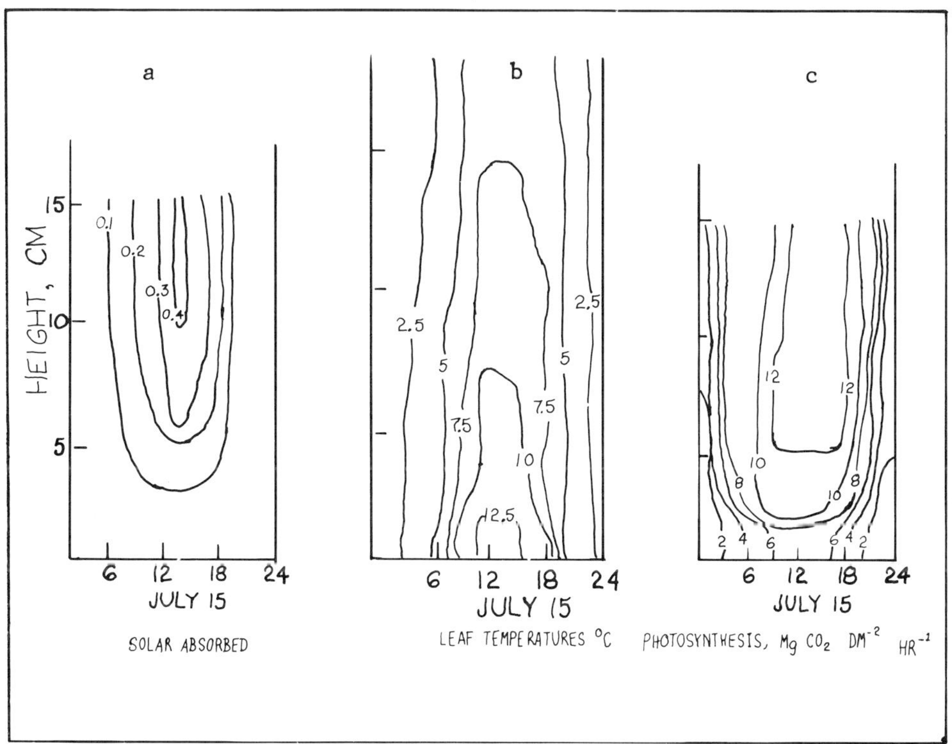

FIGURE 4 a,b,c. Isopleths of absorbed solar radiation, leaf temperature, and net photosynthesis with height in the canopy and time of day. Calculated for July 15, 1971.

If the optimum temperature were different from 15°C, the zone of maximum photosynthesis should shift. A shift in the optimum temperature of ± 10°C, shifts the zone of maximum photosynthesis from bottom to top of the canopy, with a higher temperature optimum, maximum photosynthesis is near the bottom because temperature becomes more limiting (Figure 5).

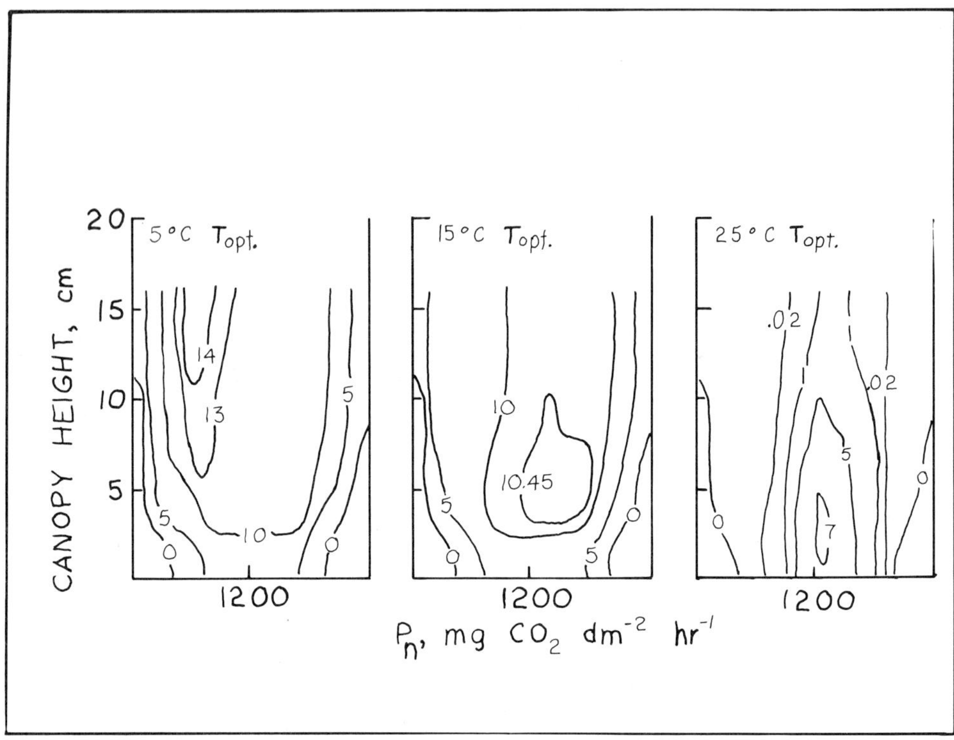

FIGURE 5. *Effect of different temprature optima on the vertical distribution of daily net photosynthesis for* Dupontia *on 15, July, 1971. (From Miller et al. 1976).*

The effect of changing the temperature optimum in vascular plants was explored to ascertain its influence on daily photosynthesis (Figure 6). The optimum temperature of 15°C is considerably above the usual daily mean air temperature of 5°C at Barrow, Alaska, during the "growing season." The simulation indicates that as the optimum temperature decreases from 25°C to 15°C, photosynthesis at the Barrow temperatures increases. However, as the optimum temperature decreases below 15°C, the increase in photosynthesis is considerably less. Selection should be relatively strong to reduce the optimum temperature to 15°C but relatively weak to reduce the optimum temperature below 15°C. Unfortunately, a cost-benefit analysis cannot be

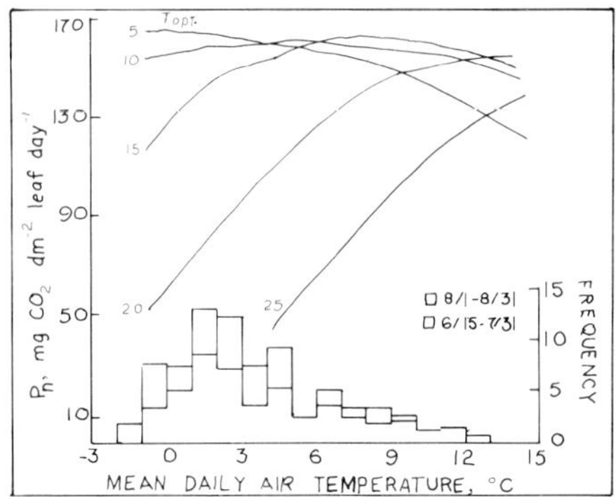

FIGURE 6. *Simulations of the effect of air temperature on daily photosynthesis at different temperature optima. Frequency distribution of mean daily temperature
(1970-1973) are also shown. (From Miller et al. 1976).*

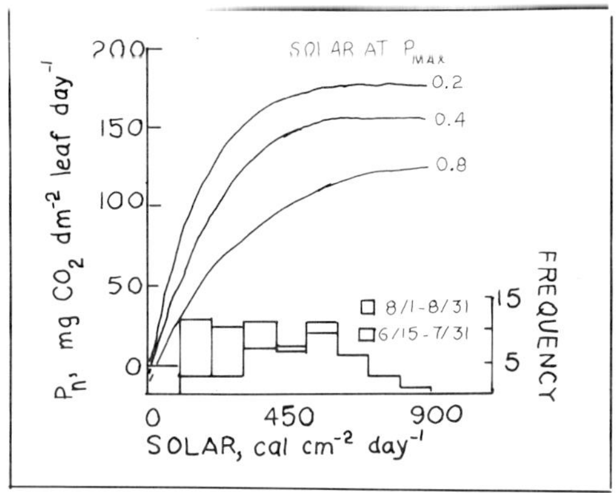

FIGURE 7. *Simulated response of photosynthesis of* Dupontia *to daily irradiance, and the frequency distribution of daily irradiances (1970-1973). Simulations of the
effects of varying the incident irradiance (cal cm^-2min^-1)
for saturation of the photosynthetic rate are also shown
(From Miller et al 1976).*

applied. Although the possible benefits of shifting the
temperature optimum are quantified, the biochemical cost
and mechanisms of this shift are unclear.

Daily photosynthesis should also be increased by
decreasing the light saturation intensity since light is
usually limiting. The magnitude of this response is indicated in simulations (Figure 7). Decreasing the light

saturation intensity from 0.4 to 0.2 cal cm $^{-2}$ min^{-1} (400–700 nm) increases photosynthesis about 15 percent. Again the benefits of this shift are quantifiable, but the biochemical costs are unclear.

Growth

Models of growth have a short heritage relative to models of canopy processes. For tundra, models of *Dupontia fisheri*, *Ledum palustre*, *Salix pulchra*, and *Eriphorum vaginatum* have been developed from research at Barrow and Meade River. The models are based on physiological assumptions about processes affecting growth and describe well the seasonal progression of leaf and stem biomass, nitrogen, phosphorus, and calcium (Figures 8 and 9). Ironically, the temperature influences on growth have not been studied and are not included.

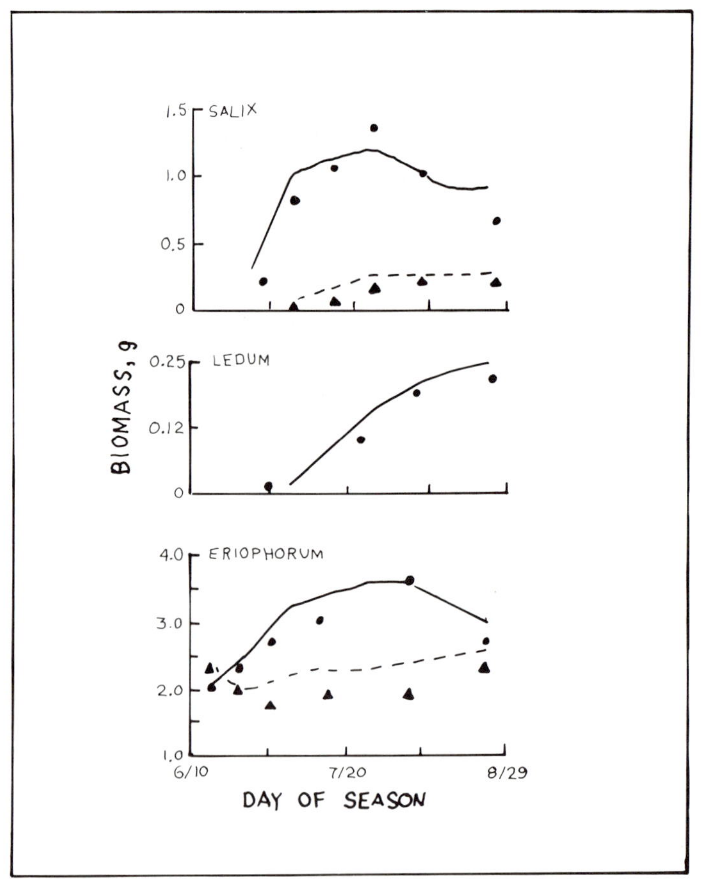

FIGURE 8. Simulated and observed seasonal progressions of biomass of leaves and stems of species of three tundra growth forms (From Stoner et al. 1977).

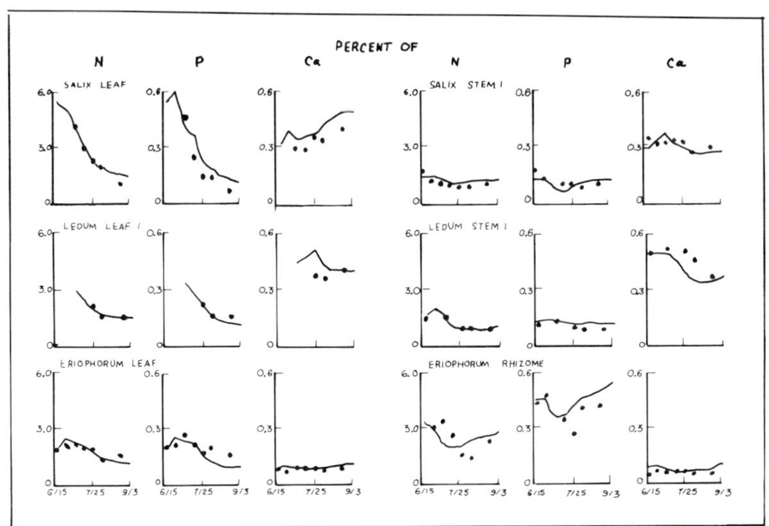

FIGURE 9.   Simulated and observed seasonal progressions
of M, P, and Ca contents of leaves and stems of species
of three tundra growth forms (From Stoner et al 1977).

The models indicate the relative advantage of the
evergreen shrub over the graminoid form with low nutrient
availability (Figure 10). The explanation for the higher
carbon gain of plants with low leaf turnover is because of
the lower growth costs associated with low leaf turnover
(Billings and Mooney 1968). An analogous pattern is observed in the field in the distribution of production among
the growth forms and nitrogen and phosphorus uptake by the
total vegetation (Figure 11). Evergreens are associated
with low nutrient availability and uptake. A critical lack
of understanding exists concerning the diffusion of nitrogen and phosphorus from the bulk soil to the root surface
and the role of mycorrhizae. This gap in our understanding
limits our ability to generalize laboratory studies of root
uptake to the field.

The growth model for *Dupontia* indicated that in the
steady state, carbon translocation in a four-tiller system
is from V_1, V_2, and V_3 tillers towards the leading rhizome
and inflorescence (Figure 12). The cost of producing a
rhizome is greater than the cost of producing an inflorescence, but the estimated survival of rhizomes vs. survival
of seed is such that reproduction via rhizomes is about 30
times more efficient in terms of phosphorus than is sexual
reproduction. In this analysis the actual survival rates
are poorly known. The pattern of carbon translocation is
not exactly that measured by Allesio and Tieszen (1975).
Some closely-controlled experiments should clear up the
disparity.

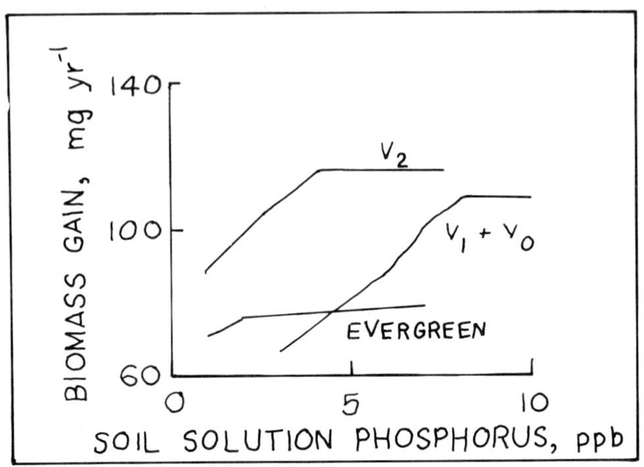

FIGURE 10. Simulated production of evergreen and two graminoid tiller age classes at different soil phosphorus levels (From Stoner et al. 1977).

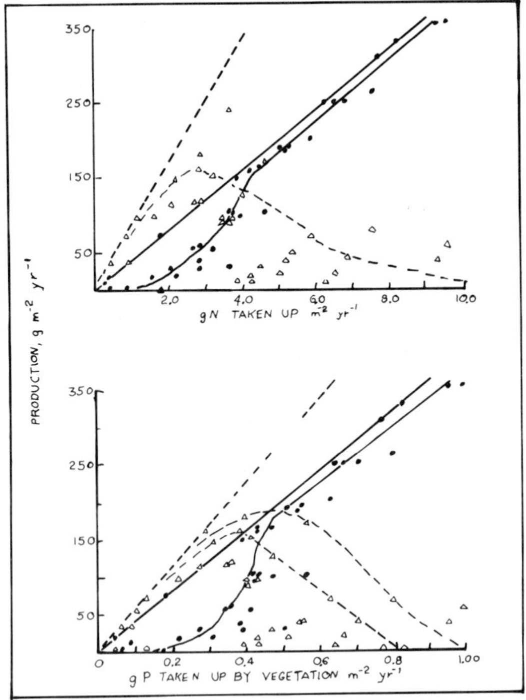

FIGURE 11. Annual productions by evergreen and deciduous shrubs and annual uptake by the vegetation of N and P (From Miller and Stoner 1978).

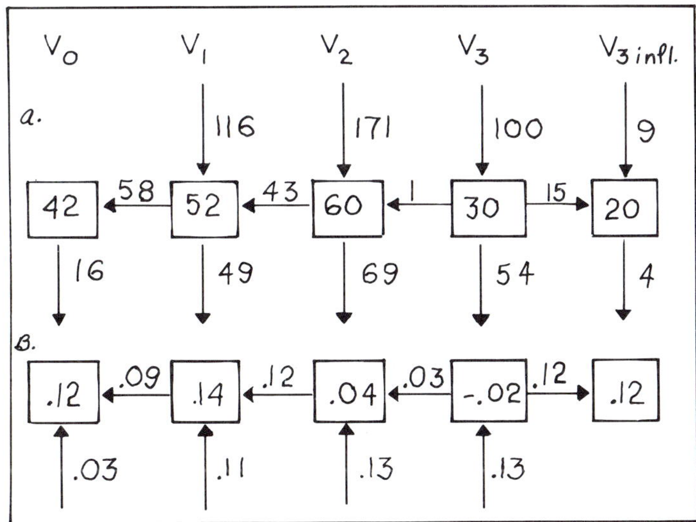

FIGURE 12. Steady state annual carbon and phosphorus fluxs simulated for a 4 tiller *Dupontia* systems (From Stoner et al. 1978).

Tundra plants are noted for their rapid growth early in the season. Their rapid growth is related to the interception of solar irradiance, the essential first step in canopy photosynthesis. As the leaf area index increases, interception increases and canopy photosynthesis increases, up to a leaf area index of about 3 (Figure 13). Above 3, leaves are sufficiently shaded that respiratory costs of some leaves are greater than photosynthetic gain. Thus, canopy photosynthesis decreases at higher leaf area indices. Two paradoxes emerge:

1. At pond margins, leaf area indices may be up to 8 (Tieszen 1972). With these leaf areas, light saturation levels or respiration rates may be lower, following the work of McCree and Troughton (1966).

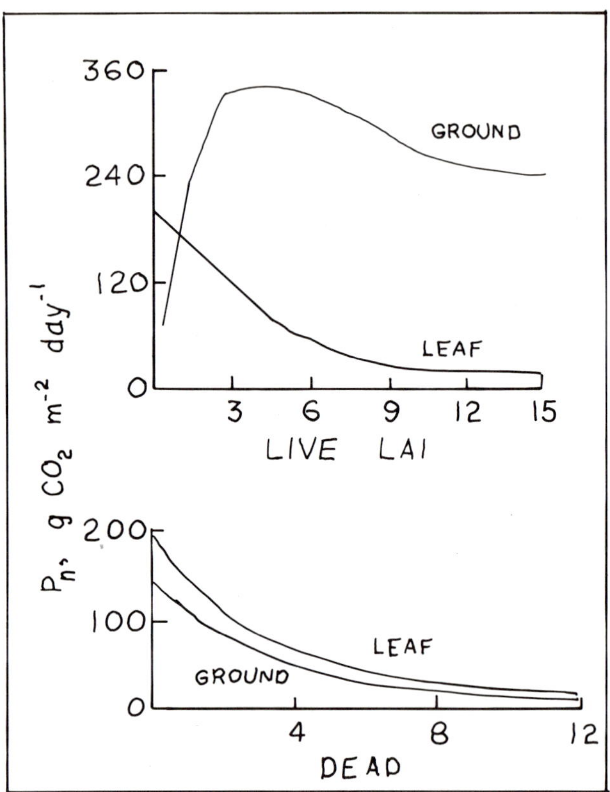

FIGURE 13.   Simulations of the effect of live and dead leaf area index (LAI) on mean leaf photosynthetic rate and community photosynthesis on a ground area basis.   In the lower graph ground area rate are for a total live LAI of 1.0 plus dead as indicated (From Miller et al. 1976).

2.   Other than at pond margins, leaf area indices are not above 2.

The latter may be due to the short growing season. The length of the growing season is one of the major features limiting seasonal production in tundra (Figure 14).   Production is increased about 2 g dry matter m⁻² with each day increase in the growing season.   From calculation of photosynthesis rates and leaf construction costs, a growing season of at least 45 days is required for positive carbon balance.   The biologically-meaningful season is apparently shorter than this since leaf senescence and retraction of nutrients must be accomplished before leaves

are lost by freezing or winter mortality. The processes
and timing involved in the senescence of leaves and the
relative merits of photosynthesizing to the end of the
season and losing more leaf material by frost and winter
mortality vs. early senescence still require study.

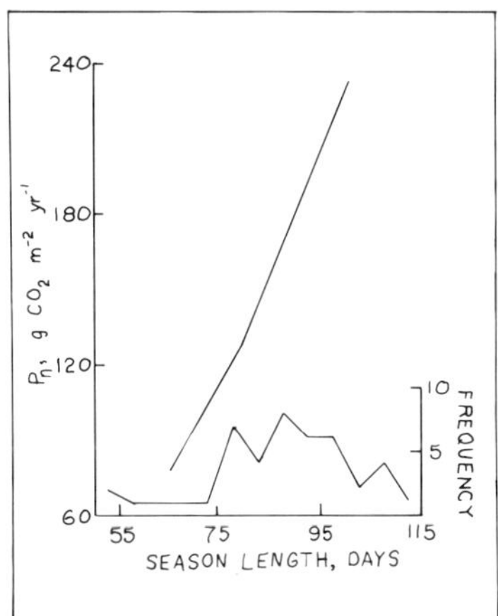

*FIGURE 14. Simulation of the effect of length of
season on the total seasonal CO_2 uptake by* <u>Dupontia</u>
*in a mixed canopy. Frequency of yearly thaw season lengths
for 50 yr (1922-1973) is also given (From Miller et al.
1976).*

Population Processes

A natural bridge between the growth models developed
from primary production process research and plant population dynamics is the analysis of shrub growing point or
tiller dynamics being developed by G. Shaver. These growing points can be treated as populations with inherent
rates of initiation and mortality which are controlled by
the environment. The pattern of initiation and mortality,
the controls by environment, the internal levels of stored
carbohydrate, nitrogen, and phosphorus, and the allocation
of stored material to growth of leaves at a growing point
or to the production of new growing points are being studied. Evidence is accumulating (McCown 1973; Shaver unpublished) that with increased nutrient availability,
tundra plants respond by increasing tillering rather than
increasing growth rates. Thus shrub and tiller system
dynamics warrant further study, since the population processes may be more sensitive than are growth or photosynthesis of a given leaf.

Population processes involving births and mortality of genets and genotypes has a long heritage of talk and a short heritage of quantification and good data. Population problems are less tractable than production processes because of their longer term duration. Additional research will be required both for theoretical and applied problems.

CONCLUSIONS

Several conclusions are apparent in this review:

1.   Plant adaptation should be considered in terms of the total environment or ecosystem and the interaction between processes and adaptation.

2.   Modeling has tied together many but not all loose ends of data, but in so doing has pointed out deficiencies in understanding.

3.   The conceptual tools for quantifying plant responses exists, but the assumption must always be questioned and improved.

4.   Previous studies emphasized physical processes, photosynthesis and respiration, growth, and population processes, in that order. Currently it is recognized that at least in relation to nutrients, the plant response is greatest in population processes, then growth, then photosynthesis.

5.   The past effort in tundra models has been involved in model development, not in testing model validity or generality. The gaming and simulation experiments must be recognized as extrapolation from data and as hypotheses for future research.

6.   The effect of temperature on plant processes may be the least studied of all influences. There is no study showing the temperature response of aboveground growth and only one showing the response of belowground growth. The recent research of the IBP was oriented toward factors influencing the distribution of species within tundra, species which were already cold-adapted.

Earlier studies distinguished adaptations of tundra species from species in other biomes. Adaptations were pointed out on a diversity of species and growth forms in a diverse ecosystem. Thus, the postulated adaptations are only loosely related to other compensating adaptations of the plant. They are seen, at present, in isolation of the specific selective forces in the habitat.

7. In a cost-benefit approach to plant adaptation, the costs or the benefits of a particular adaptation can be quantified. However, in every adaptation considered to date, information is not available to assess both the costs and benefits. What is lacking is a cellular-biochemical assessment of costs and a population-evolutionary assessment of benefits.

REFERENCES

Acock, B., J.H.M. Thornley, and J.W. Wilson. 1970. Spatial variation of light in the canopy. Pages 91-102 in Predication and measurement of photosynthetic productivity. Center for Agricultural Publishing and Documentation, Wageningen, The Netherlands.

Allesio, M.L., and L.L. Tiezen. 1975. Patterns of carbon allocations in an arctic tundra grass, *Dupontia fisheri. Am. J. Bot. 62*:797-807.

Amerman, C.R. 1971. Numerical solution of the flow equation. Pages 241-254 in D. Hillel, Soil and water. Academic Press, New York.

Anderson, M.C. 1966. Stand structure and light penetration. II. A theoretical analysis. *J. Appl. Ecol. 3*:41-54.

Anderson, M.C., and O.T. Denmead. 1969. Shortwave radiation on inclined surfaces in model plant communities. *Agron. J. 61*:867-872.

Bannister, P. 1971. The water relations of heath plants from open and shaded habitats. *J. Ecol. 59*:51-64.

Barkley, S.A., et al. 1978. Controls on decomposition and mineral release in wet meadow tundra--a simulation approach. Pages 759-778 in D.C. Adriano and I.L. Brisbin, Jr., eds. Environmental chemistry and cycling processes. Technical Information Center, U.S. Dept. of Energy, Washington, D.C.

Billings, W.D. 1974. Arctic and alpine vegetation: Plant
    adaptations to cold summer climates. In J.S. Ives and
    R.G. Barry, ed. Arctic and alpine environments. Barnes
    & Noble Books, Scranton, PA.
Billings, W.D., and P.J. Godfrey. 1967. Photosynthetic
    utilization of internal carbon dioxide by hollow-
    stemmed plants. *Science. 158*:121-123.
Billings, W.D., and H.A. Mooney. 1968. The ecology of
    arctic and alpine plants. *Biol. Rev. 43*:481-529.
Billings, W.D., G.R. Shaver, and A.W. Trent. 1973. Temper-
    ature effects of growth and respiration of roots and
    rhizomes in tundra graminoids. Pages 57-63 in L.C.
    Bliss and F.E. Wielgolaski, eds. Primary production
    and production processes, Tundra Biome. Univ. of
    Alberta Print. Services, Edmonton.
Bliss, L.C. 1962. Adaptations of arctic and alpine plants
    to environmental conditions. *Arctic. 15*:117-144.
______. 1971. Arctic and alpine plant life cycles. *Ann. Rev.
    Ecol. Syst. 2*:405-438.
Brown, J. 1971 progress report. In The structure and func-
    tion of the tundra ecosystem. Vol. 1. Progress report
    and proposal abstracts, U.S.I.B.P. Tundra Biome,
    Washington, D.C.
Brown, J., F.A. Pitelka, and H.N. Coulombe. 1970. Structure
    and function of the tundra ecosystem at Barrow, Alaska.
    Pages 41-71 in W.A. Fuller and P.G. Kevin, eds. Pro-
    ceedings of the conference on productivity and con-
    servation in northern circumpolar lands. IUCN, Morges,
    Switzerland.
Brown, J., et al. 1979. The coastal tundra at Barrow. In
    J.Brown, et al., eds. An arctic ecosystem: the coastal
    tundra at Barrow, Alaska. Dowden, Hutchinson & Ross,
    Stroudsburg, PA.
Bunnell, F.L., and P. Dowding. 1974. ABISKO--a generalized
    model for comparisons between sites. Pages 227-247 in
    A.J. Holding, et al., eds. Soil organisms and decom-
    position in tundra. IBP Tundra Biome Steering Com-
    mittee, Stockholm.
Bunnell, F.L., and K. Scoullar. 1975. ABISKO II--a computer
    simulation model of carbon flux in tundra ecosystems.
    Pages 425-448 in T. Rosswall and O.W. Heal, eds.
    Structure and function of tundra ecostems. *Ecological
    Bulletins 20.*
Callaghan, T.V., and N.J. Collins. 1976. Strategies of
    growth and population dynamics of tundra plants.
    Oikos. 27:383-388.

Chapin, F.S. et al. 1979. Control of tundra plant alloca-
    tion patterns and growth. In J. Brown, et al., eds.
    An arctic ecosystem: the coastal tundra at Barrow,
    Alaska. Dowden, Hutchinson & Ross, Stroudsburg, PA.

Courtin, G.M., and J.M. Mayo. 1974. Arctic and alpine plant
    water relations. Pages 1120-1124 in F.J. Vernberg, ed.
    Physiological adaptation to the environment. Intext
    Editions Publishers, New York.

Davidson, J.L., and J.R. Phillip. 1958. Light and pasture
    growth. Pages 181-187 in Climatology and micro-clima-
    tology. UNESCO, Paris.

Denmead, O.T. 1964. Evaporation sources and apparent dif-
    fusivities in a forest canopy. *J. Appl. Meterol.*
    3:383-389.

______. 1973. Relative significance of soil and plant eva-
    poration in estimating evapotranspiration. Pages 505-
    511 in R.O. Slatyer, ed. Plant response to climatic
    factors: proceedings of the Uppsala Symposium. UNESCO,
    Paris.

De Wit, C.T. 1965. Photosynthesis of leaf canopies. Agric.
    Res. Rep. 663, Inst. for Biol. Chem. Res. Field Crops
    Herb., Wageningen, The Netherlands.

De Wit, C.T., and Goudriaan, J. 1974. Simulation of ecolo-
    gical processes. Centre for Agricultural Publishing and
    Documentation, Wageningen, The Netherlands. 159 pp.

Duncan, W.C., et al. 1967. A model for simulating photosyn-
    thesis in plant communities. *Hilgardia*. *38*:181-205.

Ehleringer, J.R., and P.C. Miller. 1975a. A simulation model
    of plant water relations and production in the alpine
    tundra, Colorado. *Oecologia*. *19*:177-193.

______. 1975b. Water relations of selected plant species in
    the alpine tundra, Colorado. *Ecology*. *56*:370-380.

Fiscus, E.L., and P.J. Kramer. 1975. General model for
    osmotic and pressure-induced flow in plant roots. *Proc.*
    Natl. Acad. Sci. *72*:3114-3118.

Gates, D.M. 1962. Energy exchange in the biosphere. Harper
    & Row, New York. 151 pp.

______. 1965. Energy, plants and ecology. *Ecology*.
    46:1-13.

Gates, D.M., E.C. Tibbals, and F. Kreith. 1965. Radiation
    and convection for ponderosa pine. *Am. J. Bot.*
    52:66-71.

Goudriaan, J., and P.E. Waggoner. 1962. Simulating both
    aerial microclimate and soil temperature from obser-
    vations above the foliar canopy. *Noth. J. Agric. Sci.*
    20:104-124.

Hanks, R.J., and S.B. Bowers, 1962. Numerical solution of the moisture flow equation into layered soils. *Soil Sci. Soc. Am. Proc. 26*:530–534.

Harper, J.L., and J. White. 1974. The demography of plants. Pages 419–464 in R.F. Johnston, ed. Annual review of ecology and systematics. Annual Reviews, Inc., Palo Alto, CA.

Hicklenton, P.R., and W.C. Oechel. 1976. Physiological aspects of the ecology of *Dicranum fuscescens* in the subarctic. I. Acclimation and acclimation potential of CO_2 exchange in relation to habitat, light, and temperature. *Can. J. Bot. 54*:1104–1119.

Hocking, B., and C.D. Sharpin. 1965. Flower basking by arctic insects. *Nature. 206*:215.

Honert, T.H. van den. 1948. Water transport in plants as a catenary process. *Disc. Faraday Soc. 3*:146–153.

Idso, S.B., and C.T. de Wit. 1970. Light relations in plant canopies. *Appl. Opt. 9*:177–184.

Jarvis, P.G., and M.S. Jarvis. 1963. The water relations of tree seedlings. IV. Some aspects of the tissue water relations and drought resistance. *Physiol. Plant. 16*:501–516.

Kedrowski, R.A. 1976. Plant water relations in the arctic tundra near Meade River, Alaska. M.S. Thesis. San Diego State Univ., San Diego, CA.

Kleinendorst, A., and R. Brouwer. 1972. The effect of local cooling on growth and water content of plants. *Neth. J. Agric. Sci. 20*:203–217.

Klikoff, L. 1965. Photosynthetic response to temperature and moisture stress of three timberline meadow species. *Ecology. 46*:516–517.

Koranda, J.J., B. Clegg, and M. Stuart. 1978. Radio-tracer measurement of transpiration in tundra vegetation, Barrow, Alaska. Pages 359–369 in L.L. Tieszen, ed. Vegetation and production ecology of the Alaskan arctic tundra. Springer-Verlag, New York.

Kramer, P.J. 1952. Plant and soil water relations on the watershed. *J. For. 50*92–95.

________. 1969. Plant and soil water relationships: a modern synthesis. McGraw-Hill, New York. 482 pp.

Krogg, J. 1955. Notes on temperature measurement indicative of special organization in arctic and subarctic plants for utilization of radiated heat from the sun. *Physiol. Plant. 8*:836–839.

Kuiper, P.J.C. 1964. Water uptake of higher plants as affected by root temperature. *Meded. Landbouwhogesch. Wageningen. 63*:1–11.

Lange, O.L. 1965. Der CO_2 Gaswechsel von Flechten fei tiefen Temperaturen. *Planta. 64*:1-19.

Lawrence, B.A. 1974. A model of the population dynamics of *Dupontia fisheri*. M.S. Thesis. San Diego State Univ., San Diego, CA.

Lawrence, B.H., et al. 1978. A model of the population dynamics of *Dupontia fisheri*. Pages 599-619 in L.L. Tieszen, ed. Vegetation and production ecology of the Alaskan arctic tundra. Springer-Verlag, New York.

Lemon, E.R. 1967. The impact of the atmospheric environment of the integument of plants. *Internat. J. Biometerol. 3*:57-69.

Liakopoqulos, A.C. 1965. Theoretical solution of the unsteady unsaturated flow problems in soils. *Internat. Assoc. Sci. Hydrol. Bull. 10*:58-69.

Luposhinsky, W. 1969. Stomatal closure in conifer seedlings in response to leaf moisture stress. *Bot. Gaz. 130*:258-263.

Mayo, J.M., D.C. Despain, and E.M. van Zinderen Bakker, Jr. 1973. CO_2 assimulation by *Dryas integrifolia* on Devon Island, Northwest Territories. *Can. J. Bot. 51*:581-588.

McCown, B.H. 1973. Growth and survival of northern plants at low soil temperatures: growth response, organic nutrients, and ammonium utilization. Cold Regions Research and Engineering Lab., U.S. Army, Special Report 186. 15 pp.

McCree, K.J., and J.H. Troughton. 1966. Prediction of growth rate at different light levels from measured photosynthesis and respiration rates. *Plant Physiol. 41*:559-566.

McNaughton, S.J., et al. 1974. Photosynthetic properties and root chilling responses of altitudinal ecotypes of *Typa latfolia* L. *Ecology 55*:168-172.

Miller, P.C. 1972. Bioclimate, leaf temperature, and primary production in red mangrove canopies in south Florida. *Ecology. 53*:22 45.

______. 1974. Potential use of vegetation to enhance cooling in holding ponds. Pages 610-627 in J.W. Gibbons, and R.R. Sharitz, eds. Thermal ecology. AEC Symposium Series (CONF-730505), 1973.

Miller, P.C., and W.A. Stoner. 1979. Canopy structure and environmental interactions. Pages 163-173 in O. Solbrig, et al., eds. Topics in plant population biology. Columbia University Press, New York.

Miller, P.C., and L. Tieszen. 1972. A preliminary model of processes affecting primary production in the arctic tundra. *Arctic Alp. Res. 4*:1-18.

Miller, P.C., B.D. Collier, and F. Bunnell. 1975. Development of ecosystem modeling in the tundra biome. Pages 95-115 in B.C. Patten, ed. Systems analysis and simulation in ecology. Vol. 3. Part I: Ecosystem modeling in the U.S. International Biological Program. Academic Press, New York.

Miller, P.C., W.A. Stoner, and L.L. Tieszen. 1976. A model of stand photosynthesis for the wet meadow tundra at Barrow, Alaska. *Ecology. 57*:411-430.

Miller, P.C., et al. 1974. Digital simulation of potential reforestation problems in the Rung Sat. Delta, Viet Nam. Pages 100-164 in H.T. Odum, et al. eds. The effects of herbicides in South Vietnam. Part B: Working papers. Models of herbicide, mangroves, and war in Vietnam. National Academy of Sciences - National Research Council, Washington, D.C.

Monsi, M., and T. Saeki. 1953. Uber den Lichtfaktor in den Pflanzengesellschaften and sein Bedeutung fur die Stoffproduktion. *Japan. J. Bot. 14*:22-52.

Monteith, J.L. 1965. Evaporation and environment. Pages 205-234 in G.E. Fogg, The state and movement of water in living organisms. Symp. Soc. Exptl. Biol. No. XIX. Academic Press, New York.

Mooney, H.A. 1972. The carbon balance of plants. *Ann. Rev. Ecol. Syst. 3*:315-346.

Murphy, C.E., and K.R. Knoerr. 1970. Modeling the energy balance processes of natural ecosystems. U.S. IBP Anal. Ecosyst. Program Eastern Deciduous Forest Biome, Subproject Res. Rep. Final 1969-1970. Oak Ridge National Laboratory. 168 pp.

______. 1972. Modeling the energy balance processes of natural ecosystems. U.S. IBP Anal. Ecosyst. Program Eastern Deciduous Forest Biome, Subproject Res. Rep. EDFB-IBP 72-10. Oak Ridge National Laboratory. 141 pp.

Nakano, N., and J. Brown. 1972. Mathematical modeling and validations of the thermal regions in tundra soils, Barrow, Alaska. *Arctic Alp. Res. 4*:19-38.

Ng, E., and P.C. Miller. 1975. A model of the effect of tundra vegetation on soil temperatures. Pages 222-226 in G. Weller and S.A. Bowling, eds. Climate of the arctic. Proceedings of the 24th Alaska Science Conference, Fairbanks, 1973. Geophysical Institute, Univ. of Alaska, Fairbanks.

Ng, E., and P.C. Miller. 1977. Validation of a model of the effect of tundra vegetation on soil temperatures. *Arctic Alp. Res.* 9:89-104.

Oechel, W.C. 1976. Seasonal patterns of temperature response of CO_2 flux and acclimination in arctic mosses growing in situ. *Photosynthetica.* 10:447-456.

Oechel, W.C., and B. Sveinbjornsson. 1974. The microenvironment of Arctic bryophytes at Barrow, Alaska. U.S. Tundra Biome Data Report 74-25.

Oechel, W.C., et al. 1975. Temperature acclimation in *D. fuscescens* growing in situ in the arctic and sub-arctic. Pages I-131-I-144 in Proceedings of the circumpolar conference on northern ecology, 1975, Ottawa NRC with CNC/SCOPE.

Penning de Vries, F.W.T. 1972. Respiration and growth. Pages 327-347 in A.R. Rees, et al., eds. Crop processes in controlled environments. Academic Press, London.

______. 1973. Substrate utilization and respiration in relation to growth and maintenance in higher plants. Ph.D. Thesis. Agricultural Univ., Wageningen, The Netherlands.

______. 1974. Substrate utilization and respiration in relation to growth and maintenance in higher plants. *Neth. J. Agric. Sci.* 22:40-44.

Penning de Vries, F.W.T., A.H.M. Brunsting, and H.H. van Laar. 1974. Products, requirements, and efficiency of biosynthesis: a quantitative approach. *J. Theor. Biol.* 45:339-337.

Poole, D.K. 1974. Water relations of selected species of chaparral and coastal sage communities. M.S. Thesis. San Diego State Univ., San Diego, CA.

Rawlins, S.L. 1963. Resistance to water flow in the transpiration stream. Pages 69-84 in I. Zelitch, ed. Stromata and water relations in plants. *Conn. Agric. Exp. Sta. Bull. 664.*

Savile, D.B.O. 1972. Arctic adaptations in plants. Can. Dep. Agric. Mongr. 6, 81 pp.

Schultz, A.M. 1969. A study of an ecosystem: the arctic tundra. Pages 77-93 in G.M. Van Dyne, ed. The ecosystem concept in natural resource management. Academic Press, New York.

Shaver, G.R., and W.D. Billings. 1977. Effects of daylength and temperature on root elongation in tundra graminoids. *Oecologica.* 28:57-65.

Slatyer, R.O. 1967. Plant water relationships. Academic Press, New York. 366 pp.

Stewart, D.W., and E.R. Lemon. 1972. The energy budget at the earth's surface: a simulation of net photosynthesis of field corn. ECOM Atmospheric Sciences Lab. Int. Rep. 69-3. Fort Huachuca, AZ.

Stoner, W.A., and P.C. Miller. 1975. Plant water relations
     in the wet coastal tundra, Barrow, Alaska. *Arctic Alp.
     Res.* 7:109–124.
Stoner, W.A., et al. 1978. A test of a model of irradiance
     within vegetation canopies at northern latitudes.
     Arctic Alp. Res. 4:761–767.
______. 1978. Internal nutrient cycling as related to plant
     life form: a simulation approach. Pages 165–181 in
     D.C. Adriano, and I.L. Brisbin, Jr., eds. Environmental
     chemistry and cycling processes. Technical Informa-
     tion Center, U.S. Dept. of Energy, Washington, D.C.
Tibbals, E.C., et al. 1964. Radiation and convection in
     conifers. *Am. J. Bot.* 51:529–538.
Tieszen, L.L. 1972. The seasonal course of aboveground
     production and chlorophyll distribution in a wet
     Arctic tundra at Barrow, Alaska. *Arctic Alp. Res.*
     4:307–324.
______. 1973. Photosynthesis and respiration in arctic
     tundra grasses: field light intensities and temperature
     responses. *Arctic Alp. Res.* 5:239–252.
______. L.L. 1978. Photosynthesis in the principal Barrow,
     Alaska species: A summary of field and laboratory re-
     sponses. Chapter 10 in L.L. Tieszen, ed. Vegetation
     and production ecology of the Alaskan arctic tundra.
     Springer-Verlag, New York.
Tieszen, L.L., et al. 1979. Photosynthesis. In J. Brown,
     et al., eds. An arctic ecosystem: the coastal tundra
     at Barrow, Alaska. Dowden, Hutchinson & Ross, Strouds-
     burg, PA.
Tikhomirov, B.A., V.F. Shamurin, and V.S. Shtipo. 1960. The
     temperature of arctic plants. *Izv. Akad. Nauk SSR Biol.
     Ser.* 3:429–422. (In Russian)
Timin, M.E., et al. 1973. A computer simulation of the
     Arctic tundra ecosystem near Barrow, Alaska. U.S.
     Tundra Biome Report 73-1.
Waggoner, P.E., and W.E. Reifsnyder. 1968. Simulation of
     temperature, humidity, and evaporation profiles in a
     leaf canopy. *J. Appl. Meteorol.* 7:400–409.
Waggoner, P.E., G.M. Furnival, and W.E. Reifsnyder. 1969.
     Simulation of microclimate in a forest canopy. *For.
     Sci.* 15:37–45.
Weller, G., and B. Holmgren. 1974. The microclimates of the
     arctic tundra. *J. Appl. Meteorol.* 13:854–862.
Wilson, J. 1957. Arctic plant growth. *Adv. Sci.* 13:383–388.
Yim, Y., H. Ogawa, and T. Kira. 1969. Light interception by
     stems in plant communities. *Japan. J. Ecol.*
     19:233–238.

VII. NUTRIENT UPTAKE AND UTILIZATION BY TUNDRA PLANTS

F. Stuart Chapin III

Institute of Arctic Biology
University of Alaska-Fairbanks
Fairbanks, Alaska

Nutrient availability is limited in tundra soils due to many direct and indirect effects of low temperature upon nutrient cycling. Arctic plants adapt to a cold, nutrient-poor environment through: 1) a high root-to-shoot ratio which requires root growth at low temperatures and, 2) high rates of nutrient uptake. Phosphate uptake by arctic graminoids is adaptive at low temperature because of a low sensitivity to temperature (35 percent of the 20° rate being maintained at 1°C) but not through any change in temperature optimum. Tundra graminoids differ from most crop plants in having substantial tissue concentrations of nitrogen and phosphorus, even when growth is strongly limited by both elements. High nutrient concentrations appear necessary to maintain high metabolic levels necessary for rapid growth at low temperature. Tundra plants respond to nutrient deficiency by reduced growth rather than by production of nutrient-deficient tissue. Nutrient use changes substantially through the growing season. A much larger proportion of leaf nitrogen is associated with photosynthesis at mid-season than in spring or fall. Tundra deciduous shrubs translocate substantial quantities of nutrients into leaves from belowground storage each year, whereas evergreen shrubs store nutrients in leaves from one year to the next.

INTRODUCTION

All of the features that uniquely characterize arctic
tundra ecosystems derive ultimately from the small annual
input of solar energy at high latitudes. Low light intensities and low temperature are among the principal factors
limiting photosynthesis and hence growth (Miller, Stoner,
and Tieszen 1976). Yet arctic plant growth may not be
strongly energy-limited. Northern plants have unusually
high available carbohydrate levels - generally on the order
of 30 percent dry weight (McCown 1979).

The arctic climate acts both directly and indirectly
upon the processes controlling nutrient cycling and thereby
limits availability of nutrients to plants (Figure 1).
Because of low temperature and presence of permafrost,
weathering of parent material occurs at a negligible rate in
the Arctic. Furthermore, the predominance of wet or snow-
and ice-covered surfaces minimizes dust and atmospheric
input of nutrients. Nitrogen fixation occurs slowly at
prevailing temperatures. Hence, nutrient input to available
nutrient pools in the soil is extremely low in arctic tundra
systems, and most nutrients utilized by the vegetation must
be recycled by decomposition.

Recycling of organically bound nutrients by decomposition is also slow because of low temperature (Figure 1).
In many areas decomposition rates are further reduced by
poor soil aeration resulting from low evapotranspiration and
impeded drainage of soils underlain by permafrost. The soil
organic matter which accumulates as a consequence of slow
decompositiondecreases soil pH. Low pH and the presence of
abundant iron further reduces availability of some elements
such as phosphorus. In short, the arctic climate directly
and indirectly results in formation of soils with low nutrient availability (Chapin, Barsdate, and Barel 1979). Nitrogen and phosphorus tend to be particularly limiting to
plant growth. Nitrogen is present primarily as the ammonium
ion (Flint and Gersper 1974) which is not readily absorbed
by many temperate plants.

Much of the early evidence of nutrient limitation of
plant growth in the Arctic is indirect or circumstantial.
Russell (1940) found a positive correlation between total
soil nitrogen content and productivity of *Salix* in the
Canadian arctic. Warren Wilson (1957) noted that arctic
soils supported more plant growth when supplemented with
nitrogen. More recently *in situ* fertilization studies have
shown nitrogen and phosphorus to be limiting to plant growth
both in tundra (Bliss 1966; Schultz 1964; Haag 1974; Chapin

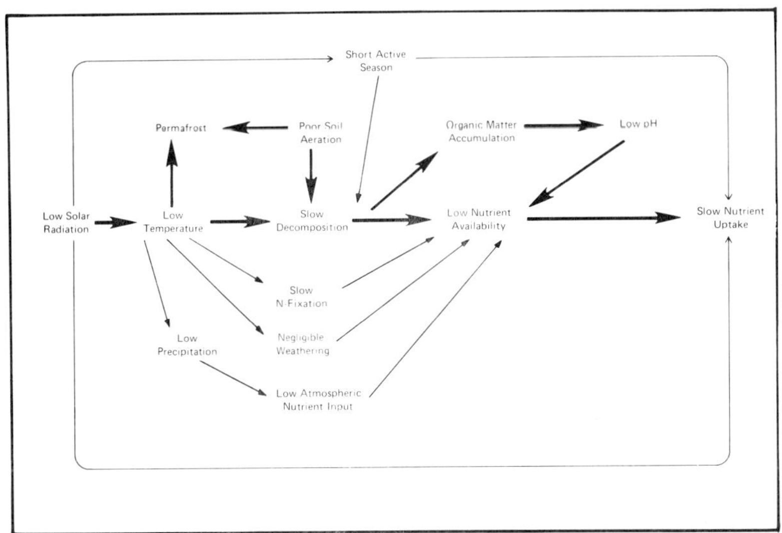

FIGURE 1. Network of cause and effect relationships linking the low radiation input of the Arctic with low nutrient availability and low nutrient uptake rates. See text for discussion (redrawn from Chapin, Miller, Billings, Coyne, and McCown in press).

Van Cleve, and Tieszen 1975; Wulgolaski, Kjelvik, and Kallio 1975; McKendrick, Ott, and Mitchell 1979) and in tundra-like moors and bogs (Tamm 1954; Gore 1961; Goodman and Perkins 1968). Plant growth has been found to be nutrient-limited in virtually every tundra community examined to date. Hence, we expect tundra plants to exhibit various adaptations for acquisition, conservative utilization, and retention of nutrients.

ROOT GROWTH

An important means by which nutrient acquisition is maximized in tundra plants is by the allocation of a large proportion of the plant biomass to roots. In Barrow graminoids 60-90 percent of the biomass occurs below-ground (Table 1) as a consequence of both substantial root longevity in many tundra species (Shaver and Billings 1975) and large belowground production. There are no direct measurements of belowground production in any tundra community, although estimates suggest a belowground production at least equal to

TABLE I.   Above- and Belowground Peak Season Biomass[1,2] *Annual Production* [3,2] *and Respiration for the Barrow Wet Meadow Tundra*

	Peak Standing Crop (g dry wt. m^{-2})	Annual Prod. (g dry wt. $m^{-2}yr^{-1}$)	Annual Respiration (gC $m^{-2}yr^{-1}$)
Aboveground	100	100	38
Belowground	660	100	40
Below/Above	6.6	1.0	1.1

[1]*Tieszen 1972;* [2]*Dennis 1977;* [3]*Chapin et al. in press.*

that determined aboveground (Shaver and Billings 1975). A latitudinal increase in root-to-shoot ratio is found when either several species or populations of a given species are compared (Chapin 1974). Little is known about the relative importance of genetic differentiation and phenotypic plasticity (Davidson 1969) in determining the high root-to-shoot ratios of arctic plants. However, McCown (1979) found that the arctic graminoid, *Dupontia fischeri*, maintained a more constant root-to-shoot ratio over a variety of root growth temperatures than did a comparable temperate graminoid. Of the total carbon fixed annually at Barrow, about 60 percent is translocated belowground. Of this, one-half is lost in respiration to maintain the substantial belowground biomass (Chapin et al. in press). It seems likely that temperate plants which experience warmer soil temperatures would not be able to sustain a comparably large root biomass unless their respiratory capacities were substantially lower.

Root growth in the Arctic is temperature-dependent as in all plants, although tundra plants have a lower temperature optimum for root initiation, elongation, and hence production, than do temperate plants (Chapin 1974). Substantial elongation is observed in arctic graminoids in growth experiments even at temperatures below 1ºC (Billings, Shaver, and Trent 1973) (Figure 2). Root elongation rate in the field appears to be controlled more strongly by photoperiod and internal factors than by soil temperature (Shaver and Billings 1975, 1977).

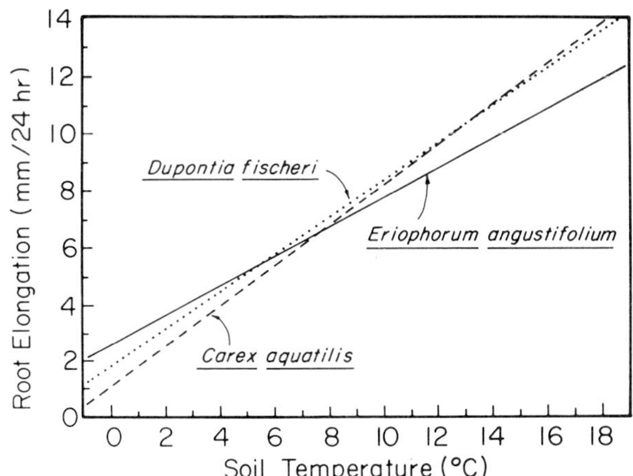

FIGURE 2. Rate of root elongation as a function of soil temperature in three Barrow graminoids grown in the phytotron (redrawn from Billings, Shaver and Trent 1973).

In the field, roots of *Eriophorum angustifolium* (Shaver and Billings 1975) and of *E. vaginatum* (Chapin 1974; Bliss 1956) follow the annual soil thaw and hence must actively grow and elongate at 0°C. Root elongation is not prevented by freezing and thawing (Shaver and Billings 1975). In fact, primary roots of *Carex aquatilis* continue elongating for 2-4 years in the field. Because spindle formation associated with mitosis is prevented at temperatures below 5°C in temperate plants (Hepler and Palevitz 1974), there is clearly some metabolic alteration of the mitotic process in arctic species.

In wet meadow tundra underlain by permafrost, anaerobic conditions may strongly influence root growth. Whereas upland moor species in England depend upon soil oxygen for root respiration, bog species have abundant aerenchyma and transport oxygen necessary for root respiration (Armstrong 1972). Many of these same species grow in tundra areas. Undoubtedly the ability of a species to transport oxygen to roots is important in determining the vegetation patterns with respect to moisture in tundra areas (Webber and Ebert-May 1979). Some species such as *Carex aquatilis* produce thick aerenchyma-containing roots in deeper anaerobic soils but also produce thinner roots that lack aerenchyma in better aerated soil horizons (Chapin 1979).

Mycorrhizal associations are strongly developed in well-drained tundra regions as in other phosphorus-deficient community types (Miller and Laursen 1979). However, they are generally less abundant in meadow tundra. Although presence of mycorrhizal associations has been documented for many species, little is known about their importance and mode of action, especially in wet anaerobic soils.

NUTRIENT UPTAKE

Nutrient absorption occurs by active transport and hence is temperature-dependent in tundra as well as temperate plants. The optimum temperature for phosphate absorption in the three principal Barrow graminoid species is at least $40^{\circ}C$ (Chapin and Bloom 1976) and does not differ from that of temperate plants (Carter and Lathwell 1967; Carey and Berry 1978) (Figure 3). Tundra plants are apparently not adapted to low soil temperature through any change in the optimum temperature for phosphate absorption. The linierity of phosphate absorption rate, even down to $1^{\circ}C$, indicates that roots of tundra plants are quite capable of absorbing phosphate from cold soils, whenever soils are thawed. For example, *Dupontia fischeri* still maintains 35 percent of its $20^{\circ}C$ phosphate absorption capacity when measured at $1^{\circ}C$. In contrast, temperate plants exhibit negligible phosphate absorption rates below $5-10^{\circ}C$ (Chapin 1974; Carey and Berry 1978; Sutton 1969). The biochemical basis of this substantial difference in ability to absorb phosphorus at low temperature is not known but is correlated with a greater degree of unsaturation of fatty acids in membrane phospholipid from cold-adapted, cold-acclimated plants (Kedrowski and Chapin 1978) (Table 2). Such unsaturation would cause membranes to remain fluid at low temperatures normally encountered during the arctic growing season.

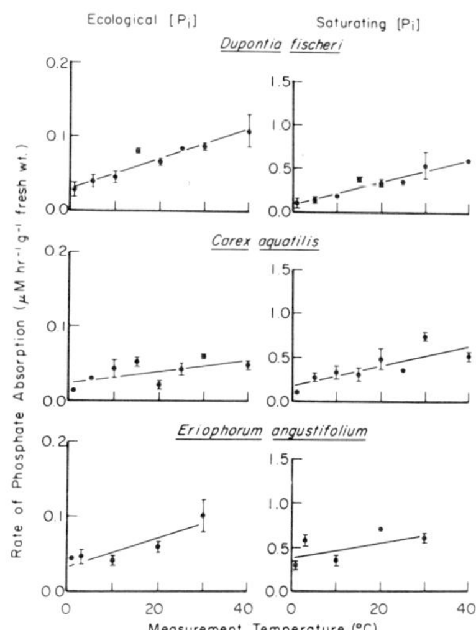

FIGURE 3. *Rate of phosphate absorption measured in the field as a function of root measurement temperature at ecological (1.0 uM) and saturating (10u M) phsophate concentrations in roots of the three principal Barrow graminoids (Chapin and Bloom 1976).*

TABLE II. *Phospholipid Double Bond Index and Phosphate Uptake Rate in Field-sampled Populations of* Carex aquatilis *from Circle Hot Springs and an Adjacent Permafrost-Dominated Site (Kedrowski and Chapin 1978).*

Population	Phospholipid Double Bond Index	Phosphate Uptake Rate[1] (nM hr⁻¹ g⁻¹ dry wt)	
		.5uM(P)	20uM(P)
Permafrost	1.18±0.02*	25±2*	420±101
Hot Springs	1.06±0.03*	15±1*	230±34

[1]*Measured at 1°C*
**Populations differ significantly at P<0.05*

Because phosphate absorption is temperature-dependent, one would expect tundra species evolving and growing in cold soil to exhibit a lower rate of phosphate absorption than a temperate species growing in its warmer soil. However, tundra species have compensated evolutionarily for this temperature effect upon phosphate absorption. When tundra and temperate plants are grown and measured together

under standard conditions, tundra species consistently
exhibit higher phosphate absorption rates (Chapin 1974). A
similar evolutionary response to low soil temperature is
observed between ecotypes of the same species. For example, *Carex aquatilis* from Barrow in northern Alaska has a
higher rate of phosphate absorption than does either a
permafrost or a hot spring ecotype when grown in a common
garden in the hot spring site in central Alaska (Figure 4).
Clearly, differences in soil temperature or associated
environmental factors have selected strongly for maintenance of differences in phosphate absorption capacity.

Acclimation of plants to low root temperature results
in an increase in phosphate absorption capacity, as measured under standard conditions (Chapin 1974). This
plastic response to the environment is compensatory in
nature and may partially explain the seasonal change in
phosphate uptake capability observed in Barrow graminoids.
Phosphate uptake capacity is high early in the season, and
declines as the soil warms (Figure 5). However, the
substantial seasonal drop in phosphate absorption capacity
is greater than would be predicted on the basis of acclimation, and may also result from root aging. The ability
of tundra species to acclimate is much less strongly developed than that of species from more thermally fluctuating environments (Chapin 1974).

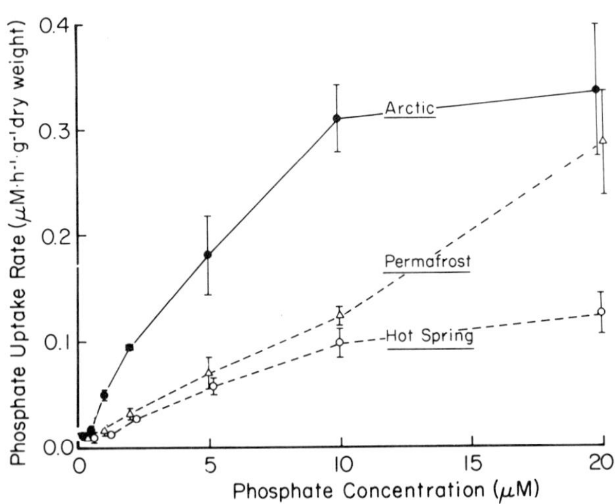

FIGURE 4. *Rate of phosphate absorption at 15°C by roots of various* Carex aquatilis *populations grown in a common garden at Circle Hot Spring, central Alaska. Mean ± standard error, n=4.*

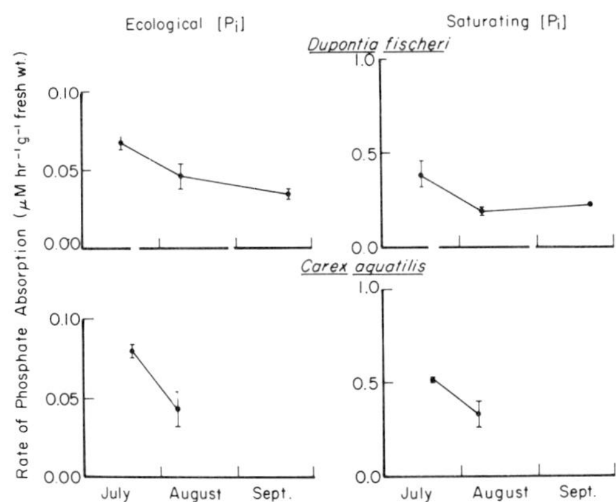

FIGURE 5. Rate of phosphate absorption measured in the field as a function of time of season in Dupontia fischeri *and* Carex aquatilis *measured at 10°C at ecological (1.0uM) and saturating (20uM) phosphate concentrations (Chapin and Bloom 1976).*

Tundra and temperate grasses also differ dramatically in their ability to absorb nitrogen (Figure 6). *Dupontia fischeri* grown at low root temperature responds much more to added nitrogen than does *Bromus pumpillianus*. At warmer root temperatures, both species show a similar responsiveness to added nitrogen. There are indications that *Dupontia* absorbs ammonium as well as or better than nitrate, whereas the temperate *Bromus* exhibits considerably more growth when grown with nitrate (McCown 1979). Such differences in utilization would make sense in light of the predominance of ammonium in Barrow wet meadow soils and of nitrate in temperate grasslands.

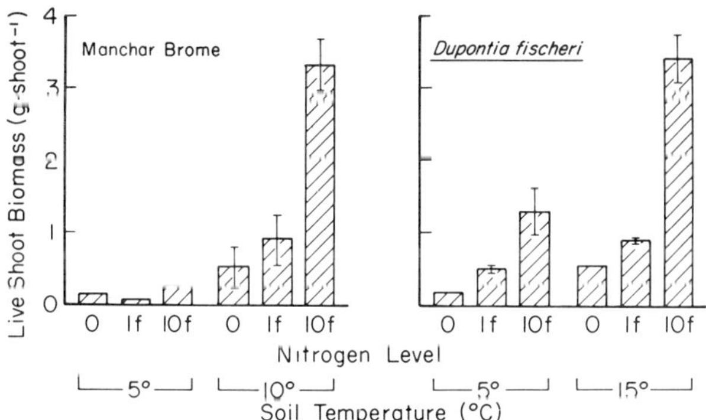

FIGURE 6. Shoot biomass of Manchar brome and Dupontia fischeri *grown at differing nitrogen levels and soil temperatures (redrawn from McCown 1979).*

NUTRIENT UTILIZATION AND RETENTION

The general principles developed to explain plant
growth and nutrition are based almost exclusively upon
studies of agricultural species. Such plants generally
evolved from weeds growing in disturbed sites around human
habitations. These habitats tended to have high nutrient
availability so that light was frequently more limiting to
growth than were nutrients. Hence, rapid shoot growth
would have been strongly selected for. Given this basic
genetic stock, agronomists have selected for high aboveground yield in high nutrient environments. In such situations, there is little selective premium for efficient
utilization of nutrients. In short, generalities of plant
nutrition are based almost entirely upon plants that are
selected to grow rapidly, and where efficient nutrient
utilization is of little selective advantage. In closed K-
selected competitive communities, selection may not be so
much for rapid growth, but rather for efficient utilization
of nutrients, and the general ground rules governing plant
nutrition may be somewhat different from those developed in
agriculture.

As discussed above, fertilization studies have shown
nitrogen and/or phosphorus availability to be among the
factors limiting plant production in every tundra community
thus-far examined. Likewise, Ulrich and Gersper (1979)
consider Barrow tundra plants nutrient-limited, on the
basis of the extremely low levels of inorganic nitrogen and
phosphorus present in shoots. Yet leaves of tundra plants
generally have quite substantial concentrations of total
nitrogen and phosphorus (Chapin, Van Cleve, Tieszen 1975;
Chapin, Johnson, and McKendirck in press; Dowding et al. in
press). Perhaps plants evolving in the arctic where the
growing season lasts only 8 weeks, cannot "afford" to
produce nutrient-deficient tissues but require high nitrogen and phosphorus concentrations in order to support their
observed rapid growth. This is in strong contrast to crop
plants that frequently continue to grow, even if nutrients
are unavailable, and consequently produce nutrient-deficient
leaves with low photosynthetic rates.

Plants adapted to cold climates and tissues of plants
acclimated to cold conditions, generally have high respiratory capacities and high enzymatic activities (Mooney and
Billings 1961; McNaughton 1972; Billings et al 1971). This
may be largely a consequence of high concentrations of
enzymes. If so, high tissue protein nitrogen concentrations are a prerequisite for the high metabolic potential

and rapid growth observed in tundra plants in the field. Tissue activities and nitrogen concentrations in populations of *Carex aquatilis* growing in cold and warm soils, are consistent with this hypothesis. In cold soils, roots and rhizomes have higher phosphate uptake rates and respiration rates and higher tissue nitrogen concentrations than in the hot spring plants. A large proportion of the nitrogen in leaves is associated with the photosynthetic enzyme ribulase-1,5-bisphosphate carboxylase-oxygenase (RuBP), the concentration of which is strongly affected by factors such as light. Hence, nitrogen concentration may show less of a correlation with ambient temperature in leaves than in other tissues.

Active biosynthesis associated with rapid growth requires a proliferation of membranes, which are largely constructed from phospholipid. This phosphorus invested in phospholipid plus that required for sugar phosphates and nucleotide phosphates involved in cellular energy transfer must result in high phosphorus concentrations in metabolically active tissues. *Carex aquatilis* roots growing in cold soils have high total phosphorus concentrations, high phospholipid concentrations, and a higher proportion of phosphorus bound in phospholipid (Table 3). Hence, on the basis of known high metabolic capacities of tundra plants we should expect the observed high concentrations of tissue nitrogen and phosphorus.

TABLE III. Phosphate Uptake Rate, Nitrogen and Phosphorus Concentrations in Roots of Field-Sampled Populations of Carex aquatilis from Circle Hot Springs and and Adjacent Permafrost-Dominant Site (Kedrowski and Chapin 1978).

Population	P uptake Rate[1] (nM hr^{-1} g^{-1} dry wt.)	Nitrogen (% dry wt.)	Phosphorus (% dry wt.)	Lipid - P (% total P)
Permafrost	1750±240**	0.83±.02	.200±.004	18.4±.7*
Hot Springs	480±90**	0.78±.07	.200±.004	12.1±0.0*

[1]*Measured at 23ºC, 20 uM Phosphats.*
**, ** Population differe significantly at P<.05, .01, respectively.*

The seasonal changes in nutrient concentrations of tundra plants generally parallel those observed in other community types. In deciduous species leaf nitrogen concentration is high at snowmelt and increases slightly as buds break and new leaves are formed (Figure 7). Then there is a gradual decrease in leaf nitrogen through the remainder of the growing season; first as structural tissue is added to leaves effectively diluting the nitrogen present, and later as nitrogen is actually translocated out of leaves back into stems. Leaves continue to decline in nitrogen concentration in fall, but newly produced stems retain or accumulate nitrogen. The nitrogen concentrations of older stems and roots decrease during early summer when leaves are rapidly expanding and accumulating nitrogen, but increase again from mid-summer onward. These seasonal patterns are shown here for dwarf birch, *Betula nana*, (Figure 7), but are basically similar in other deciduous shrubs, graminoids, and forbs. Phosphorus and potassium show similar patterns (Chapin, Van Cleve, Tieszen 1975; Chapin, Johnson, and McKendrick in press). Evergreen shrubs exhibit the same basic seasonal changes in nitrogen concentration in new growth and old stems as are observed in the deciduous species, except that young leaves do not continue to lose nitrogen in the fall. Nitrogen concentration of older leaves generally parallels that of stems, decreasing slightly in late summer when new leaves are produced. This parallel seasonal change in concentration suggests that leaves and stems may both represent small storage pools of nitrogen.

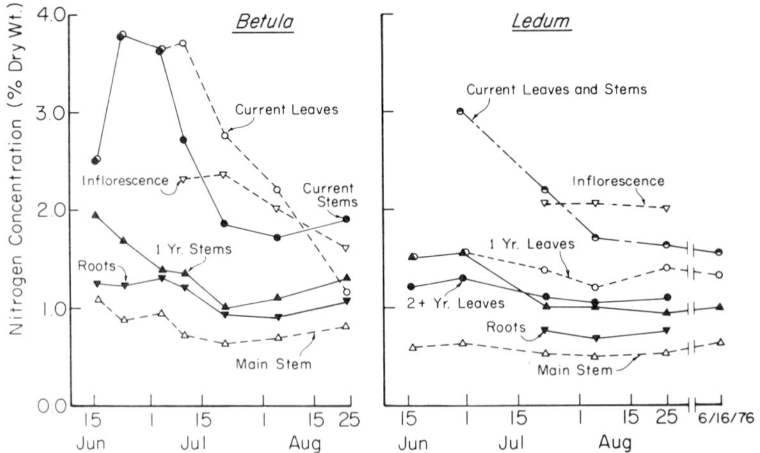

FIGURE 7. Nitrogen concentration of various plant parts of Betula nana and Ledum palustre sampled through the 1975 growing season at Atkarook, Alaska (Chapin, Johnson, and McKendrick in press).

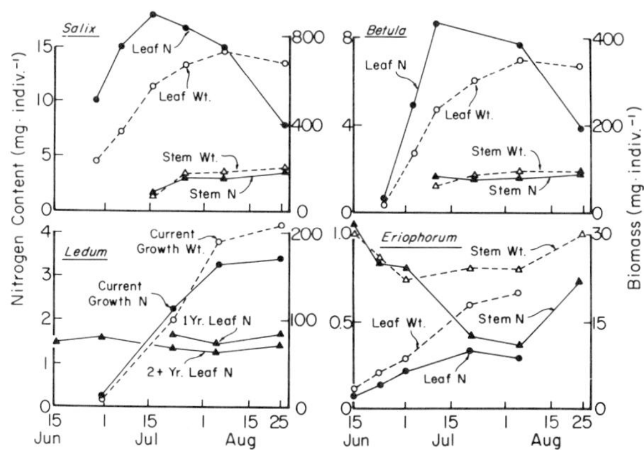

FIGURE 8. Quantity of nitrogen and biomass contained in leaves and current years stems of *Salix pulchra* and *Betula nana* sampled through the 1975 growing season at Atkasook, Alaska. In *Ledum palustre* current growth was lumped and 1-year old and 2-year-and-older leaves are shown. In *Eriophorum vaginatum* leaf and total stem (rhizome) are shown for the same study site (Chapin, Johnson, and McKendrick in press).

The magnitude of the seasonal movement of nitrogen into and out of leaves is impressive. In *Salix*, the equivalent to 12 percent of the total aboveground nitrogen pool is translocated into leaves in the first week following bud break (Figure 8). By mid-July 41 percent of the aboveground nitrogen pool is present in leaves. At no time is more than 19 percent of the aboveground carbon pool invested in leaves. Moreover, maximal leaf nitrogen content per individual is achieved within 3-4 weeks, whereas leaves continue to accumulate biomass over a 6-week period. The nitrogen requirements for leaf production coincide with the time of minimal soil thaw, minimal soil temperature in the thawed zone, and minimal current root biomass and hence, must be supplied largely by nitrogen reserves rather than by concurrent uptake. At Barrow (Tieszen 1972), and presumably at Atkasook where these data were collected, the biomass accumulation in leaves, parallels or closely follows, the seasonal development of peak photosynthetic competence, so leaf production may not greatly deplete carbonhydrate stores. Hence, it appears that the rapid flush of leaf production by tundra deciduous shrubs strains nutrient reserves of the plant more severely than carbon reserves.

The nitrogen balance of the evergreen shrub is quite different (Figure 8). Leaf nitrogen requirements become substantial only in mid-season. Moreover, nitrogen is

translocated into leaves over a much longer time interval
in *Ledum* than in deciduous shrubs. Hence, this evergreen
shrub is more likely to meet its annual nitrogen requirements from concurrent uptake and to depend less upon
storage. Nutrient remobilization from senescing leaves
should be selected for in any environment where nutrients
are among the factors limiting plant growth. In Barrow and
Atkasook tundras, 40-60 percent of the peak leaf nitrogen,
phosphorus, and potassium standing stocks, are back-trans-
located out of leaves into stems or rhizomes. This back-
translocation begins several weeks before peak leaf biomass
is achieved and presumably represents reutilization of the
leaf nutrients for biosynthesis and metabolism in other
plant parts as well as for overwinter storage (Chapin, Van
Cleve, Tieszen 1975; Chapin, Johnson, and McKendrick in
press). The ability of plants to back-translocate and
reutilize nutrients is clearly evident in a greenhouse
study of *Carex aquatilis*. Barrow populations of this species continued to grow and reproduce vegetatively in sand
in the greenhouse with no phosphorus addition for a period
equivalent to three growing seasons. The phosphorus capi-
tal for growth must have come primarily from senescing
leaves (Shaver, Chapin, and Billings in press). Nothing is
known of the carbon cost of nutrient back-translocation.

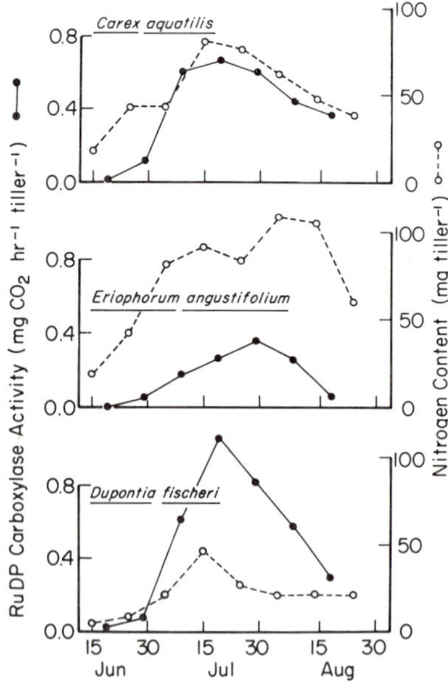

FIGURE 9. *Seasonal course of nitrogen content and
RuDP carboxylase activity (Tieszen 1979) of the three principal Barrow graminoids sampled in the field through the
growing season.*

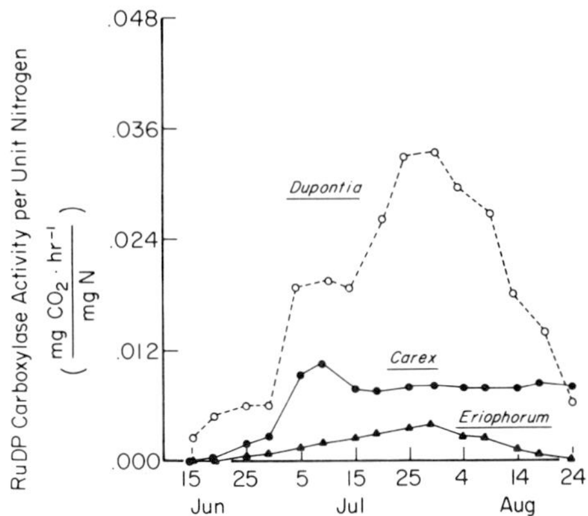

FIGURE 10. *RuDP carboxylase activity per unit nitro-*
gen in the three principal Barrow graminoids sampled in
the field through the growing season (reprinted from
Chapin et al. in press).

Seasonal nutrient patterns have now been examined in
a variety of tundra species and sites (Chapin, Van Cleve,
Tieszen 1975; Wielgolaski, Kjelvik, and Kallio 1975;
Chapin, Johnson, and McKendrick in press; Rodin and Bazilevich 1967). What does this tell us? This information is
obviously necessary for studies of herbivore nutrition and
plant-animal interaction. In the area of plant nutritional
ecology it provides a context in which to examine plant
function. For example, Tieszen found that ribulose-1,
5-bisphosphate carboxylase-oxygenase (RuBP) activity, as
measured under standard conditions, follows the same
general seasonal pattern as does leaf nitrogen concentration (Figure 9). However, when activity is expressed on
a unit nitrogen basis, it becomes clear that early and late
in the season much of the leaf nitrogen is not directly
associated with photosynthesis (Figure 10). Presumably the
high nitrogen concentrations found in leaves early in the
season, are largely associated with growth and biosynthesis.
During mid-season, photosynthetic enzymes constitute a much
more important part of leaf nitrogen. In the fall structural nitrogen and catabolic enzymes predominate. These
patterns fit with the pattern of lipid changes observed by
McCown (1979). Shoots show extremely high lipid concentrations during early spring growth and then fall to a
lower, more stable level (Figure 10). The high spring
lipid concentration may reflect the well developed systems
of membranes associated with biosynthesis and rapid growth
rather than a pool of storage lipid as suggested by earlier
work (McNair 1945; Bliss 1962).

In the past five years, the concept of carbon balance
has been used effectively to quantify patterns of allocation and provide a measure of the "adaptiveness" of a given
growth pattern. Yet in situations such as tundra where
nutrients play a key role in regulating plant growth and
allocation, the nutrient cost of growth must also be considered. As shown above, carbon and nitrogen are not
allocated in parallel. Calcium and other nutrients show
still different patterns of utilization (Chapin, Van Cleve,
Tieszen 1975; Chapin, Johnson, and McKendrick in press).
There may well be certain times of year or certain years
when nitrogen or phosphorus limits growth or reproduction
more strongly than does carbon. Under such circumstances
growth patterns that conserve nutrients but are expensive
in terms of carbon, may be selectively advantageous. The
shifting limitation between water, carbon, and various
nutrients may well result in a changing allocation pattern
that is developmentally fixed or changes in response to the
environment, depending upon environmental predictability.
The control over plant nutrient and carbon allocation is a
key area of future research in arctic plant nutrition.

ACKNOWLEDGMENTS

Emanuel Epstein and Brent McCowan critically reviewed
the manuscript. Ideas and new information presented were
developed through work at Circle Hot Springs (NSF grant
BMS74-12776), at Atkasook in the program for Research on
Arctic Tundra Environments (RATE:NSF grant DPP76-80642),
and at Barrow during the International Biological Programme, Tundra Biome (NSF grant GV29342). The latter two
programs received logistic support from the Naval Arctic
Research Laboratory at Barrow, Alaska.

REFERENCES

Armstrong, W. 1972. A re-examination of the functional
    significance of aerenchyma. *Physiol. Plant.*
    27:173-177.

Billings, W.D., G.R. Shaver, and A.W. Trent. 1973. Temperature effects of growth and respiration of roots and rhizomes in tundra graminoids. Pages 57–63 in L.C. Bliss and F.E. Wielgolaski, eds. Primary production processes. Tundra Biome. Univ. of Alberta Print. Services, Edmonton.

Billings, W.D., et al. 1971. Metabolic acclimation to temperature in arctic and alpine ecotypes of *Oxyria digyna*. *Arctic Alp. Res.* *3*:277–289.

Bliss, L.C. 1956. A comparison of plant development in microenvironments of arctic and alpine tundras. *Ecol. Monogr.* *26*:303–337.

______. 1962. Caloric and lipid content in alpine tundra plants. *Ecology*. *43*:753–757.

______. 1966. Plant productivity in alpine microenvironments on Mt. Washington, New Hampshire. *Ecol. Monogr.* *36*:125–155.

Carey, R.W., and J.A. Berry. 1978. Effects of low temperature on respiration and uptake of rubidium ions by excised barley and corn roots. *Plant Physiol.* *61*858–860.

Carter, O.G., and D.J. Lathwell. 1967. Effects of temperature on orthophosphate absorption by excised corn roots. *Plant Physiol.* *42*:1407–1412.

Chapin, F.S., III. 1974. Morphological and physiological mechanisms of temperature compensation in phosphate absorption along a latitudinal gradient. *Ecology*. *55*:1180–1198.

______. 1979. Phosphate uptake and nutrient utilization by Barrow tundra vegetation. Pages 483–507 in L.L. Tieszen, ed. The ecology of primary producer organisms in the Alaskan arctic tundra. Springer-Verlag, New York.

Chapin, F.S., III, and A. Bloom. 1976. Phosphate absorption: Adaptation of tundra graminoids to a low temperature, low phosphorus environment. *Oikos*. *26*:111–121.

Chapin, F.S., III, R.J. Barsdate, and D. Barel. 1979. Phosphorus cycling in Alaskan coastal tundra: A hypothesis for the regulation of nutrient cycling. *Oikos*.

Chapin, F.S., III, D.A. Johnson, and J.D. McKendrick. In press. Seasonal nutrient allocation patterns in various tundra plant life forms in northern Alaska: implications for herbivory. *J. Ecol.*

Chapin, F.S., III, K. Van Cleve, and L.L. Tieszen. 1975. Seasonal nutrient dynamics of tundra vegetation at Barrow, Alaska. *Arctic Alp. Res.* *7*:209–226.

Chapin, F.S., III., et al. In press. Control of tundra
    plant allocation patterns and growth. In J. Brown, et
    al., eds. An arctic ecosystem: The coastal tundra of
    northern Alaska. U.S. Tundra Biome Terrestrial Syn-
    thesis Volume. Dowden, Hutchinson & Ross, Stroudsberg,
    PA.

Chapin, F.S., III., et al. In press. Carbon and nutrient
    budgets and their control in the coastal tundra at
    Barrow. In J. Brown, et al., eds. An arctic ecosystem:
    The coastal tundra of northern Alaska. U.S. Tundra
    Biome Terrestrial Synthesis Volume. Dowdwn, Hutchinson
    & Ross, Stroudsberg, PA.

Davidson, R.L. 1969. Effect of root/leaf temperature
    differentials on root/shoot ratios in some pasture
    grasses and clover. *Ann. Bot. 33*:561-569.

Dennis, J.G. 1977. Distribution patterns of belowground
    standing crop in arctic tundra at Barrow, Alaska.
    Arctic Alp. Res. 9:113-127.

Dowding, P., et al. In press. Nutrients in tundra eco-
    systems. In J.J. Moore and J. Cragg, eds. Tundra:
    Comparative analysis of ecosystems. Cambridge Univ.
    Press, Cambridge, England.

Flint, P.S., and P.L. Gersper. 1974. Nitrogen nutrient
    levels in arctic tundra soils. Pages 375-387 in A.J.
    Holding, et al., eds. Soil organisms and decomposition
    in tundra. IBP Tundra Biome Steering Committee, Stock-
    holm.

Goodman, G.T., and D.F. Perkins. 1968. The role of mineral
    nutrients in *Eriophorum* communities. III. Growth
    response to added inorganic elements in two *E. vag-
    inatum* communities. *J. Ecol. 56*:667-683.

Gore, A.J.P. 1961. Factors limiting plant growth on high-
    level blanket peat. *I. Calcium and phosphate. J.
    Ecol. 49*:399-402.

Haag, R.W. 1974. Nutrient limitations to plant production
    in two tundra communities. *Can. J. Bot.
    52*:103-116.

Hepler, P.K., and B.A. Palevitz. 1974. Microtubules and
    microfilaments. *Ann. Rev. Plant Physiol.
    25*:309-362.

Kedrowski, R.A., and F.S. Chapin, III. 1978. Lipid pro-
    perties of *Carex aquatilis* from hot spring and perma-
    frost-dominated sites in Alaska, implications for
    nutrient requirements. *Physiol. Plant. 44*:231-237.

McCown, B.H. 1979. Observations on the interaction of
    organic nutrients, soil nitrogen, and soil temperature
    on plant growth and survival in the arctic environ-
    ment. In L.L. Tieszen, ed. The ecology of primary
    producer organisms in the Alaskan arctic tundra.
    Springer-Verlag, New York.

McKendrick, J.D., V.J. Ott, and G.A. Mitchell. 1979. Ef-
    fects of nitrogen and phosphorus fertilization on the
    carbohydrate reserve and seasonal nitrogen and phos-
    phorus levels in *Dupontia fisheri* and *Arctagrostis
    latifolia* shoots and rhizomes near Barrow, Alaska. In
    L.L. Tieszen, ed. The ecology of primary producer
    organisms in The Alaskan arctic tundra. Springer-
    Verlag, New York.

McNair, J.B. 1945. Plant fats in relation to environment
    and evolution. *Bot. Rev. 11*:1–59.

McNaughton, S.J. 1972. Enzymic thermal adaptations: the
    evolution of homeostasis in plants. *Amer. Natur.
    106*:165–172.

Miller, O.K., Jr., and G.A. Laursen. 1979. Ecto- and
    endomycorrhizae of arctic plants at Barrow, Alaska. In
    L.L. Tieszen, ed. The ecology of primary producer
    organisms in the Alaskan arctic tundra. Springer-
    Verlag, New York.

Miller, P.C., W.A. Stoner, and L.L. Tieszen. 1976. A model
    of stand photosynthesis for the wet meadow tundra at
    Barrow, Alaska. *Ecology. 57*:411–430.

Mooney, H.A., and W.D. Billings. 1961. Comparative phy-
    siological ecology of arctic and alpine populations of
    Oxyria digyna.   *Ecol. Monogr. 31*:1–29.

Rodin, L.E., and N.I. Bazilevich. 1967. Production and
    mineral cycling in terrestrial vegetation. Translated
    from Russian by Oliver and Boyd, London. (First
    published in Moscow, 1965 by Izdatel'stvo Nauka).

Russell, R.S. 1940. Physiological and ecological studies on
    an arctic vegetation. II. The development of vege-
    tation in relation to nitrogen supply and soil micro-
    organisms on Jan Mayen Island. *J. Ecol. 28*:269–288.

Schultz, A.M. 1964. The nutrient recovery hypothesis for
    arctic microtine cycles. II. Ecosystem variables
    in relation to arctic microtine cycles. Pages 57–68 in
    D.J. Crisp, ed. Grazing in terrestrial and marine
    environments. Blackwell Scientific Publ., Oxford,
    England.

Schultz, A.M. 1969. A study of an ecosystem: the arctic
    tundra. Pages 77–93 in G.Van Dyne ed. The ecosystem
    concept in natural resource management. Academic
    Press, New York.

Shaver, G.R., and W.D. Billings. 1975. Root production and
    root turnover in a wet tundra ecosystem, Barrow,
    Alaska. *Ecology. 56*:401-409.
Shaver, G.R. and W.D. Billings. 1977. Effects of day length
    and temperature on root elongation in tundra gramin-
    oids. *Oecol. 28*:57-65.
Shaver, G.R., F.S. Chapin, III, and W.D. Billings. In press.
    Ecotypic differentiation in *Carex aquatilis*
    as related to ice-wedge polygonization in the Alaskan
    coastal tundra. *J. Ecol.*
Sutton, C.D. 1969. Effect of low soil temperature on
    phosphate nutrition of plants - A review. *J. Sci. Food
    & Agric. 20*:1-3.
Tamm, C.O. 1954. Some observations on the nutrient turnover
    in a bog community dominated by *Eriophorum vaginatum
    L.   Oikos. 5*:189-194.
Tieszen, L.L. 1972. The seasonal course of aboveground pro-
    duction and chlorophyll distribution in a wet arctic
    tundra at Barrow, Alaska. *Arctic Alp. Res. 4*:307-324.
______. 1979. Photosynthesis in the principal Barrow, Alaska
    species: A summary of field and laboratory responses.
    In L.L. Tieszen, ed. The ecology of primary producer
    organisms in the Alaskan arctic tundra. Springer-Ver-
    lag, New York.
Ulrich, A., and P.L. Gersper. 1979. Plant nutrient limita-
    tions of tundra plant growth. In L.L. Tieszen, ed. The
    ecology of primary producer organisms in the Alaskan
    arctic tundra. Springer-Verlag, New York.
Warren Wilson, J. 1957. Arctic plant growth. *Adv. Sci.
    13*:383-388.
Webber, P.J., and D.C. Ebert. 1979. Spatial and temperal
    variation of the vegetation and productivity of the
    US-IBP tundra biome sites. In L.L. Tieszen, ed.
    The ecology of primary producer organisms in the Alas-
    kan arctic tundra. Springer-Verlag, New York.
Wielgolaski, F.E., S. Kjelvik, and Pl. Kallio. 1975.
    Mineral content of tundra and forest tundra plants in
    Fennoscandia. Pages 316-332 in F.E. Wielgolaski, ed.
    Fennoscandia tundra ecosystems. Vol. 16 part I. Plants
    and micro-organisms ecological studies. Springer-
    Verlag, New York.

VIII. MECHANISM OF THERMAL TOLERANCE

A. Ian de la Roche

Research Station
Agriculture Canada
Ottawa, Canada

The plasmalemma is likely the primary site of injury in plant cells that are killed by extracellular freezing. However, it is still not known what constitutes freezing damage to the membrane, nor is it known how the plasmalemma of resistant cells are able to accommodate this stress. The increase in membrane lipid unsaturation observed in winter cereals during cold hardening is clearly not involved in the process of low temperature acclimation. Finally, the relative insensitivity of other organellar membranes to extracellular freezing damage implies that the augmentation of these other organelles in herbaceous plants during acclimation is not a factor in the mechanism of freezing tolerance.

Injury to plants from low temperature can conveniently be divided into two broad areas--chilling or cold injury which occurs to most tropical and many temperate species when exposed to low temperatures in the range of 0°C to 13°C, and freezing or frost injury which occurs when plants are exposed to sub-freezing temperatures. During chilling, a sensitive species will suffer marked dysfunction of normal metabolic processes which over time will lead to structural lesions and eventually to cell death. In contrast, injury sustained by freezing is more direct and results in immediate physical damage to the cell and its constituents. Two types of freezing injury can occur and these are mainly a function of the rate of freezing. With rapid decreases in temperature, ice crystals can form intracellularly and cause complete disruption of the membranes and components of the protoplasm. With slower rates of freezing, ice initially forms extracellularly and be-

cause of the resulting drop in vapor pressure outside the
cell, water is withdrawn from the cell into the extracellular spaces. In the latter case, injury to the cell is
believed to be due to the dehydrative stresses imposed on
the cell and its constituents when the water is withdrawn.

Visual evidence of cell damage by intra- and extracellular ice formation comes from an early report by Siminovitch and Scarth (1938) on *Catalpa* and *Cornus* cortical
cells and epidermal cells of red cabbage and more recently
from freezing studies (Siminovitch, Singh, and de la Roche
1978) on freshly isolated protoplasts from winter rye
(de la Roche et al. 1977). In the later study, free protoplasts were subjected to either fast or slow rates of
freezing on a microscope-adapted thermoelectric stage.
During rapid freezing to $-12^{o}C$, ice formation was observed
inside the protoplasts as indicated by flashing in the
cells which occurred when the refractive index of the
protoplasm changed as a result of water crystallization.
Upon thawing, complete destruction of the plasmalemma and
coagulation of protoplasm was observed. During slow freezing, ice crystallization spread across the field entrapping
and encasing the protoplasts. As the temperature continued
to decrease towards $-12^{o}C$, it was apparent that the protoplasts were undergoing progressive dehydration and contraction due to the movement of water out of the protoplasm. At $-12^{o}C$, the shrunken protoplasts had completely
lost their spherical shape and were barely visible between
the crevices of ice lenses. Upon thawing, the shrunken
protoplasts slowly rehydrated and after one hour most had
resumed their original spherical shape and volume. In
contrast, when protoplasts were slowly frozen to lower
temperatures that were lethal, they exhibited even further
dehydration. Upon thawing, they usually rehydrated to
50-75 percent of their original volume at which time the
plasmalemma ruptured and the contents of the cytoplasm were
expelled into the liquid medium (Singh 1977a).

Since intracellular ice formation is always lethal
under field conditions, there is no primary tolerance to
this stress. In contrast, a considerable range of resistance to the dehydration stress of extracellular freezing
can be found among various species of plants. This can
range from as little as two to three degrees below zero in
the cells of very tender species to greater than $-196^{o}C$ in
a very tolerant species such as the bark cells of black
locust trees (Levitt 1972).

EVIDENCE THAT FREEZING INJURY IS DUE TO MEMBRANE DAMAGE

Scarth and coworkers (Scarth, Levitt, and Siminovitch 1940; Siminovitch and Levitt 1941; Scarth 1941) were the first to demonstrate a direct relationship between the level of freezing resistance in herbaceous and woody plants and changes in the permeability and viscosity of the plasmalemma during acclimation and freezing. Later, Greenham (1966) conducted electrical resistance studies on alfalfa and determined that the integrity of the plasmalemma was lost at the moment of freezing injury. Frey et al. (1977) incorporated a spin label (TEMPO) into wheat leaves and monitored the temperature-dependent partitioning of TEMPO between the hydrophobic lipid environment of the membranes and the outside aqueous phase as the temperature was lowered to the killing point of the tissue. Van't Hoff analysis of the data indicated that an abrupt phase transition occurs in the membranes just prior to the lethal temperature. These results, while not conclusively proving, certainly suggest that the plasmalemma is a primary site of freezing damage and that membrane lesions occur before or simultaneously with cell death.

More direct studies on the chemical and physical properties of plasmalemma during acclimation and freezing have been impaired by difficulties in isolating these membranes from plant systems. As a consequence, several laboratories (Heber and Santarius 1974; Steponkus et al. 1977; Pomeroy 1974) have turned their attention to other organellar membrane systems of the cell which are more easily isolated. Heber and Santarius (1964, 1974) examined the loss of function and integrity of mitochondrial and chloroplast membranes subjected to fast freezing and osmotic dehydrative stresses *in vitro*. On the basis of these results, they proposed that the major cause of injury under *in vivo* conditions was due to loss of membrane semipermeability brought about by certain intracellular solutes that accumulate to toxic levels during dehydration. Steponkus and coworkers (1977) identified three sites of injury in spinach chloroplasts that had been subjected to slow rates of freezing to -29°C. As well as confirming the loss in membrane permeability, they also demonstrated a loss of chloroplast coupling factor (CF$_1$) and plastocyanin. The loss of plastocyanin appeared to be a result of physical leakage associated with a change in thylakoid morphology from a tubular to a vesicular structure. However, one must be cautious, however, in extrapolating the results of *in vitro* studies on isolated organelles to conditions *in vivo*. Since in both these studies, injury may have been due to mechanical disruption of the isolated organelles by the ice crystals that are formed during freezing *in vitro* rather than to freeze-dehydration injury which occurs with extracellular freezing *in vivo*. Further, the freezing experiments *in vivo* were carried out at temperatures (-25°C and -29°C) which were much greater than the killing temperature of the cells *in vivo*. Finally, Heber and Santarius (1964) could find no difference in the response to freezing of mitochondria and chloroplasts isolated from plants widely differing in freezing tolerance.

CHANGES IN FATTY ACID UNSATURATION DURING ACCLIMATION

During the low temperature acclimation of herbaceous species there is a dramatic increase in the unsaturation of cellular lipids (Redshaw and Zalik 1968; de la Roche et al. 1972; Smith 1968; Gerloff, Richardson, and Stahmann 1966; Grenier et al. 1972; Smolenska and Kuiper 1977). In wheat and rye shoots an increase in linolenic acid (18:3) is observed (Redshaw and Zalik 1968; de la Roche et al. 1972), while in alfalfa roots an increase in linoleic acid (18:2) is observed (Grenier et al. 1972). De la Roche et al. (1975) have shown that the increased proportion of unsaturated fatty acids observed during low temperature acclimation of cereals is the result of both altered desaturase activity and shifts in the proportions of major lipid classes possessing unique fatty acid compositions.

Willemot (1977) applied an inhibitor of linolenic acid synthesis to 12 day-old plants of Kharkov wheat and then subjected them to low temperature hardening conditions for two weeks. He found that this compound inhibited not only the accumulation of linolenic acid but also the development of freezing tolerance (Table 1). From these data he concluded that "increased unsaturation of fatty acids is therefore probably an important part of the mechanism of cold adaption in winter wheat."

TABLE I. *Freezing Tolerance of Whole Plants and Fatty Acid Composition of Roots During Hardening of Winter Wheat Treated With BASF 13-338 (From Willemot 1977).*

Treatment	Control		Treated
Days of Hardening	0	14	14
Fatty Acids[1]			
16:0	7.4	16.7	19.0
18:0	0.6	0.4	0.9
18:1	3.8	4.6	3.9
18:2	44.5	36.0	55.0
18:3	33.8	42.3	19.7
LD_{50}[2]	> - 5.0	-17 7	> 6.0

[1]*Palmitic, stearic, oleic, linoleic, and linolenic acids respectively.*
[2]*Temperature at which 50 percent of the test population was killed.*

Chemical and physical studies (Miller, de la Roche, and Pomeroy 1974) of membranes of hardened and unhardened wheat have also suggested a direct link between unsaturation, fluidity of membranes, and cold acclimation. As seen in Table 2, the mitochondrial and microsomal membranes of cold hardened Kharkov wheat possess approximately twice the proportions of linolenic acid as do the membranes of unhardened plants grown to the same stage of development. Spin labelling studies (Miller, de la Roche, and Pomeroy 1974; Miller 1977) indicated that the more unsaturated membranes from hardened plants possessed greater fluidity, particularly at 1.5°C. From the electron spin resonance spectra (Figure 1), Miller (1977) calculated order parameters S of 0.7 and 0.5 respectively for 24°C and 2°C-grown wheat seedlings when measured at 1.5°C.

*TABLE II. Fatty Acid Composition of Lipid Isolated
From Mitochondrial and Microsomal Membranes of Wheat.
Young Epicotyls were Excised from Seedlings of Kharkov
Grown at 24ºC and 2ºC for 48 hours and 4 weeks, Respectively.
(From de la Roche unpublished; Miller, de la Roche, and
Pomeroy 1974).*

	Fatty acid composition (mole %)				
	16:0	18:0	18:1	18:2	18:3
Mitochondria					
48 hours at 24ºC	24.4	1.7	6.2	37.6	30.1
4 weeks at 2ºC	22.8	1.0	4.9	17.2	54.2
Microsomes					
48 hours at 24ºC	27.6	0.5	8.6	34.7	28.5
4 weeks at 2ºC	27.4	0.5	6.5	17.2	48.0

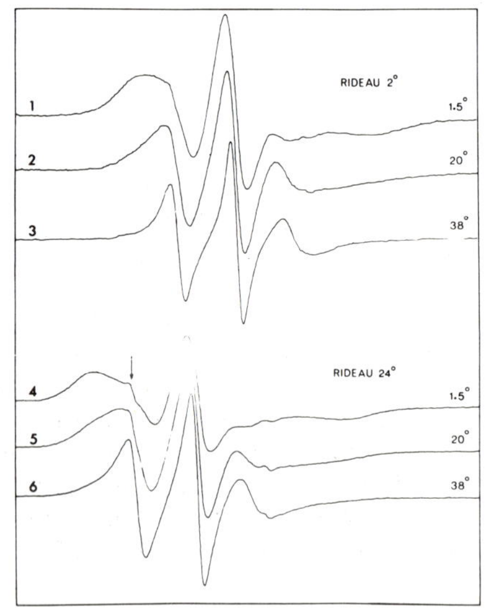

*FIGURE 1. Electron spin resonance spectra of 12
nitroxyl stearic acid label in mitochondrial preparations
isolated from Rideau wheat seedlings grown at 2ºC and 24ºC
respectively. Spectra were taken at 1.5ºC, 20ºC and 38ºC
(From Miller, de la Roche, and Pomeroy 1974).*

TABLE III. Fatty Acid Composition of the Total Membranes from Excised Embryos of Seedlings Germinated and Grown at 2°C and 24°C for Five Weeks and 50 Hours respectively (From de la Roche, Pomeroy, and Andrews 1975).

Cultivar	Growth Temperature (°C)	LD_{50}[1]	Fatty acid composition[2] (mole %)						
			16:0	18:0	18:1	18:2	18:3	20:0	20:4
Kharkov	2	-18	21.4	0.4	4.9	30.9	42.0	0.1	0.4
	24	- 2	24.6	0.5	7.4	47.5	19.8	0.1	0.1
Rideau	2	-13	21.5	0.4	4.7	33.3	39.2	0.2	0.7
	24	- 2	24.6	0.4	6.8	49.8	18.3	T	T
Cappelle-Desprez	2	- 6	22.8	0.3	4.6	32.1	39.8	0.1	0.4
	24	- 1	24.7	0.5	5.5	48.8	20.2	0.1	0.1
Marquis	2	- 5	22.0	0.3	4.5	32.3	40.5	0.1	0.3
	24	- 2	22.8	0.4	7.2	52.0	17.4	0.1	0.1
$S\overline{X}$			0.5	0.1	0.2	0.4	0.8	0.1	0.1

[1]*Temperature at which 50 percent of the test population is killed.*
[2]*Palmitic, stearic, oleic, linoleic, linolenic, arachidic and arachidonic acids, respectively.*

The first direct evidence to show that increased unsaturation of membranes was not directly involved in the mechanism of freezing tolerance came from a study on several cultivars of wheat that were genetically different in freezing tolerance. De la Roche et al. (1975) showed that although low temperatures stimulated the synthesis of linolenic acid in membranes of wheat seedlings, the level of increase was similar in all cultivars and did not bear any relationship to the level of freezing tolerance that

they were capable of developing (Table 3). Thus an increase in membrane unsaturation could not be responsible
for the differences observed in hardiness (LD_{50} of $-5^{\circ}C$
to $-18^{\circ}C$) among four cultivars. These conclusions were
later confirmed by Willemot et al. (1977) using two cultivars of winter wheat which were at a more advanced stage
of development.

Physical studies of membrane lipids from hardened and
unhardened tissues of *Secale cereale* have also been carried
out (Singh, de la Roche, and Siminovitch 1977). Figure 2
compares the differential scanning calorimeter (d.s.c.)
cooling curves of total lipid and isolated phospholipids
from both the total epicotyl tissue and the isolated phospholipids of non-hardy winter rye with that of hardy winter
rye. The phase transition of the lipids from all samples
was broad and extended over a range of $40^{\circ}C$. Although
there was a large difference in unsaturation, the onset and
midpoint temperatures of the d.s.c. exotherms were only
slightly lower when hardy tissue was used (Table 4) and did
not reflect the ability of hardened cells to withstand
freezing temperatures $23^{\circ}C$ lower than above unhardened
cells (Table 4). The above chemical and physical data
demonstrated clearly that differences in membrane unsatur-

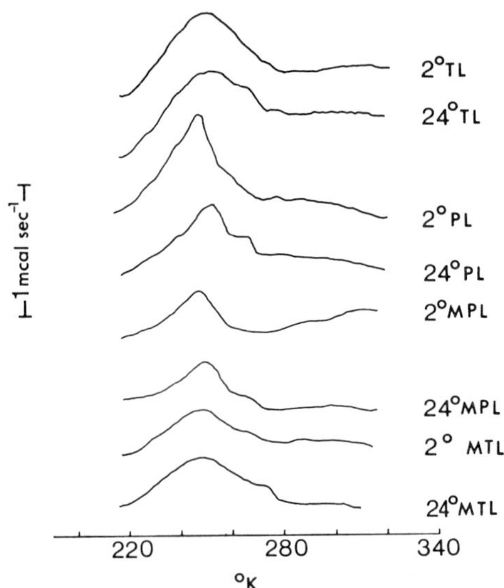

FIGURE 2. *Differential scanning calorimeter cooling
curves for total lipids and phospholipids isolated from
winter rye seedlings. 24°C, winter rye grown for 50 hours
at 24°C; 2°C, winter rye grown for five weeks at 2°C; TL,
total lipids; PL, total phospholipids; M, 105,000 x g
fraction (From Singh, de la Roche, and Siminovitch 1977).*

TABLE IV. *Fatty Acid Composition and Transition Temperatures of Lipids Isolated from Winter Rye Seedlings Grown at 2°C and 24°C for five Weeks and 50 Hours Respectively. The Hardened and Non-Hardened Rye Respectively had Freezing Tolerances (LD$_{50}$ of -28°C and -5°C*

Source of Extract	Growth Temperature (°C)	Fatty acid composition (mole %)					Phase transition temp. (°C)	
		16:0	18:0	18:1	18:2	18:3	Onset	Midpoint
Excised embryos	2	19.0	0.3	5.9	19.6	54.9	+ 4	-22
Total lipid	24	20.1	0.7	7.3	28.4	43.4	+ 3	-20
Phospholipid	2	21.6	0.3	6.8	23.0	48.1	- 4	-25
	24	22.9	0.4	7.7	32.7	36.3	- 3	-21
105,000g Membrane fraction	2	21.6	0.1	7.0	17.4	53.3	+ 5	-21
Total lipid	24	24.4	0.1	7.1	32.3	35.8	+ 5	-25
	2	23.6	0.1	6.8	18.6	50.1	-11	-25
Phospholipid	24	26.1	0.1	6.6	33.1	33.5	- 7	-22

(*From Singh, de la Roche and Siminovitch 1977*)

ation between hardy and unhardy cells were anot involved in
the mechanism of freezing resistance. It should be noted,
however, that none of the above studies precluded the
possibility that increased unsaturation was a prerequisite
to cold hardening and that the differences observed after
cold hardening among cultivars were due to other unidentified factors.

Recently de la Roche (1977a) and Siminovitch (1977)
were able to induce freezing tolerance in wheat and rye
seedlings without low temperature acclimation by germin-
ating seed under reduced water potential at 24°C for 12
days. When Kharkov wheat seedlings were treated in this
manner, then rehydrated for 24 hours and subjected to
artificial freezing tests, the desiccated seedlings were
able to survive exposure to -13°C whereas the unhardened
control (24°C) was killed at -7°C. The fatty acid composition of the phospholipids from epicotyls of unhardened
(24°C), cold hardened (2°C), and desiccation hardened
(24°C) seedlings of Kharkov wheat are presented in Table 5.
The stimulated synthesis of linolenic acid which is always
observed with low temperature hardening of cereals was not
apparent in tissues hardened by desiccation. In fact the
proportion of linolenic acid in the desiccated material was
less than the proportion in the unhardened control grown at
24°C. Similar results were also obtained for two other
cultivars of wheat.

*TABLE V. Fatty Acid Composition of Phospholipids
from Eipcotyls of Unhardened (24°C), Cold Hardened (2°C)
and Desiccation Hardened (24°C) Seedlings (From de la
Roche 1977a).*

	Fatty acid composition (mole %)				
Treatment	*16:0*	*18:0*	*18:1*	*18:2*	*18:3*
24°C	*23.9*	*0.4*	*9.3*	*27.0*	*39.4*
2°C	*22.2*	*0.6*	*7.2*	*17.8*	*52.2*
90% R.H., 24°C	*21.2*	*1.1*	*25.2*	*21.0*	*31.5*

TABLE VI. Fatty Acid Composition of Phosphilipids from Eipcotyls of Unhardened (24ºC) and Cold Hardened (2ºC) Wheat in the Presence and Absence of inhibitors (From de la Roche 1977b).

	Fatty acid composition (mole %)				
Treatment	16:0	18:0	18:1	18:2	18:3
24ºC	23.1	0.7	9.6	31.1	35.5
2ºC	21.4	0.3	9.2	17.2	51.9
SANDOZ[1] 2ºC	18.5	0.6	12.6	35.3	33.0
BASF[2] 2ºC	19.3	0.2	11.1	47.1	22.3

[1] *4-chloro-5-(dimethylamino)-2-(-trifluoro-m-tolyl)-(2H)-pyridazinone.*
[2] *4-chloro-5-(dimethylamino)-2-phenyl-3-(2H)-pyridazinone.*

In another experiment, de la Roche (1977b) investigated the effects of BASF 12-338 and SANDOZ 6706 on cold hardening of wheat and rye seedlings. It is recalled that the first compound was also used by Willemot (1977) in his studies with Kharkov wheat. Seeds were imbibed with the two pyridazinone inhibitors and then germinated at 2ºC for five weeks in the dark. The results from lipid analyses on treated seedlings and untreated controls grown at 2ºC and 24ºC are presented in Table 6. It is apparent that low temperature stimulation of linolenic acid is completely inhibited in young seedlings treated with either BASF 13-338 or SANDOZ 6706. In fact the level of linolenic acid from these treatments is even lower than the 24ºC grown controls. Similar results were obtained for root tissue and for rye. These lipid results agree well with those published by Willemot (1977) for root and leaf tissues of the same cultivar but at a more advanced stage of development. De la Roche (1977b) then measured the freezing tolerance of Kharkov wheat seedlings and isolated tissues. Survival was determined by the amount of regrowth after the freezing treatment. No significant difference in freezing resistance was found among 2ºC-grown seedlings treated with inhibitors or the untreated control. All these seedlings

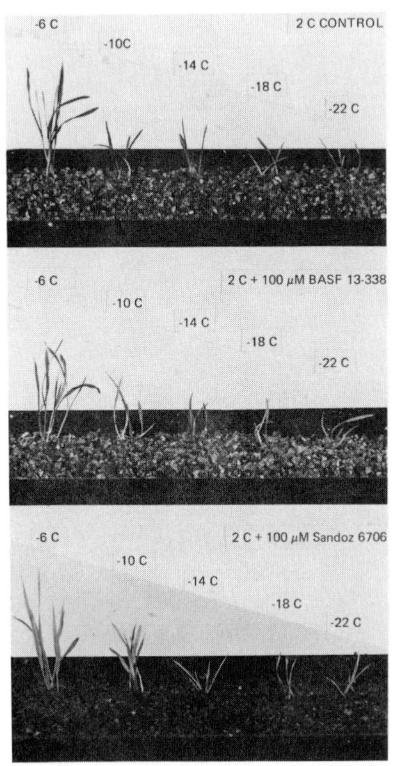

FIGURE 3. Survival of Kharkov seedlings after exposure to a range of freezing temperatures. The seedlings were grown in the dark under the indicated conditions for five weeks before freezing. Photographed after one week regrowth in a controlled environment (24°C light, 18°C dark with a 16 hour photoperiod). All seedlings were killed by freezing to -24°C. Seedlings grown at 24°C were killed by freezing to -4°C.

survived -18°C and exhibited some damage at -22°C (Figure 3), while unhardened seedlings grown at 24°C were all killed at -4°C.

The freezing data are in direct contrast to the freezing results obtained previously by Willemot who found that plants treated with BASF 13-338 were incapable of developing any freezing tolerance during cold exposure (Table 1). The discrepancies in freezing results between ferences in the physiological state of the tissues utilized.

Willemot (1977) observed that BASF 13-338, as well as inhibiting linolenic acid biosynthesis and cold hardening, also inhibited the usual increase in dry weight and phospholipid content associated with low temperature hardening of roots. These observations coupled with the considerable evidence (Hilton et al. 1971; St. John 1976; Vaisberg and Schiff 1976) indicating that the pyridazinone herbicides inhibit normal chloroplast functions suggest that lack of photosynthetic carbon and not the absence of 18:3 accumulation is the casual factor inhibiting hardening in 14 day-old plants. The requirement of photosynthesis for cold hardening of such tissues has been well documented in several studies (Levitt 1972). In the study by de la Roche (1977b) this complication was avoided because very young seedlings (dark-grown) were utilized which derived their fixed carbon from endospermal reserve and not from photosynthesis. The inhibitor studies by de la Roche thus have shown that a greater degree of unsaturation is not a prerequisite for the hardening process at lower temperatures, since Kharkov wheat grown in the presence of the inhibitors was equally able to harden off.

The difference in membrane fluidity for cereals grown at 2°C and 24°C which were mentioned earlier merit further consideration at this time. Although fluidity as measured by electron spin resonance can bear no relationship to the development of freezing tolerance, it has also been suggested that greater membrane fluidity at low temperature above freezing derived from increased unsaturation may be an important adaptive mechanism in cereals for metabolism at low temperature. This apparently is not the case, at least with mitochondrial membranes, since it has recently been demonstrated that mitochondria isolated from hardened and unhardened tissues of wheat and rye exhibit no significant difference in function at lower temperatures (Pomeroy and Andrews 1975).

There have been relatively few studies on the fatty acid composition of lipids in woody plants during acclimation. Van der Schans (1965) observed that there were small but significant increases in the quantity of linolenic acid in the bark and roots of red osier dogwood after cold acclimation. At approximately the same time Ketchie (1966) reported that the ratio of unsaturated to saturated fatty acids in total lipids of bark samples of Halehaven peach trees increased from 0.56 to 0.91 during two weeks in the fall. However, he obtained only a small increase in hardiness during this period. Yoshida and Sakai (1973) have reported that there was no significant change in total fatty acid unsaturation during cold acclimation of cortical

TABLE VII. Fatty Acid Composition and PC:PE Ratios of Summer and Winter Living Bark Tissues of Black Locust Trees (From Singh, de la Roche, and Siminovitch 1975).

| Tissue | Phospholipid | Fatty acid composition (mole %) | | | | | | |
		16:0	18:0	18:1	18:2	18:3	Δ /mol[1]	PC:PE[2]
Summer	Total	21.4	1.7	5.8	62.8	8.3	1.6	1.5
$(LD_{50}^{3}$ -10ºC	PE	23.7	1.5	4.8	64.1	6.9	1.5	
	PC	15.7	2.2	6.7	63.1	12.1	1.7	
Winter	Total	19.8	0.4	2.3	69.7	7.7	1.6	1.6
$(LD_{50}$ -196ºC)	PE	23.2	0.4	1.8	69.1	5.5	1.6	
	PC	12.8	0.5	2.2	74.5	9.9	1.8	

[1]Δ/mol = 1.0 (%monoene)/100 + 2.0 (%diene)/100 + 3.0 (%triene)/100.

[2]*by mol*

[3]*Temperature at which 50% of the cells were killed.*

bark cells of poplar. More recently a study was carried out on the changes in membrane lipids in black locust bark cells during the period when freezing tolerance increased from -10ºC to greater than -196ºC (Singh, de la Roche, and Siminovitch 1975; Siminovitch, Singh, and de la Roche 1977). During this time no significant change occurs in the fatty acid unsaturation of total membrane phospholipid or in either of the two major phospholipids, phosphatidyl-choline and phosphatidylethanolamine (Table 7).

AUGMENTATION OF CELLULAR MEMBRANES DURING ACCLIMATION

Increases in phospholipid content of tissues during acclimation at low temperatures have been demonstrated in several plant species (Redshaw and Zalik 1968; de la Roche et al. 1972; Ketchie 1966; Yoshida and Sakai 1973; Singh, de la Roche and Siminovitch 1975; Siminovitch, Singh, and de la Roche 1977; Grenier and Willemot 1974; Willemot 1975;

TABLE VIII.   Augmentation of Phospholipid and Protein in Insoluble Homogenates of Summer and Winter Black Locust Bark Tissues (From Singh, de la Roche, and Siminovitch 1975).

Tissue	Temperature at LD_{50}	Phospholipid/DNA	Protein/DNA
Summer	- 10ºC	4.7	14.0
Winter	-196ºC	11.1	33.0

LD_{50} *refers to the temperature at which 50 percent of the cells were killed.*

Values presented represent means of three replicates.

Willemot et al. 1975). The studies (Singh, de la Roche, and Siminovitch 1975) with black locust bark tissues are particularly noteworthy because there it was clearly demonstrated that the increases in total phospholipid during acclimation was a direct measurement of the increase in total membrane content of the cells. As seen in Table 8, both the phospholipid and protein contents of the insoluble homogenate containing almost all of the cellular membranes increased by more than 100 percent during the transformation of cells from the unhardy (killed at -10ºC) to hardy (survives -196ºC) condition. Ultrastructural studies have shown that during the period of maximal hardening in locust bark cells large increases are observed in organellar membranes, lipoprotein vesicles, and plasmalemma, the latter being manifested as invaginations at the plasmalemma surface (Pomeroy and Siminovitch 1971). Nevertheless, the significance of these increases with respect to freezing tolerance in woody tissues is not known. Studies with the herbaceous species have been complicated by the fact that the cells undergo considerable division during hardening, and consequently the increases in phospholipid content reported of such tissues may have been a result of new cell division rather than an actual increase within individual cells. Using DNA content as a measure of cell number, de la Roche and Singh (unpublished) were recently able to measure and express phospholipid content in herbaceous tissue on a "per cell" basis (e.g. The validity of using DNA content as a measure of cell number in tissues grown at 2ºC (hardened) and 24ºC (unhardened) was established by preparing protoplasts from these tissues and determining their cell number and DNA content. It was found that the

DNA content per cell was identical in 2°C and 24°C grown tissue). They found that the individual cells of tissues grown and hardened at 2°C did indeed have a higher content of phospholipid than comparable cells from unhardy tissues grown at 24°C (Table 9). However, the increase in phospholipid content was not as great as had been reported earlier for wheat and rye using a "per plant" or dry weight basis. Willemot et al. (1975) recently showed that these increases occurred to the same extent in a hardy and less hardy strain of winter wheat during low temperature acclimation. Redshaw and Zalik (1968) obtained similar results with a spring and a winter cultivar of rye. Thus it would appear that phospholipid augmentation in cereals during acclimation may be merely a response to low temperature growth conditions and not related to the development of freezing tolerance.

Yoshida and Sakai (1973) reported that the augmentation of phospholipids in poplar cortex during acclimation was primarily due to a selective increase in the rate of synthesis of phosphatidylcholine and phosphatidylethanolamine. In contrast, Singh et al. (1975) observed no such preferential synthesis during acclimation of black locust bark cells. Studies on herbaceous plants suggested that the increase in membrane phospholipids during low temperature hardening is only of a quantitative nature (Willemot 1975; Thompson and Zalik 1973; de la Roche, Andrews, and Kates 1973).

TABLE IX. Phospholipid Content of Cells from Tissues of Hardy and Non-hardy Secale cereale *L. cv. Puma (From de la Roche and Singh, unpublished).*

Tissue	Growth Temperature[1] (°C)	Cell Survival[2] (°C)	Phospholipid/ DNA	% Increase in Phospholipid
Epicotyl	2	-22	.22	35
	24	- 4	.16	
Coleoptile	2	-18	.08	61
	24	- 4	.05	

[1]*seedlings were grown in the dark at 2°C and 24°C for five weeks and 64 hours respectively.*

[2]*minimum temperature for 100 percent cell survival.*

TABLE X.   Sterol Composition of Total Membrane from Four Cultivars of Wheat Differing in Freezing Resistance (From de la Roche, unpublished).

Cultivar	Growth Temperature (ºC)	Sterol/ Phospholipids	Sterol composition (mole %)[1]	
			Campesterol	Sitosterol
Kharkov	2	.14[2]	21.0	79.0
	24	.14	32.2	67.8
Rideau	2	.17	23.6	76.4
	24	.14	29.9	70.1
Cappelle-Desprez	2	.16	23.4	76.6
	24	.16	33.5	64.5
Marquis	2	.16	23.8	76.2
	24	.15	32.4	67.6

[1]*Free sterols were isolated from a total membrane extract by thin-layer chromatography and analysed by gas-liquid chromatography (Siminovitch 1977).   Cholestane was used as an internal standard for quantification purposes. Cholesterol was also present in trace quantities.*

[2]*Mean of three replicates.*

The free and esterified sterol content of cortical and xylem tissues of poplar exhibit little seasonal changes, unlike the change in phospholipids (Yoshida and Sakai 1973). Grenier et al. (1975) found no increase in the proportion of ^{14}C-incorporation into free sterols of alfalfa roots during hardening. De la Roche (unpublished) analyzed the free sterols of membranes from four cultivars of wheat grown under acclimating (2ºC) and non-acclimating (24ºC) conditions. He found that growth temperature had no effect on the ratio of sterol to phospholipid content but had a significant influence on the proportion of the two major components, β-sitosterol and campesterol (Table 10). These differences, however, did not appear to be directly related to the mechanism of freezing tolerance, as the same changes were found in the four cultivars which differed significantly in hardiness. These few studies suggested that qualitative changes in the lipid components of the membrane do not play a significant role in the development of cold acclimation.

DIFFERENT TOLERANCES OF CELLULAR MEMBRANES
TO FREEZE-DEHYDRATION

Although the plasmalemma has been identified as a site
of freezing injury in several systems (Siminovitch, Singh,
and de la Roche 1978; de la Roche et al. 1977; Singh 1977a;
Levitt 1972; Scarth, Levitt, and Siminovitch 1940; Siminovitch and Levitt 1941; Scarth 1941; Greenham 1966; Frey et
al. 1977), there is little evidence for implicating other
cellular membrane systems. Closer examination of these
other membrane systems reveals that they probably do not
share the same sensitivity to freezing as does the plasmalemma.

Chambers and Hale (1932) were the first to suggest
that vacuolar membranes (tonoplasts) were more resistant
than the plasmalemma to freezing stress. Recently Singh
(1977b) has observed that when cereal protoplasts are
subjected to lethal treatments of extracellular freezing,
the plasmalemma invariably ruptures during thawing before
complete turgor is regained and frequently the vacuole is
released with its tonoplast intact. De la Roche (unpublished) studying protoplast behavior in hypotonic solutions
has found that the tonoplast is considerably more elastic
than the plasmalemma in hardened wheat and rye.

Singh et al. (1977) have recently shown that mitochondrial membranes *in situ* are immune to freeze-dehydration stresses that are clearly lethal to the cell. They
reported that there were no significant differences in
respiratory functions (membrane intactness) of mitochondria
rapidly isolated from coleoptile cells that had been non-
lethally and lethally frozen. In contrast, isolated
mitochondria subject to the same freezing conditions *in
vitro* were severely damaged. The latter results were
consistent with the earlier *in vitro* studies of Heber and
Santarius (1974) and reaffirm earlier comments as to the
invalidity of extrapolating *in vitro* results to conditions
in vivo.

Indirect evidence suggests that chloroplast membranes
are similarly more resistant to freeze-dehydration than the
plasmalemma. Leaf cells of cereal seedlings that are dark-
grown and therefore not possessing fully developed chloroplasts are considerably less hardy than cells from comparable light-grown seedlings which possess fully developed
chloroplasts (Andrews, Pomeroy, and de la Roche 1974).
Recently Senser and Beck (1977) showed that freezing isolated chloroplasts of spruce needles to temperatures that

were lethal to the intact cell did not markedly alter their
photochemical activities. This observation indicated that
contrary to the conclusion drawn from similar experiments
with spinach (Heber and Santarius 1974; Steponkus et al.
1977; Heber and Santarius 1964), the chloroplast membranes
of spruce are not the most sensitive membranes of the cell.
Steponkus and coworkers (1977) in a review of their earlier
studies on the cold acclimation and freezing of chloroplasts have also concluded that "it is highly unlikely that
the primary site of freezing injury is on the chloroplasts."
While it may be argued that all cellular membrane systems
are undergoing the same acclimating processes as the plasmalemma during low temperature incubation, it is clear that
they do not respond to freeze-dehydration stresses in the
same manner as does the plasmalemma.

CONCLUSIONS

It is apparent from the foregoing discussion that the
mechanism of resistance of plant cells to extracellular
freezing is still far from being solved. The evidence
showing that the plasmalemma are primary targets of injury,
although considerable, is still not unequivocal. For
instance, it is still not known what constitutes freezing
damage to these membranes or whether there are chemical and
structural differences between plasmalemma of freezing
resistant and susceptible cells. Answers to these questions await the availability of a reliable method for
isolating plasmalemma from cells of higher plants.
The complexity of the plasmalemma suggests that a
consideration of only one component, phospholipid unsaturation, may be too simple an approach. Future studies
will also have to consider the structural proteins of the
membrane and their interactions with the lipid moiety.
The numerous reports concerned with changes in other
cellular constituents during low temperature acclimation
have been extensively reviewed in previous articles (Levitt
1972; Olien 1967; Alden and Hermann 1971). Nevertheless,
it should be pointed out that in all the reports of chemical differences between hardened and unhardened tissues,
it was not demonstrated whether the differences were directly
involved in the mechanism of freezing tolerance or were
merely general metabolic responses to low temperature
growth conditions. In this respect, the recent studies on
lipid unsaturation considered in this review take on added
significance.

ACKNOWLEDGMENT

The author thanks Dr. Michael Burke for his encouragement and assistance during the symposium and Drs. Jasbir Singh, Alan Wilson, and Nancy Long for discussion and helpful criticism during the preparation of the manuscript.

REFERENCES

Alden, J., and R.K. Hermann. 1971. Aspects of the cold hardiness mechanism in plants. *Bot. Rev. 37:37*.

Andrews, C.J., M.K. Pomeroy, and A.I. de la Roche. 1974. The influence of light and diurnal freezing temperatures on the cold hardiness of winter wheat seedlings. *Can. J. Bot. 52:*2539.

Chambers, R., and H.P. Hale. 1932. The formation of ice in protoplasm. *Proc. Royal Soc. Lond. B. 110:*337.

de la Roche, A.I. 1977a. Development of freezing tolerance in wheat by water stress: changes in lipid unsaturation. *Plant Physiol. Suppl. 59:*36.

______. 1977b. Development of freezing tolerance in wheat without changes in lipid unsaturation. In Winter hardiness in woody perennial, First International Symposium, Posnan, Poland, September 1977.

de la Roche, A.I., C.J. Andrews, and M. Kates. 1973. Changes in phospholipid composition of a winter wheat cultivar during germination at 2°C and 24°C. *Plant Physiol. 51:*468.

de la Roche, A.I., M.K. Pomeroy, and C.J. Andrews, 1975. Changes in fatty acid composition in wheat cultivars of contrasting hardiness. *Cryobiology. 12:*506.

de la Roche, A.I., et al. 1972. Lipid changes in winter wheat seedlings (*Triticum aestivum*) at temperatures inducing cold hardiness. *Can. J. Bot. 50:*2401.

______. 1977. Isolation of protoplasts from unhardened and hardened tissues of winter rye and wheat. *Can. J. Bot. 55:*1181.

Frey, L., et al. 1977. Spin label studies of plant cell membranes transformed using TEMPO. *Plant Physiol. Suppl. 59:*35.

Gerloff, E.D., T. Richardson and M.A. Stahmann. 1966. Changes in fatty acids of alfalfa roots during cold hardening. *Plant Physiol. 41:*1280.

Greenham, G.G. 1966. The stages at which injury occurs in alfalfa. *Can. J. Bot. 44:*1971.

Grenier, G., and C. Willemot. 1974. Changes in roots of
    frost hardy and less hardy alfalfa varieties under
    hardening conditions. *Cryobiology*. *11*:324.

Grenier, G., et al. 1972. Changements dans les lipides de
    la lucerne en conditions menant a l'endurcissement au
    froid. *Can. J. Bot. 50*:1681.

Grenier, G., H.J. Hope, C. Willemot, and H.P. Therrien. 1975.
    Sodium $-1,2-^{14}C$ acetate incorporation in roots of
    frost hardy and less hardy alfalfa varieties under
    hardening conditions. *Plant Physiol. 55*:906.

Heber, U., and K.A. Santarius. 1964. Loss of adenosine tri-
    phosphate synthesis caused by freezing and its re-
    lationship to frost hardiness problems. *Plant Physiol.
    39*:712.

______. 1974. Water stress during freezing. In O.L. Lange,
    L. Kappen, and D. Schulze, eds. Ecological studies,
    water and plant life - problems and modern approaches.
    Vol. 19. Springer-Verlag.

Hilton, J.L., et al. 1971. Interactions of lipoidal mater-
    ials and a pyridazinone inhibitor of chloroplasts.
    Plant Physiol. 48:171.

Ketchie, D.O. 1966. Fatty acids in the bark of Halehaven
    peach as associated with hardiness. *Proc. Am. Soc.
    Hort. Sci. 88*:201.

Levitt, J. 1972. Response of plants to environmental stresses
    Academic Press, New York. 93-95 pp.

Miller, R.W. 1977. Personal communication.

Miller, R.W., A.I. de la Roche, and M.K. Pomeroy. 1974.
    Structural and functional responses of wheat mitochon-
    drial membranes to growth at low temperatures. *Plant
    Physiol. 53*:426.

Olien, C.R. 1967. Freezing stresses and survival. *Ann. Rev.
    Plant Physiol. 18*:387.

Pomeroy, M.K. 1974. Studies on the respiratory properties
    of mitochondria isolated from developing winter wheat
    seedlings. *Plant Physiol. 53*:653.

Pomeroy, M.K., and C.J. Andrews. 1975. Effect of temper-
    ature on respiration of mitochondria and shoot segments
    from cold-hardened and non-hardened wheat and rye
    seedlings. *Plant Physiol. 56*:703.

Pomeroy, M.K., and D. Siminovitch. 1971. Seasonal cyto-
    logical changes in secondary phloem parenchyma cells
    in *Robinia pseudoacacia* in relation to cold hardiness.
    Can. J. Bot. 49:787.

Redshaw, E.S., and S. Zalik. 1968. Changes in lipids in
    cereal seedlings during vernalization. *Can. J. Biochem.
    46*:1093.

St. John, J.B. 1976. Manipulation of galactolipid fatty
     acid composition with substituted pyridazinones. *Plant
     Physiol.* *57*:38.
Scarth, G.W. 1941. Dehydration injury and resistance. *Plant
     Physiol.* *16*:171.
Scarth, G.W., J. Levitt, and D. Siminovitch. 1940. Plasma-
     membrane structure in the light of frost hardening
     changes. In Cold Spring Harbour Symposia on Quanti-
     tative Biology. 8:102.
Senser, M., and E. Beck. 1977. On the mechanism of frost
     injury and frost hardening of spruce chloroplasts.
     Planta. *137:195.*
Siminovitch, D. 1977. Induction of cold hardiness and plas-
     molysis injury resistance in coleoptile and epicotyls
     of winter rye seedlings in the dark at room temper-
     ature by desiccation. *Plant Physiol. Suppl. 59*:4.
Siminovitch, D., and J. Levitt. 1941. The relationship
     between frost resistance and the physical state of
     protoplasm. II. The protoplasmic surface. *Can. J. Res.
     C. 19*:9.
Siminovitch, D., and G.W. Scarth. 1938. A study of the
     mechanism of frost ism of frost injury to plants. *Can.
     J. Res. C. 16*:467.
Siminovitch, D., J. Singh, and A.I. de la Roche. 1977.
     Studies on membranes in plant cells resistant to
     extreme freezing. 1. Augmentation of phospholipids and
     membrane substance without changes in unsaturation of
     fatty acids during hardening of black locust bark.
     Cryobiology. 12:144.
______. 1978. Freezing behavior of free protoplasts of
     winter rye. *Cryobiology. 15*:205.
Singh, J. 1977a. Freezing tolerance of rye protoplasts.
     Plant Physiol. Suppl. 59:4.
______. 1977b. Personal communication.
Singh, J., A.I. de la Roche, and D. Siminovitch. 1975.
     Membrane augmentation in freezing tolerance of plant
     cells. *Nature. 257*:669.
______. 1977a. Differential scanning calorimeter analysis of
     lipids isolated from hardy and unhardy black locust
     bark and from winter rye seedling. *Cryobiology.
     14*:620.
______. 1977b. Relative insensitivity of mitochondria in
     hardened and non-hardened rye coleoptile cells to
     freezing *in situ.*   *Plant Physiol. 60*:713.
Smith, D. 1968. Varietal chemical differences associated
     with freezing resistance in forage plants. *Cryobiology.
     5*:148.

Smolenska, G., and P.J.C. Kuiper. 1977. Effect of low temperature upon lipid and fatty acid composition of roots and leaves of winter rape plants. *Physiol. Plant.* 41:29.

Steponkus, P.L., et al. 1977. Effects of cold acclimation and freezing on structure and function of chloroplast thylakoids. *Cryobiology.* 14:303.

Thompson, L.W., and S. Zalik. 1973. Lipids in rye seedlings in relation to vernalization. *Plant Physiol.* 52:268.

Vaisberg, A.J., and J.A. Schiff. 1976. Events surrounding the early development of Euglena chloroplasts. 7. Inhibition of carotenoid biosynthesis by the herbicide SAN 9789 (4-chloro-5-(methylamino)-2-(α,α,α, -trifluro-$\underline{m}$-tolyl)-3-(2H) pyridazinone) and its developmental consequences. *Plant Physiol.* 57:260.

Van der Schans, C. 1965. Cold acclimation of bark and roots of *Cornus stolonifera*. Ph.D. Thesis, Univ. of Minnesota.

Willemot, C. 1975. Stimulation of phospholipid biosynthesis during frost hardening of winter wheat. *Plant Physiol.* 55:356.

__________. 1977. Simultaneous inhibition of linolenic acid synthesis in winter wheat roots and frost hardening by BASF 13-338; a derivative of pyridazinone. *Plant Physiol.* 60:1.

Willemot, C., et al. 1975. Effet de l'endurcissement a la gelee sur la teneur en matiere seche, en phosphore lipidique et en proteines solubles et insolubles des ble d'hiver. *Can. J. Plant Sci.* 57:555.

Willemot, C., et al. 1977. Changes in fatty acid composition of winter wheat during frost hardening. *Cryobiology.* 14:87.

Yoshida, S., and A. Sakai. 1973. Phospholipid changes associated with the cold hardiness of cortical cells from poplar stem. *Plant Physiol.* 14:353.

DISCUSSION: WATER IN PLANTS:
THE PHENOMENON OF FROST SURVIVAL[1]

M.J. Burke

Department of Horticulture
Colorado State University
Fort Collins, Colorado

Many plants freeze and survive. This is unusual in living systems, as most organisms are killed if frozen, including most higher plants if frozen in summer. However, in winter many plants "cold acclimate" and survive freezing and exposure to +4°K. Often when freezing injury occurs, it is associated with a general lysis of the tissue cells; the lysing takes place after brief exposure below the killing temperature. In some cases the general cell lysis occurs at temperatures well below the freezing of most of the tissue water and long after the tissue water has frozen. In other cases a freezing event terminating deep undercooling is associated with cell lysis. In this review the freezing of plant water and the associated membrane changes leading to cell lysis are discussed.

INTRODUCTION

This review will focus on a very small part of the general frost hardiness phenomenon in plants. There are a number of recent reviews that provide more detailed and comprehensive treatments of the subject (Levitt 1972; Levitt 1956; Burke et al. 1976, 1975; George and Burke 1976; Olien 1967; Mayland and Cary 1970; Siminovitch et al. 1968; Parker 1963; Weiser 1970).

[1]*These studies were supported in part by grants from the National Science Foundation (BMS 74-23137), the Petroleum Research Fund (PRF 9702-AC1, 6), the Research Corporation, the Horticultural Research Institute and the Colorado Experiment Station.*

Here the discussion is limited to several plant species that may serve as model systems for future physiological study of frost hardiness in arctic plants.

Plants overwintering in arctic, alpine, and temperate regions generally contain a considerable fraction of freeze-able water. These plants must survive subfreezing temperatures in winter; therefore, most of these plants must avoid (by undercooling) or tolerate freezing of their tissue water. As a general statement one could say that most living tissues are killed if subjected to freezing temperatures and most plants are killed if frozen during periods of active growth in summer; however, in winter the same plant cold acclimates and may survive very low temperatures.

Levitt (1972) points out that plants survive freezing temperatures by three general means: The first means is dehydration of the freezable tissue water before winter comes; such tissues are cold in winter but they don't freeze. Lichens are a good example. The second means is undercooling (supercooling) of the plant tissues to below the "normal" freezing point, and to as low as the homogeneous ice nucleation temperature (about $-41^{\circ}C$ for plant solutions). Again such undercooling plant tissues do not freeze and the tissues survive by avoiding tissue ice. Good examples are the living xylem tissues in many hardwood trees (oak, elm, etc.) (George and Burke 1976). The third means is extracellular freezing of the cell water. This means of frost survival is found in many intermediately hardy and very hardy plants. Such plants simply partition the ice away from the interior of living cells. The ice growth is restricted to the extracellular air spaces and as the ice grows, the cells dehydrate and tolerate the "physiological drought".

The plants which have the above three means of frost survival occur with various levels of cold hardiness ranging from $-1^{\circ}C$ to lower temperatures. Ideally, the frost killing phenomenon is a well defined, easily described event. In such ideal cases, a critical low temperature (the killing temperature) is involved. If this killing temperature threshold is exceeded, i.e., if the temperature drops below the killing point, then almost complete lysis of tissue cells is observed. For example, hardy Kharkov winter wheat leaves show no injury at $-15^{\circ}C$ but show 90 percent cell lysis after brief exposure to $-18^{\circ}C$. In hardy turf leaves this occurs between $-35^{\circ}C$ (no lysis) to $-41^{\circ}C$ (90 percent cell lysis). Here the discussion is limited to such ideal cases. In the ideal situation, freezing of

water leads to disruption of the cell membrane; therefore,
discussion of new and unique means of studying water and
membranes will be the starting point.

NUCLEAR MAGNETIC RESONANCE (NMR) AND ELECTRON SPIN RESONANCE
(ESR) AS MEANS FOR THE STUDY OF WATER AND MEMBRANES IN
WHOLE LIVING PLANTS

There are very few methods which can be used to characterize water and membranes in whole living biological
tissues, particularly when they are partially frozen.
Pulse proton NMR and ESR spin probe methods provide some of
the best tools for this purpose. Based on NMR relaxation
times, it is possible to measure the amount of water which
is not ice and also to characterize the time averaged
environment of the liquid water in partially frozen plant
tissues. ESR spin labeling techniques can be used to study
"phase separations" in plant cell membranes (e.g., see
Raison et al. 1971 and Fey et al. 1977). In many crop
plants these phase separations occur between $+5^o$ and $+20^oC$
and result in a type of injury called chilling injury.
There is now evidence that such membrane phase separations
occur at subfreezing temperatures in undercooled plants
(Fey, et al. 1977).
The NMR relaxation times make it possible to measure
the amount of water which is not ice because of the very
large change in these properties which accompanies freezing. Details of the NMR methods for distinguishing between
ice, liquid, and organic proton containing substances are
discussed by Burke et al. (1975) and by Burke and Bryant
(in preparation). For example, the spin-spin relaxation
time, T_2, for biological solids is 10^{-5} seconds or less,
for ice in hardy plants it falls into the 10^{-6} second range
while for liquid aqueous phases in tissues, it ranges
between 1 second and 10^{-4} seconds. Since the relaxation
times for the liquid phases are distinctly different, these
properties can be used to determine the fraction of liquid
water at subfreezing temperature. The relaxation time
properties can also be used to characterize the unfrozen
water in partially frozen samples (Burke, George, and
Bryant 1975). An interesting result of relaxation time
measurements on plant tissues is that T_2 (spin-spin relaxation time) is shortened and T_1 (spin-lattice relaxation
time) is changed after frost injury has occurred. Examples
are in Table 1.

TABLE I.   NMR Relaxation Times for Living and Dead Plant Tissues

Sample	T_1 (slow) (msec)		T_2 (slow) (msec)	
	alive	dead	alive	dead
Red osier dogwood twig	400	550	39	23
Azalea floral primordium[14]	47	26	25	6
Shagbark hickory twig[15]	60	33	43	7

[14]*George and Burke 1977a.*

[15]*George and Burke 1977b.*

Two types of ESR spin labels have been used in the
study of membrane structural transitions. One type including 12-nitroxide steric acid provides information
regarding membrane fluidity. The second type spin label
includes TEMPO (2, 2, 6, 6 – tetramethylpiperidinooxyl).
This spin-label has been used extensively by McConnel and
coworkers (e.g., see Shimshick and McConnell 1973). TEMPO
partitions between the aqueous phases and membrane (lipid)
phases of the sample. The temperature dependence of the
partitioning of this label is used to obtain information
about the membrane. An example of results of ESR-TEMPO
labeling of whole tomato leaf is in Figure 1. Tomato leaf
has a well characterized membrane phase separation between
12º and 15ºC. Note the break in the Van't Hoff plot at
13ºC. The Van't Hoff analysis is used because the ESR
results provide information on the temperature dependence
for the equilibrium constant of TEMPO partitioning between
the lipid and aqueous fractions of the leaf. TEMPO partitioning is determined according to the methods of Shimshick and McConnell (1973). Briefly, the ESR spectrum of
TEMPO has two high field lines, one from TEMPO in lipid and
the other from TEMPO in water. These two lines can be
deconvoluted and the fraction of the TEMPO in the lipid, ℓ,
and the water, ω, determined. The equilibrium constant is
the ratio of the concentrations of TEMPO in lipid and water
according to equation 1. Here $\underline{W}$ and $\underline{L}$ are the number of
moles of liquid water and lipid in the leaf, respectively.
Therefore, if $\underline{W}$ and $\underline{L}$ are held constant by preventing
change in $\underline{W}$ by freezing, etc., then the equilibrium constant can be easily obtained from the ESR spectrum and the

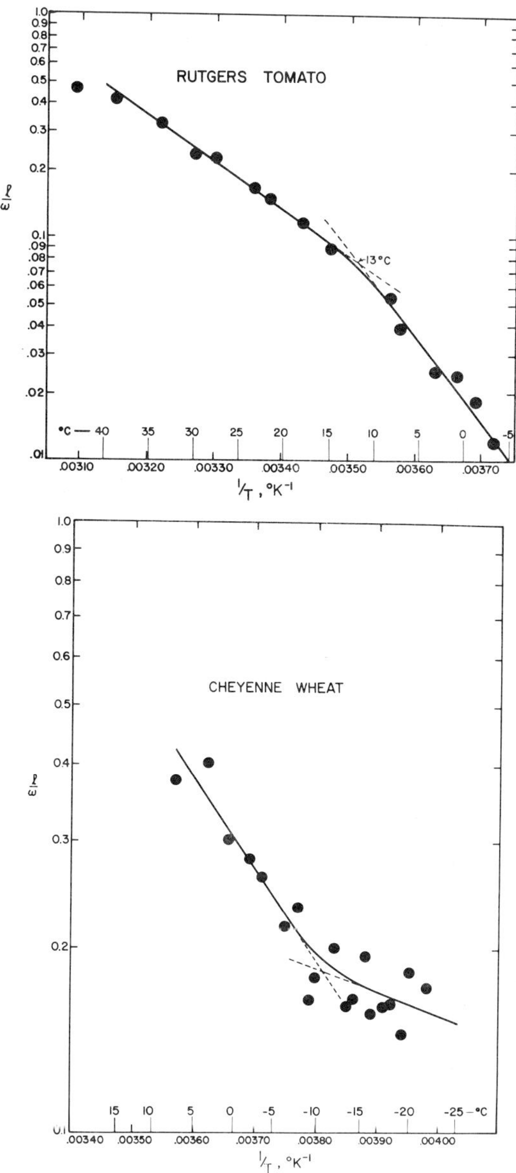

FIGURE 1. Van't Hoff plots of TEMPO partitioning between lipid and water fractions in mascerated tomato leaf and whole wheat leaf. ℓ is the high field ESR line intensity of the TEMPO label in lipid and w is the intensity for the TEMPO in water. The ratio is proportional to the equilibrium constant. Breaks in the plots indicate structural transitions in the leaf, presumably in the leaf membranes. Wheat leaves used in these studies were undercooled to -27°C (Fey 1977).

Van't Hoff analysis made. Results for undercooled wheat
leaf are in Figure 1b. A possible break is at $-10^{\circ}C$ (Fey
1977).

$$K = \frac{\ell}{\omega} \quad \times \quad \frac{W}{L}$$

(Equation 1)

PHYSICAL CONSIDERATIONS IN THE FREEZING OF PLANTS:
EXTRACELLULAR FREEZING

As discussed above, plant tissues and cells survive by
avoiding ice growth within the cell by mechanisms including
undercooling and extracellular freezing (Levitt 1972). Here
the discussion revolves around plants such as wheat and
turfgrass which survive freezing temperatures by toleration
of extracellular freezing. Winter hardy tissues that survive by extracellular freezing generally have ice initiated
outside their living cells near -1° or $-2^{\circ}C$. A single ice
nucleation somewhere within or outside the generally wet
plant can start the whole process rolling and the initial
ice growth in most hardy plants is restricted to water
already in the extracellular spaces. Unlike animal tissues,
extracellular water in plants is quite low in solutes and it
generally has a small melting point depression ($\sim 0.01^{\circ}C$)
compared to the cell solution ($\sim 1.0^{\circ}C$). During slow freezing, free energy restrictions make it necessary that water
from cell solutions move to the extracellular spaces and
freeze. This leaves the cells in a progressively dehydrated
condition. The cell is then essentially under drought
stress and the cold hardy plants survive the "drought".
Thermodynamics of freezing of dilute aqueous solutions
in a semipermeable plant cell is discussed elsewhere (Mazure
1963; Dusta et al. 1975). To a first approximation, the free
energy considerations of importance for equilibrium freezing
are that the melting point depression of the cell solutions
(colligative property) must equal the ambient temperature
of the tissue, i.e., the temperature of the tissue ice. A
relation (equation 2) for liquid water as a function of temperature was derived by Gusta et al. (1975). "f" is the
weight fraction of water which remains liquid at temperature,

T(^{o}K). "f" is determined useing the NMR methods described above. "b" is the portion of absorbed water which does not freeze or have solvent properties. ΔTm (thaw) is the melting point depression of the thawed cell solution at zero turgor pressure. Equation 2 predicts a straight line with slope – ΔTm (thaw), and intercept "b" should be obtained when plotting "f" vs. $(T - 273)^{-1}$. The equation predicts the freezing to be a continuous gradual process with most of the freezing occurring above –10°C (+263°K). Typical results are in Figure 2. Note the rather continuous behavior of the freezing curves at the killing temperatures (–18°, –9°, and –3°C). Similar results are obtained for crown and leaf tissues of turf and wheat plants.

$$f = (\Delta Tm \text{ (thaw)} / (T - 273) + b$$

(Equation 2)

Although the liquid water content is a gradual function of temperature near the killing point, the fractional cell lysis (fractional electrolyte leakage) abruptly changes at the killing temperature. Results for several plant species are in Figure 3. Note particularly the abrupt changes in cell lysis for Kharkov wheat at –18°C and Merion Blue turfgrass at –38°C. At these killing temperatures, we have not been able to demonstrate, using NMR or differential thermal analysis procedures, that anomalous behavior in quantity of liquid water as a function of temperature occurs. In fact, at –18° or –38°C, respectively, there is no significant change in liquid water over the killing temperature region. Additional studies using NMR and ESR membrane probes will be necessary to research the killing temperature region more thoroughly. As is noted in Figure 1, membrane structural transitions at subfreezing temperatures may occur.

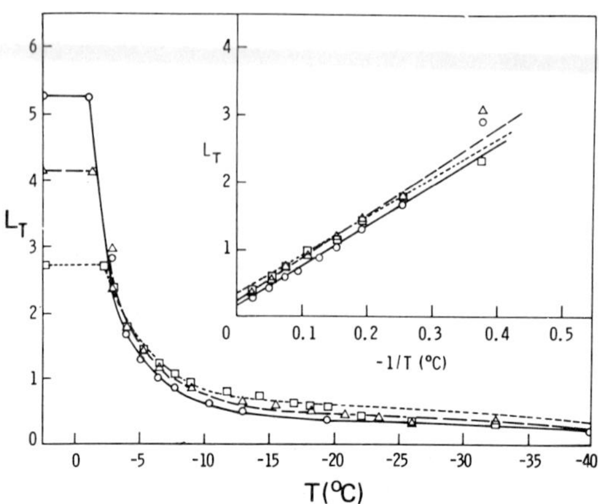

FIGURE 2. *Freezing curves for hardy (□), intermediately hardy (△), and tender (o) Kharkov winter wheat. The
hardy, intermediately hardy and tender Kharkov wheat was
killed at -18° ±1°, -9° ±1°, and -3° ±1°C, respectively.
The liquid water content, L_t, is expressed in grams of liquid water per gram dry sample. L_t is plotted versus
temperature and the reciprocal of temperature. The lines
drawn are the best fitting hyperbolae in the L_t versus T
plots and the best fitting strait line in the L_t versus 1/T
plots for the results between -2.5° and -40°C (Gusta et
al. 1975).*

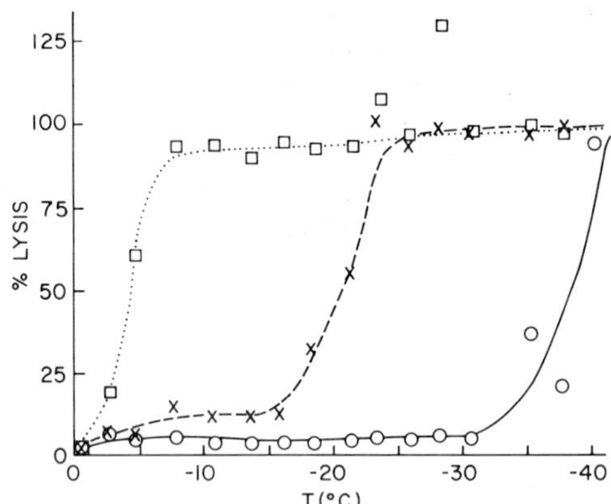

FIGURE 3. *Temperature dependence of percentage cell
lysis as determined by fractional electrolyte loss measurements. Data are for leaves of Inia 66 spring wheat (□),
Kharkov winter wheat (X), and Merion Blue turfgrass (o).
Plants were 7 weeks old and were grown for the last 4 weeks
at 1°C with a 12 hour light period. Two cm long leaf cuttings were nucleated with ice at -1.5°C, held 12 hours then
cooled at a rate of 1°C/hour. Note the strong temperature
dependence of electrolyte leakage at -4° (Inia 66) -18° (Khar
kov) and -38°C (Merion Blue).*

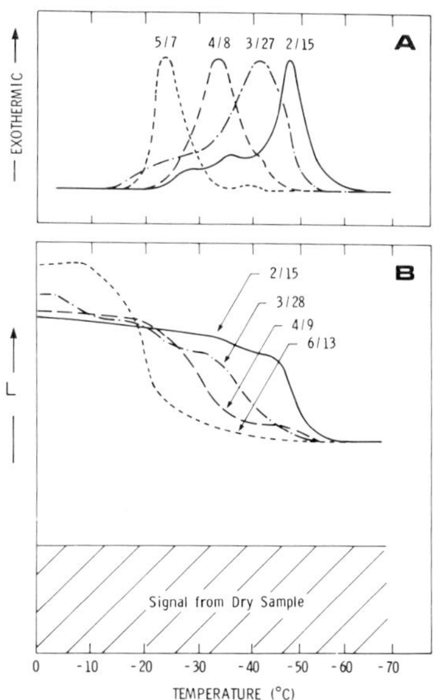

FIGURE 4. Changes in the differential thermal analysis (A), and freezing curves obtained by nuclear magnetic resonance (B), of shagbark hickory twigs during deacclimation from winter. In the DTA's above the peaks or "exotherms" indicated freezing points moving from -45°C in midwinter to -2°C in early spring. Samples were cooled at about 1°C/min. In the lower freezing curves liquid water, L is plotted on the left axis. This plant is clearly in control of it's freezing (Burke et al. 1975).

PHYSICAL CONSIDERATIONS IN THE FREEZING OF PLANTS: DEEP UNDERCOOLING

In addition to the plants that survive frost by restricting ice to the extracellular spaces, trees and shrubs like hickory, oak, elm, etc., have tissues which deep under cool and survive by avoiding freezing. This part of the discussion deals only with those tree and shrub tissues which undercool to survive subfreezing temperatures.

Many of the plants which undercool reach the homogenous ice nucleation temperature limit, about -41°C, before freezing begins and as can be seen in Figures 4 and 5, they control undercooling during cold acclimation and deacclimation (Burke, George, and Bryant 1975; George and Burke 1976, 1977; George, Burke and Weiser 1974). A plant structural feature that always accompanies deep undercooling is that only small volumes are frozen per ice nucleation. The tissues are either partitioned into many small isolated groups of cells which freeze independently or the whole

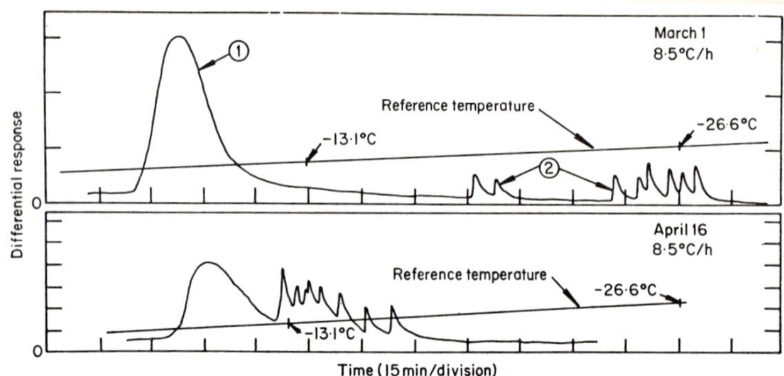

FIGURE 5. Differential thermal analysis recordings of whole azalea flower buds collected in late winter (March 1) and early spring (April 16) during deacclimation from cold temperatures. The exotherms of type 1 correspond to freezing of water in the bud scales which are dead and the stem axis upon which the primordia are attached. The stem axis survives freezing. Exotherms of type 2 indicate freezing of individual primordia. The primordia are killed when they freeze. Note the change in the primordia exotherm positions during deacclimation from cold. In mid-winter primordia exotherms are near -40°C (Burke et al. 1975).

tissue itself is very small. Considerable work has been done on deep undercooling in the tree shagbark hickory and a shrub rhododendron (George and Burke 1977; George, Burke and Weiser 1974).

Undercooling is important in trees and shrubs chiefly because when the undercooled fraction freezes, a large portion of the plant stem cells lyze. Figure 6 gives percent stem cell lysis as a function of temperature. Note that for this pear tree twig, the abrupt increase in cell lysis at -34°C occurs at the low temperature freezing point for the twig.

Considerable information is available on the physical aspects of undercooling in liquids. Undercooled pure water can only exist between the freezing point of water, 0°C and the homogenous nucleation point of water, -37.1°C (Rasmussen and MacKenzie 1972). Undercooling of dilute aqueous solutions, such as cellular cytoplasm, occurs over a slightly lower range. The freezing point depression and homogenous nucleation point, show an apparent relation with the homogenous nucleation point, acting like a colligative property of the solution. The homogenous nucleation point depression (below -37.1°C) is about twice as large as the freezing point depression (below 0°C) (Rasmussen and MacKenzie 1972). For higher plant cells with freezing point depressions between -2° and -4°C, this means the homogeneous nucleation point is between -39° and -45°C. This is precisely the

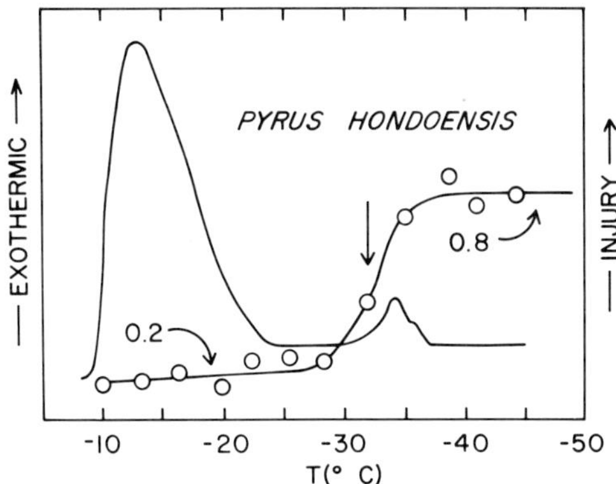

FIGURE 6. Differential thermal analysis (solid curve) and fractional electrolyte loss or cell lysis (0 connecting solid curve) for a twig of the perar species Pyrus hon- doensis. *The peaks or exotherms in the differential thermal analysis tracing indicated freezing points. Note the almost all or nothing change in electrolyte loss that coincides with the low temperature exotherm. The estimated killing temperature based on browning of stem tissue is indicated by the arrow. This sample was collected in February 1977 at Oregon State University by Dr. M. Westwood.*

temperature range observed for freezing and killing of the living wood cells, xylem ray parenchyma, of most hardwood tree plants in winter (George et al. 1974).

There are two simple means of undercooling water to $-37^{\circ}C$. These are using small water droplets in a cloud chamber or homogenizing water into droplets in hydrocarbon solvents (Rasmussen and MacKenzie 1972; Rasmussen, MacAulay, and MacKenzie 1974). These solvents include heptane, sili- cone oil, and safflower oil. Some of the droplets freeze due to ice nucleation from heterogeneous impurities; this is an almost undectable fraction of the droplets, if the drop- lets are sufficiently numerous relative to the heterogeneous impurities. The essence of the droplet technique is the partitioning of a sample of liquid into a large number of small volumes, thereby isolating heterogeneous ice nuclea- tors in the original sample to only a small fraction of the volumes; the great majority of the remaining liquid is thus capable of homogeneous nucleation. Ice growth is restricted to the nucleated droplets. This model system of Rasmussen and MacKenzie (1972) is surprising in its similarity to undercooling in plants, as will be seen in the next section.

Theoretical and experimental studies of ice nucleation from the undercooled state have been made and there is a considerable literature on the subject (Hobbs 1974; Dufour and DeFay 1963; Fletcher 1970). Turnball and Fisher (1949) derived an approximate relation giving the homogeneous nucleation rate, J, as a function of temperature, T, and undercooling, ΔT. This was modified by Hoffman (1958) and Buckle (1961) to give:

$$J = \Omega \, \exp(- K/T^3 \Delta T^2)$$

(Equation 3)

$K/T^3 \, \Delta T^2$ is the total free energy change in forming ice-like clusters of critical size for ice growth and is a function of the ice-water interfacial tension cubed. This "activation energy" is obviously temperature dependent so typical Arrhenius behavior is not expected. Ω is usually considered constant over the narrow temperature range (-30° to -40°C) where nucleation rates are rapid enough to make experimental measurements possible. Wood and Walton (1970) measured Ω (10^{48} nucleations/ml sec) and K(+1.63 x 10^{12} °K^5) (Figure 7). Similar relations for heterogeneous ice nucleation have been derived (Fletcher 1970). For a heterogeneous impurity, Fletcher's treatment suggests that forming ice-like clusters of critical size for ice growth depends on the surface tension and size of the catalytic heterogeneous nucleator. In general large heterogeneous nucleators with low interfacial surface tension (with ice) are more effective nucleators. Of course, ice itself is a good nucleator. K is generally smaller for heterogeneous ice nucleation. Therefore, in both homogeneous and heterogeneous ice nucleation, log J is expected to be a function of $1/T^3 \, \Delta T^2$.

The temperature dependence of ice nucleation rate has been measured for water, yeast, and hickory. The droplet method described above was used for yeast and water (Rasmussen, MacAulay, and MacKenzie 1974; Wood and Walton 1970) and the methods used for hickory are described by George and Burke (1977). In Figure 7, the results are compared graphically and the values of Ω and K in equation 3 are given (Table 2). These results suggest substantial variation of undercooling ability for biological samples. It is clear that hickory which survives winter by deep undercooling has favorable ice nucleation kinetics (i.e., they are slow) when compared to yeast which does not survive in winter by undercooling.

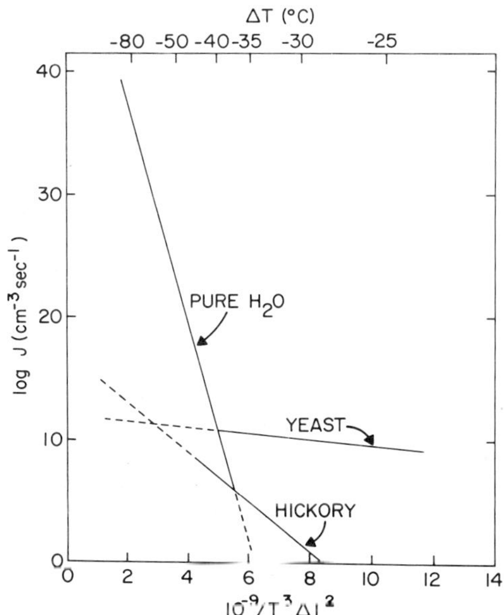

FIGURE 7. The logarithm of the ice nucleation rate J (log J) plotted v.s. $1/T^3\Delta T^2$ for pure water (Wood and Walton 1970), hickory (George and Burke 1977b), and yeast (Rasmussen et al. 1974). The amount of undercooling ΔT, is plotted on the upper axis. T is in $^\circ K$. From the figure it is clear the ice nucleation rate, J, at $-30^\circ C$ differs by more than 10 orders of magnitude for hickory and yeast. It differs by 30 orders of magnitude for yeast and pur water at $-30^\circ C$.

TABLE II. The Ice Nucleation Paramters of Equation 3 for Water in Yeast, Water in Hickory Xylem and Pure Water.

	Ω	K
yeast	10^{12}	5×10^{10}
hickory	10^{17}	2×10^{12}
water	10^{48}	1.6×10^{12}

To gain some physical insight into the meaning of these
above numbers it is helpful to take as an example, equal
size droplets (1mm^3) of packed cells or water and then
calculate, using equation 3, the freezing temperature of
the droplet. Such cavalier use of equation 3 may not be
justified in view of sample heterogeniety, temperature dependence of Ω and K, etc. In any case, equation 3 predicts
the yeast will freeze at -12°C, the hickory at -19°C, and
the pure water at -33°C. It also predicts freezing in a
threshold manner. The freezing probability at the freezing
temperatures change by 100-1000 times per degree C. This
strong temperature dependence of heterogeneous ice nucleation was predicted and observed by Vali and Stansbury
(1966). Their results show that the freezing probability
changes so rapidly with temperature, that normally, heterogeneous ice nucleation will occur within an interval of
only 1/4°C around the threshold temperature regardless of
cooling rate.

The biological samples most closely approaching the
homeogeneous ice nucleation limit of water (i.e., the Ω and
K of water) may be the floral primordia described in Figure
5. In Figure 5, the floral primordia are in the process of
deacclimating from winter. In mid-winter they typically
undercool 34°C (i.e., freeze at -36°C) (Graham and Mullin
1976a, 1976b). They contain approximately 1 mg of a dilute
aqueous solution. Undercooling such a large particle without a single ice nucleation requires Ω and K values in
equation 3 very close to those obtained for pure water.
Unlike the floral primordia which can tolerate only 1 ice
nucleation, the hickory requires many ice nucleations to
freeze and this allows it to undercool to -40°C even though
the Ω and K in equation 3 are less favorable. This is
discussed in the next section.

Of the above biological samples none could be considered good ice nucleators (i.e., those that freeze water
above -5°C). However, ny biological ice nucleators are
known to exist. Bacteria and other biogenic ice nucleators
may be of importance in droplet freezing in clouds (Schnell
and Vali 1972, 1973). Some bacteria nucleate ice above
-2°C (Vali et al. 1976; Maki et al. 1974). Bacterial ice
nucleators grow predominately on plant leaves. Some cry-
stalline organic compounds are well known as ice nucleators.
Among these are phloroglucinol (-2°C), trichlorobenzene
-12°C), ℓ-leucine (-4.5°C), ℓ-tryptophane (-5.5°C), etc.
(Hobbs 1974).

BOTANICAL CONSIDERATIONS IN DEEP UNDERCOOLING OF PLANTS

Generally for undercooling to be of importance in plant protection from cold, it is necessary that the tissues be protected from external sources of heterogeneous ice nucleation and that the undercooled system be stable for long periods of time (hours-days). This is true for many woody plant species, and it is now apparent that undercooling is a major factor in the winter survival of plants and some overwintering insects which survive in plants (George and Burke 1976; Salt 1961)

The cells which deep undercool are generally only a small fraction of the cells of the host plant. These undercooling cells include whole tissue units such as the ray parenchyma cells in the living wood of hardwood tree species, and the floral primordia in the flower buds of many of the same tree species plus others (George and Burke 1976; George et al. 1974; Graham and Mullin 1976; Quamme et al. 1975). In the winter these cells are killed when they freeze. Cold injury to the ray parenchyma cells causes "blackheart damage" and eventually leads to the death of the host plant, usually several years after exposure. Flower bud kill is less serious for the whole plant, but it does lead to loss of reproductive capacity for a year, and loss of the horticulturally important fruit product. In the winter, the other cells of the host plant survive freezing which usually occurs near $0^{o}C$, and they even survive immersion in liquid nitrogen (Burke, George, and Bryant 1975).

The Development and Structure of Plants Which Undercool

How and why plants and animals developed the capacity of undercooling is an open question. Certainly many plants have developed other means of providing frost protection, even to below $-269^{o}C$ in winter. This is adequate on a planet where the temperature only rarely approaches $-90^{o}C$. However, deep undercooling has a lower limit of only $-40^{o}C$. This is not adequate - a large part of the land surface of the northern hemisphere is subject to winter minimum below $-40^{o}C$. The ecological implications of this $-40^{o}C$ lower hardiness limit will be discussed later.

Deep undercooling for frost protection in plants has both surprising and disadvantageous facets. First, it is extremely difficult to undercool water to the homogeneous ice nucleation point to gain the frost protection; however, plants seem to have accomplished this (Figures 4,5, and 6).

Second, the undercooled state is a metastable state at best,
and in plants it is actually a transient condition. In
plant undercoolers there is a finite heterogeneous ice
nucleation rate although it is slow (Figure 7) (George and
Burke 1977). The undercooled water also is expected to have
a higher vapor pressure than ice and therefore, the liquid
should evaporate (it does not) (George and Burke 1977).
Present evidence is that its free energy is lowered *just
enough* to bring it into vapor phase equilibrium with ice
at -40°C (George and Burke 1977). On the surface it seems
surprising that plants have this capacity to deep undercool;
however, upon deeper investigation the surprise abates.

The cells which deep undercool in plants are in isolated groups throughout most of the winter and during much
of cold acclimation and deacclimation (George and Burke
1977). The groups of cells are isolated, in that there are
no continuous columns of water linking the groups of cells
together. So, in winter one ice nucleation is necessary for
every group of cells which freezes (Axelrod 1966; George and
Burke 1977; George, Burke, and Weiser 1974). The ice cannot
propagate from one group to the next. The plant structures
which make undercooling possible are quite similar to the
deep undercooling model systems which were discussed earlier
(droplets in clouds and emulsions).

Geographic Plant Distribution and Undercooling

In North America, the species which deep undercool are
all confined to the deciduous forest which covers most of
the eastern one half of the United States and parts of
southeastern Canada. Among the 77 species so far observed
to deep undercool are the oaks, elms, maples, beeches,
ashes, walnuts, hickories, roses, rhododendrons, apples,
pears, peaches, plums . . . (George and Burke 1976; George
et al. 1974). Willow species, white birch, trembling aspen
and the other deciduous species of the more northern boreal
forest, the forest which covers Canada, Alaska and northern
Eurasia, do not deep undercool and they survive much lower
temperatures. Also, the conifers which are the dominant
trees of the boreal forest region do not deep undercool.
The relation between geographic distribution and undercooling is depicted in Figure 8. Interestingly, the northern geographic limit of the tree undercoolers corresponds
with the region where -40°C midwinter minimum air temperatures become probable (Figure 9) (George et al. 1974).

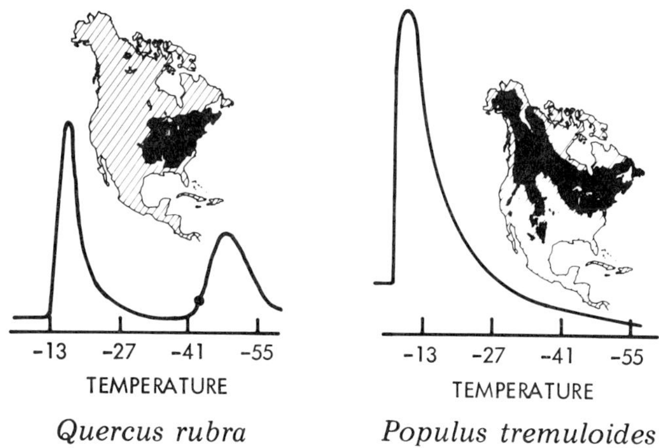

FIGURE 8. Differential thermal analyses of the living wood and the natural geographic ranges of red oak and trembling aspen. The peaks in the differential thermal analyses indicate freezing points. The solid circle in the differential thermal analyses of red oak at -42ºC indicates the tissue killing temperature (George et al. 1974b).

While the geographic distribution of tree undercoolers is limited to southern regions, this rule does not hold for shrubs. More thorough investigations of arctic shrubs regarding deep undercooling occurrence are not yet complete; however, arctic shrubs such as kinnikinick *(Arctostaphylos uva-ursi (L.)* Spreng.) and wild rose *(Rosa acicularis* Lindl.) do have deep undercooling of xylem and have cold hardiness limits of -40°C (Rajashekar pers. comm.). Such shrubs must survive arctic winters under snow, other temperature avoidance means, or by toleration of winter injury to exposed tissues. Detailed study of winter hardiness and occurrance of deep undercooling in arctic and other shrubs will be reported elsewhere.

Probably the plants of today that are capable of deep undercooling evolved and were well-established, in much warmer regions of a warmer era, the late mesozoic and early cenozoic (about 40 to 100 million years ago) (Axelrod 1966; Gongquist 1968; Stebbins 1965). It is likely that such plants were occasionally exposed to subfreezing temperatures in winter; however, it is unlikely they were exposed to temperatures as low as the deep undercooling limit, -40°C. At this time these trees were distributed throughout the land mass of the northern hemisphere (Axelrod 1966; Cronquist 1968). As time passed, the earth cooled gradually until in the pleistocene epoch (about 2 million years ago) and glacial conditions began. Figure 10 depicts the temperature changes of the earth through these times (Frenzel

FIGURE 9. *Map of North America divided into three
regions depending on extreme minimum winter temperatures.
Region A has average annual minimums below -40ºC and extreme minimums much lower. Region B has finite probability of -40ºC minimums. Region C has very little probability of -40ºC. Plants exhibiting deep undercooling
have not been observed to occur very far north of region B
(George et al. 1974b).*

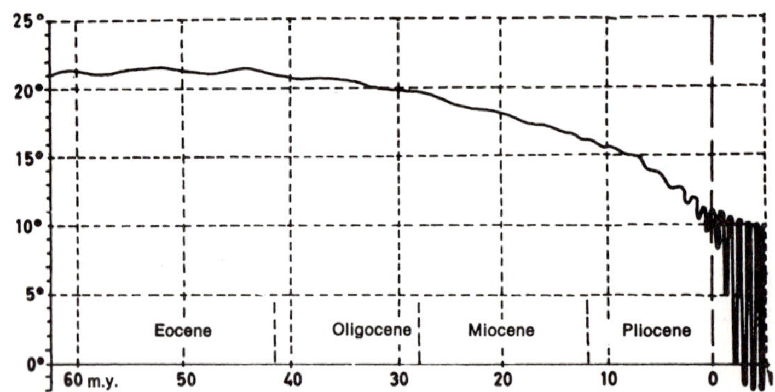

FIGURE 10. *Climatic fluctuations during the times of
higher plant (angiosperm) evolution in Central Europe. The
curve gives the average annual temperature of Central Europe
in ºC. Plants capable of deep undercooling evolved and
were well established 40 to 100 million years ago. The
lower axis is in millions of years. Clearly the earth is
cooling. Also an increase in temperature fluctuations is
recognizable (to resolve the temperature fluctuations on
the right of the figure, the time scale is exaggerated five
times) (Frenzel 1973).*

1973). During these past several million years, it is
clear -40°C midwinter minimum temperatures were common on a
large part of the land mass of the northern hemisphere
(Frenzel 1973). As shown in Figure 9, today most of Canada
and the U.S.S.R. are exposed to -40°C winter minimums, and
the tree undercoolers are not found in the cultivated or
native flora of these regions. They are dominant in the
flora of the more southern regions. This discussion suggests a dynamic relation between climate and the northern
advance of the deciduous forest, and the important climatological factor is the occurrence of -40°C midwinter minimum temperatures. The discussion also suggests that deep
undercooling was once a fully adequate frost hardiness
mechanism that has become less adequate in northern regions.

Plant Control of Undercooling

Plants come with varying abilities to undercool and
their ability to undercool changes with changing season
(Figures 4 and 5). Briefly, as most hardwood trees begin to
cold acclimate in autumn, two freezing points become apparent, both just below 0°C (Burke, George, and Bryant 1975;
George and Burke 1977). Hickory as seen in figure 4, is an
exception, for it only has the low temperature freezing
point. As autumn progresses, the low temperature freezing
point advances slowly over a period of several weeks to -
40°C. The reverse process occurs in spring.

There are two types of explanations for this changing
freezing behavior in the living wood samples like hickory.
First, the ice nucleation rate changes with cold acclimation
and deacclimation, and the number of freezing volumes
remains the same. The alternate explanation is that the ice
nucleation rate does not change, but during cold acclimation
and deacclimation, there are changes in the ease of ice
propagation from one freezing volume to the next. There
have been no other studies of these controlled undercooling
changes in plants such as hickory. In floral primordia,
only the first explanation above is reasonable because the
primordia can be artificially isolated and their freezing
volumes adjusted, and yet the ability to undercool changes
with season just as it does in the intact flower buds
(George, Burke, and Weiser 1974).

CONCLUSIONS

     There are subtle and important differences between the
general areas of plant stress physiology (which is discussed
here) and plant ecophysiology. So far, much of the physiological study of arctic plants falls into the area of
ecophysiology, and such study is not included above.
Ecophysiology is an important physiological discipline
within the general area of plant ecology concerning environmental influences on plants. For example, in ecophysiology
of arctic plants, interest is in such things as how low
growing temperatures effect plant growth, and possible competition potential. Plant stress physiology, on the other
hand, is a discipline within plant physiology. It attempts
to understand how stresses produce injurious effects on
plants, and how plants physiologically defend themselves
against stresses. In the arctic, an important stress is low
temperature or freezing stress,, and the stress physiological approach determines the plant killing temperature
and the physiological cause of injury. Unfortunately, at
this point very little is known about the stress physiology
of arctic plants. For example, what are the killing temperatures for arctic plants during the summer growing
season and later in winter? Are arctic plants appreciably
hardier than temperate plants? Do arctic plants generally
survive by avoidance or tolerance of extreme environmental
temperatures?
     In this review, several timely methods for stress
physiological study are discussed, and several applications
regarding non-arctic plants are given as examples. It is
hoped that the future will bring the application of stress
physiological principles to the study of arctic plants.

REFERENCES

Axelrod, D.I. 1966. Origin of deciduous and evergreen
     habits in temperate forests. *Evol.* *20*:1-15.
Bryant, R.G., C. Rajashekar, and M.J. Burke. In preparation.
     Ordinary garden variety NMR: application to plant frost
     hardiness. Critical reviews in environmental control.
Buckel, E.R. 1961. Studies on the freezing of pure liquids.
     II. The kinetics of homogeneous nucleation of super-
     cooled liquids. *Proc. Royal Soc.* *A261*:189-196.

Burke, M.J., M.F. George, and R.G. Bryant. 1975. Water in
    plant tissues and frost hardiness. Pages 111-135 in
    R.B. Duckworth, ed. Water relations of foods. Academic
    Press, London, England.
Burke, M.J., et al. 1976. Freezing and injury in plants.
    Ann. Rev. Plant Physiol. 27:507-528.
Cronquist, A. 1968. The evolution and classification of
    flowering plants. Houghton Mifflin, Boston, MA.
Dufour, L., and R. DeFay. 1963. Thermodynamics of clouds.
    Academic Press, New York.
Fey, R.L. 1977. Membrane studies using spin labeling of
    plant tissue. M.S. Thesis, Colorado State University.
Fey, R.L., et al. 1977. Spin label studies of plant cell
    membrane transformations using TEMPO. *Plant Physiol.
    59*:197 (Abstr.)
Fletcher, N.H. 1970. The chemical physics of ice. Cambridge
    University Press, Cambridge, England.
Frenzel, B. 1973. Climatic fluctuations of the ice age. Case
    Western Reserve University Press, Cleveland, OH.
George, M.F., and M.J. Burke. 1976. The occurrence of deep
    supercooling in cold hardy plants. *Current Adv. in
    Plant Sci. 22*:349-360.
______. 1977a. Cold hardiness and deep supercooling in xylem
    of shagbark hickory. *Plant Physiol. 59*:319-325.
______. 1977b. Supercooling in overwintering azalea flower
    buds: additional freezing parameters. *Plant Physiol.
    59*:326-328.
George, M.F., M.J. Burke, and C.J. Weiser. 1974a. Super-
    cooling in overwintering azalea flower buds. *Plant
    Physiol. 54*:29-35.
George, M.F., et al. 1974b. Low temperature exotherms and
    woody plant distribution. *Hort. Sci. 9*:519-522.
Graham, P.R., and R. Mullin. 1976a. The determination of
    lethal freezing temperatures in buds and stems of
    deciduous azalea by a freezing curve method. *J. Am.
    Soc. Hort. Sci. 101*:3-7.
Graham, P.R., and R. Mullin. 1976b. The study of flower bud
    hardiness in azalea. *J. Am. Soc. Hort. Sci. 101*: 7-10.
Gusta, L.V., M.J. Burke, and A.C. Kapoor. 1975. Determin-
    ation of unfrozen water in winter cereals at subfreez-
    ing temperatures. *Plant Physiol. 56*:707-709.
Hobbs, P.V. 1974. Ice physics. Clavendon Press, Oxford,
    England.
Hoffman, J.D. 1958. Thermodynamic driving force in nuclea-
    tion and growth processes. *J. Chem. Phys. 30*:1192-1193.
Levitt, J. 1956. The hardiness of plants. Academic Press,
    New York. 278pp.

Levitt, J. 1972. Responses of plants to environmental
    stresses. Academic Press, New York. 697 pp.
Maki, L.R., et al. 1974. Ice nucleation induced by *Pseu-
    domonas syringae.*   *Appl. Microbiol. 28*:456–459.
Mayland, H.F., and J.W. Cary. 1970. Frost and chilling
    injury in growing plants. *Adv. Agron. 22*:203–234.
Mazur, P. 1963. Kinetics of water and ion movement from
    cells at subzero temperatures and the likelihood of
    intracellular freezing. *J. Gen. Physiol. 47*:347–69.
Olien, C.R. 1967. Freezing stresses and survival. *Ann. Rev.
    Plant Physiol. 18*:387–408.
Parker, J. 1963. Cold resistance in woody plants. *Bot. Rev.
    29*:124–201.
Quamme, H.A., et al. 1975. An improved exotherm method for
    measuring cold hardiness of peach flower buds. *Hort.
    Sci. 10*:521–523.
Raison, J.K., et al. 1971. Temperature-induced phase changes
    in mitochondrial membranes detected by spin labeling.
    J. Biol. Chem. 246:4036–4040.
Rasmussen, D.H., and A.P. MacKenzie. 1972. Effect of solute
    on ice-solution interfacial free energy; calculation
    from measured homogeneous nucleation temperatures.
    Pages 126–145 in H.H.G. Jellinek, ed. Water structure
    at the water ploymer interface. Plenum Press, New York.
Rasmussen, D.H., M.N. MacAulay, and A.P. MacKenzie. 1974.
    Supercooling and nucleation of ice in single cells.
    Cryobiology. 12:328–339.
Salt, R.W. 1961. Principles of insect cold-hardiness. *Ann.
    Rev. Entomol. 6*:55–74.
Schnell, R.C., and G. Vali. 1972. Atmospheric ice nuclei
    from decomposing vegetation. *Nature. 236*:163–165.
______. 1973. Worldwide source of leaf-derived freezing
    nuclei. *Nature 246*:212–213.
Shimshick, E.J., and H.M. McConnell. 1973. Lateral phase
    separation in phospholipid membranes. *Biochem. 12*:2351–
    2359.
Siminovitch, D., et al. 1968. Phospholipid, protein, and
    nucleic acid increases in protoplasm and membrane
    structures associated with development of extreme
    freezing resistance in black locust tree cells. *Cryo-
    biology. 5*:202–225.
Stebbins, G.L. 1965. The probable growth habit of the
    earliest flowering plants. *Ann. Mo. Bot. Gard. 52*:457–
    468.
Turnbull, D., and J.C. Fisher. 1949. Rate of nucleation of
    condensed systems. *J. Chem. Phys. 17*:71–73.

Vali, G., and E.J. Stansbury. 1966. Time-dependent charac-
    teristics of the heterogeneous nucleation of ice. *Can.
    J. Phys.* 44:477-502.
Vali, G., et al. 1976. Biogenic ice nuclei. Part II. Bac-
    terial sources. *J. Atmos. Sci.* 33:1565-1570.
Weiser, C.J. 1970. Cold resistance and injury in woody
    plants. *Science.* 169:1269-1278.
Wood, G.R., and A.G. Walton. 1970. Homogeneous nucleation
    kinetics of ice from water. *J. Appl. Phys.* 41:3027-
    3036.

IX. METABOLIC AND ENZYMATIC ADAPTATIONS
TO LOW TEMPERATURE

Brian F. Chabot

Section of Ecology and Systematics
Cornell University
Ithaca, New York

Most of the research on metabolism of tundra plant species at low temperatures concerns whole-leaf or whole-plant gas exchange. Optimum and compensation point temperatures for net photosynthesis occur at lower temperatures and absolute metabolic rates below 10°C are usually higher for tundra species and ecotypes than for species or ecotypes from warmer habitats. Metabolic acclimation in response to seasonal changes in temperature is an important adjunct to the evolutionary shift in metabolic function. Limited work on cellular mechanisms of temperature adaptation for tundra and non-tundra species is reviewed. Improved metabolic performance at low temperatures appears to be primarily a function of resistance of chloroplast and mitochondrial membranes to low temperature phase changes and quantitative increases in enzyme levels. Changes in stomatal conductance and enzyme kinetic properties occur in some species, but are of secondary importance. Based upon existing fragmentary evidence, there appear to be no qualitative differences between tundra and non-tundra species in their adaptations for low temperature metabolism.

INTRODUCTION

Arctic plants make a number of metabolic adjustments to the low temperature which they experience during the growing season. However, this is only part of the story. At the outset, it should be remembered that, in addition to metabolic mechanisms of low temperature tolerance, plants can significantly modify the thermal conditions to which they

283

are exposed. Many plant species are capable of short-term
behavior that achieves some degree of thermoregulation.
Examples are leaf nyctinasty (Darwin and Darwin 1896;
Schwintzer 1971; Smith 1974) solar tracking in flowers
(Kevan 1975), and diurnal opening and closing of flowers.
A few species of plants possess thermogenic tissue capable
of generating metabolic heat. On a slightly longer time
scale, the production of leaves of different sizes or
degrees of pubescence modifies thermal energy balance
(Ehleringer, Björkman, and Mooney 1976; Mooney, Björkman,
and Troughton 1974). Overall growth form can place leaves,
and other tissues, in markedly different micro-climates.
This is particularly important in tundra plant species whose
morphology places them within a high temperature zone near
the soil surface (Billings and Mooney 1968). Dormancy in
plants is a phenomenon ecologically similar to hibernation
in animals in that the organisms face only a selected fraction of the annual environmental regime. Thus, differences
between plants and animals are not necessarily extreme.
More importantly, we must be wary of considering temperature
effects on metabolism without some knowledge of the ecology
of the species. In particular, we need to know the environmental conditions actually faced at each phase of the growth
cycle. This has not always been done.

This review will focus on metabolism in arctic plant
species at low, but above freezing, temperatures. Most of
the literature on metabolism of arctic plants concerns whole
leaf gas exchange. There are only a small number of studies
dealing with organelle and enzyme systems at low temperatures. As a result, it is necessary to include research
results from both arctic and non-arctic species. This is
done in part, for comparative purposes in order to point up
the unique capabilities of arctic species. Also, adaptation
to low temperature appears to be one end of a continuum of
temperature adaptation so that information on behavior at
higher temperatures will be employed in understanding response to low temperature. Freezing tolerance in plants
will not be covered, as this topic is considered elsewhere
in this publication, and has recently been reviewed by
others (Larcher 1973; Bauer, Larcher, and Walker 1975).

GENERAL TRENDS IN WHOLE PLANT METABOLISM

A useful starting point for describing cellular and
enzymatic adaptations to low temperature is a brief sum-
marization of temperature effects on whole organ gas exchange. Evidence for the ecological value of specific
cellular modifications usually comes from correlation with
higher level metabolic functions. Processes such as photosynthetic and respiratory gas exchange integrate complex
cellular activity, but they also interface readily with
growth and reproduction, which have directly interpretable
ecological relevance. Only a brief review of leaf gas exchange in tundra plants will be given here since the subject
has been treated by Billings and Mooney (1968) and Bliss
(1971).

Net Photosynthesis

Tundra species usually have higher rates of net photosynthesis at low temperatures (below 10°C) than non-tundra
species (Figure 1). This is not to say that non-tundra
species are incapable of photosynthesis at temperatures near
freezing. Indeed, *Atriplex confertifolia* of the Great Basin
Desert has been shown to continue photosynthesis to -5°C
(Caldwell 1972). However, tundra species as a group have
their optimum and low-temperature compensation points for
net photosynthesis at lower temperatures than is true for
other major floristic assemblages. These differences are
for the most part, genetically based. Plants have evolved
to maximize carbon uptake under temperature conditions most
like those of their native habitats. This genetic fixation
of metabolic performance has occurred between species in
different habitats as well as for ecotypes of widely ranging
species (Billings and Mooney 1968; Mooney and Shropshire
1967; Mächler and Nösberger 1977; Slatyur and Ferrar 1977a).
Photosynthetic performance is also a phenotypically
plastic character that can be modified by growth regime.
Plants grown at low temperature shift both the optimum and
compensation point temperatures for photosynthesis to lower
temperatures (Figure 2). This phenotypic pattern of acclimation follows the genetic pattern of adaptation in that
carbon gain usually increases at the growing temperatures as
a result of the acclimation process (Mooney and Shropshire
1967; Strain, Higginbotham, and Mulroy 1976; Chabot in press
1978). Acclimation has been observed in both field and lab

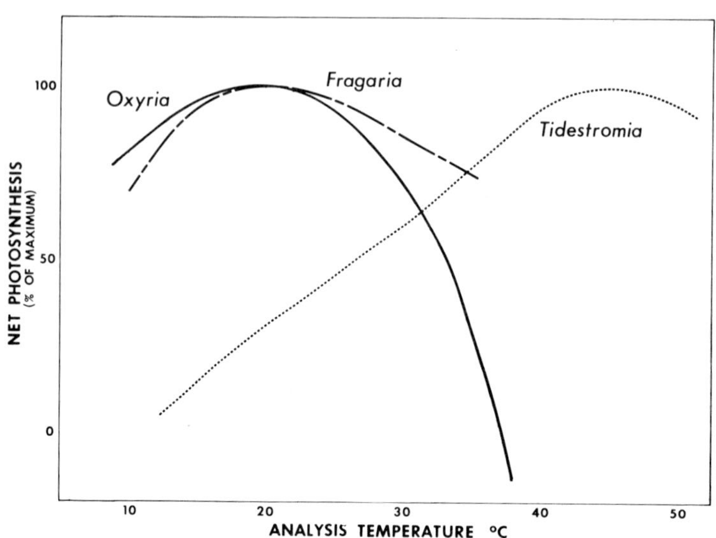

FIGURE 1.   *Relative net photosynthesis in three species from diverse climatic regions: Oxyria digyna--alpine tundra (Chabot and Billings 1972), Fragaria vesca--temperature woodland (Chabot 1978), Tidestromia oblongifolia--hot desert (Pearcy et al. 1971).*

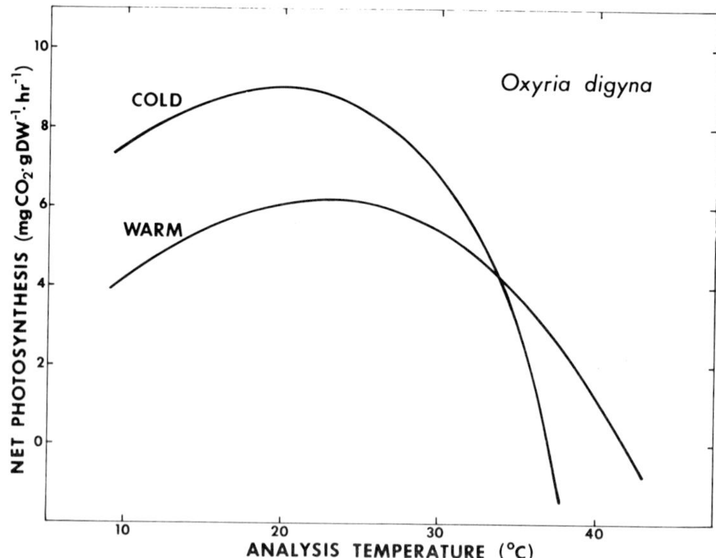

FIGURE 2.   *Acclimation of net photosynthesis in Oxyria digyna from Ellesmere Island when grown under warm and cold conditions (adapted from Billings et al. 1971).*

grown plants and appears to be a mechanism for compensating
for temperature changes during the growing season (Strain,
Hibbinbotham, and Mulroy 1976; Moore, Miller, Ehleringer,
and Lawrence 1973; DePuit and Caldwell 1973; Lange, Schulze,
Evenari, Kappen, and Buschbom 1974; Larson and Kershaw 1975;
Kershaw 1977a; Oechel 1976; Pearcy 1976; Slatyer and Morrow
1977). Absolute rates of photosynthesis do change with
acclimation. The highest absolute rates are usually developed under conditions similar to the natural habitat.
For tundra species, this means that growth at high temperatures usually depresses the absolute rate of photosynthesis
over that achieved at low growth temperatures (Billings and
Mooney 1968; Oechel 1976) (Figure 2). By contrast, warm
climate species reach their highest absolute rates of
carbon uptake under high temperatures (Mooney and Shropshire
1967; Chabot 1978; Sawada and Miyachi 1974; Slatyer and
Ferrar 1977a). There are limits on this process of course.
At very low growth temperatures even arctic species show
declines in absolute net photosynthesis (Hicklenton and
Oechel 1976; Kershaw 1977a).

Dark Respiration

In contrast with photosynthesis, less attention has
been paid to dark respiration in plants. Limited comparisons suggest that at any given temperature, tundra species
may have higher respiratory rates than species from warmer
habitats (Billings and Mooney 1968; Forward 1960; Hartgerink
and Mays 1976). Again, this parallels some results from
acclimation studies which have demonstrated that cold-grown
plants may have higher respiratory rates at comparable
analysis temperatures than warm-grown plants (Chabot and
Billings 1972). In both cases it may be argued that such
compensation for the expected depressing effects of low
temperatures leads to a form of metabolic homeostasis.

Photorespiration

Photorespiratory CO_2 release as a function of temperature is still something of a mystery. Several studies indicate that photorespiration reaches an optimum point within
normal biological temperatures (Pearson and Hunt 1972;
Wilson 1972) although other studies show a steady increase
between 13 to 20 C in wheat (Keys, Sampaio, Cornelius, and

Bird 1977) and 3 to 24 C in *Trifolium repens* (Mächler, Nösberger, and Erismann 1977). It is suggested that in alpine clones of *Trifolium*, higher photorespiration rates account for decreased sensitivity of net photosynthesis to temperature and the eventual decline of net photosynthesis at high temperatures (Mächler, Nösberger, and Erismann 1977). The acclimation response of photorespiration sometimes corresponds to that for dark respiration (Mooney and Harrison 1970). However, in alfalfa, photorespiration is not affected by growth temperature (Pearson and Hunt 1972). In yet other cases there is an inverse behavior; low growth temperatures depress rates of photorespiration beyond that due to thermal kinetics of enzyme systems (Wilson 1972; McNaughton 1973). This inverse response is unusual for biological systems and, in general, difficult to explain functionally. There are methodological difficulties in measuring photorespiration and it has not been studied in many species or over a great range of environmental conditions. It will not be considered further in this review.

Time Course Studies

The rate of acclimation of gas exchange capacity to temperature changes has been studied for a small number of species, including several from the Arctic. The results reveal something of the possible cellular mechanisms for low temperature acclimation. For both photosynthetic and respiratory acclimation there is typically a time lag of several hours befor acclimation begins and the process may be completed in as little as 24 hours (Chabot and Billings 1972, Sawada and Miyachi 1974; Kershaw 1977b). The rate of acclimation is dependent on temperature or the physiological state of the tissue, requiring significantly longer time periods at low versus high temperature (Chabot and Billings 1972; Slatyer and Ferrar 1977b). At certain phases of the growth cycle, photosynthesis does not acclimate to temperature changes (Kershaw 1977b). These facts argue that biochemical or stomatal processes play a primary role in rapid acclimation. However, acclimatory adjustments may continue for several days to weeks (Mooney and Harrison 1970; Hicklenton and Oechel 1976; Slatyer and Ferrar 1977b) suggesting that some structural reorganization at the cellular level may also occur.

MECHANISMS OF METABOLIC ADAPTATION LOW TEMPERATURE

The question addressed here is: 'How is the active metabolism of tissues maintained at low temperatures?' For the most part the answers for plants appear to parallel those arrived at for animal systems. Studies in plants are less complete, but suggest a primary role for both enzyme systems and cellular membranes. Dark respiration has received less attention in plants than has photosynthesis. However, the process is most likely to bear a close resemblance to the behavior of animal respiratory metabolism and will be considered first.

Dark Respiration

Acclimation of mitochondrial activity parallels that for whole-gas exchange. The rate of mitochondrial oxidation at a given temperature increases either when a genotype is grown at progressively lower temperature in the laboratory (Billings, et al. 1971; Geronimo and Beevers 1964; Pomeroy and Andrews 1975) or in field populations collected from successively higher elevations (Klikoff 1966, 1968). That is, there exists a tendency in both genetic differentiation and phenotype modification for mitochondrial oxidation to increase under lower growth temperatures. Such changes appear to be intrinsic to the mitochondria rather than involving an alteration of the concentration of mitochondria in the tissue such as have been described in other situations (Simon, et al. 1971; Weintraub 1972).

In several studies, changes in the activity of respiratory enzymes have been noted to result from exposure to low temperatures. Quantiatative increases in the amount of soluble protein at low temperatures have been observed in many (Parker 1962; Kanaryan and Gasparyan 1966; Gerloff, Stahmann, and Smith 1967; Chabot, Chabot, and Billings 1972; Krasnuk, Jung, and Witham 1976) but not all studies (Pomeroy and Andrews 1975). Reports of increases in the activity of specific enzymes (Sysoev and Krasnaya 1967; McNaughton 1972) may involve changes in the amount of the enzyme rather than its kinetic properties. It is likely that the major part of increased mitochondrial and tissue respiration at low temperatures is due to this quantitative augmentation of enzyme levels.

Much work has been done on electrophoretic patterns of proteins following cold-hardening under the hypothesis that isozymic substitutions should contribute to greater catalytic efficiency (Roberts 1969). Isozymic variants may increase, decrease, or remain the same depending upon the specific enzyme and plant species (Gerloff, Stahmann, and Smith 1967; Chabot, Chabot, and Billings 1972; Roberts 1967; Davis and Gilbert 1970; Hall, et al. 1970; Krasnuk, Witham, and Jung 1976). No general patterns of isozymic change appear from such studies. It is often not clear whether hardening conditions have induced a dormant state or have led to acclimation of active metabolism to low temperatures. Additionally, it is not clear as to the real functional significance for changes in isozymic variants at low temperatures since kinetic studies on these variants have not been performed. Kinetic studies usually have been done on bulk protein extracts. Roberts (1967) found a slight decrease in the activation energy of invertase following cold-hardening in wheat. McNaughton (1972) described complex changes in the activation energy (E_a) of malate dehydrogenase as a function of genotype and growth environment. E_a was high in a coastal population of *Typha latifolia* grown at cool temperatures and decreased when this population was grown at warmer, and more widely ranging temperatures. A reverse pattern was found for a continental population. McNaughton suggested that his data supported a trend toward increasing temperature insensitivity in a thermal range most like that of the native habitat. Alexandrov (1977) has recently reviewed evidence for the importance of enzyme structural changes in high temperature metabolism.

A most promising area of investigation concerns temperature effects on membrane composition. The physical state and chemical composition of mitochondrial membranes affects rates of oxidative metabolism. This has been reviewed by Lyons (1972), and is covered by de la Roche in this volume.

Photosynthesis

Mechanisms of adaptation of photosynthesis to temperature have focused on three areas: CO_2 transport resistances, carboxylase enzyme activity, and photochemical processes. The thermal sensitivity of ribulose-1,5-bisphosphate carboxylase-oxygenase (RuBPcase) has been regarded by several individuals to control both the maximum rates of photosynthesis and the temperature optimum for net photosynthesis. Treharne and Cooper (1969) found that the thermal optima for RuBPcase was 10°C higher in tropical than in temperate grass species corresponding to differences in their photosynthetic performance. Phillips and McWilliam (1971) correlated thermal optima of carboxylase enzyme activity with net photosynthesis in three plant species from contrasting habitats. The cold adapted alpine species, *Caltha intraloba*, proved to have an enzyme with a lower activation energy and greater sensitivity to heat denaturation than the other species. By contrast, other studies have shown that there is no thermal optima for carboxylase activity below 35-40°C in several arctic species (Chabot, Chabot, and Billings 1972; Tieszen and Sigurdson 1973) or below 30°C in a group of non-arctic species (Björkman and Pearcy 1971). Additionally, the activation energy did not change with growth condition or ecotype. Thus, the shape of the net photosynthesis vs. temperature curve cannot be interpreted simply as a result of thermal properties of the carboxylase enzyme. More recently, Huner and Macdowall (1976) have found that cold-hardened rye plants produced a form of RuBPcase which had different kinetic and stability properties than the enzyme from non-hardened plants. In the unhardened enzyme, there was a decrease in activation energy above 30°C while the hardened enzymes showed a decrease in E_a below 10°C. These changes tend to make the enzyme more efficient in the temperature range corresponding to growth conditions. Such differences in the kinetic properties of RuBPcase have not been found consistently. Assay procedures for RuBPcase have changed substantially in the past 10 years. Whether existing discrepancies are the result of true species differences or methodological problems must await further work.

Quantitative differences in carboxylase enzyme concentration appear when plants are grown at different temperatures. In spite of the greater thermal efficiencies, Huner and Macdowall (1976) report three times less activity in the low temperature hardened plants. Sawada et al. (1974), working with wheat, also found that absolute activity was reduced following exposure to low temperatures. This corresponded to a reduction in maximum rates of photosynthesis for these hardened plants. In arctic and alpine populations of *Oxyria digyna*, both the activity of RuBPcase and rates of net photosynthesis were highest in cold grown plants (Chabot, Chabot, and Billings 1972). The increased enzyme activity appeared to be due in part to quantitative changes in the RuBPcase enzyme. A similar result was found in *Dactylis glomerata* (Jensen and Bahr 1977). Such evi-

dence suggests that the light saturated rates of photosynthesis at low temperatures depends directly upon the
catalytic capacity of carboxylase enzyme which in turn
relates closely to enzyme concentration. Maintenance of
enzyme concentration would be a combined function of protein synthesis and degradation processes. Little work has
been done on protein synthesis at low temperatures. However, the limited data presented here suggests that tundra
species may be distinguished by their capacity to maintain
enzymatic function at low temperatures where other species
have shown a loss of function.

The work cited here on RuBPcase activity with respect
to temperature needs to be repeated and extended to a
larger group of species. Understanding of this key enzyme
has changed greatly within the past few years (see Jensen
and Bahr 1977). Maximum activity under analytical conditions is obtained following a specific set of incubation
requirements which have generally not been used in existing
ecological studies. These procedural difficulties are
compounded by the competing carboxylase and oxygenase
functions of this enzyme. We are probably also going to
need a better understanding of the cell environment in
which this enzyme functions. Concentrations of cofactors
and pH, which are known to influence the enzyme *in vitro*,
are likely to change over a wide range of temperatures.

Photochemical processes are also known to be temperature sensitive. Hill reaction activity shows a temperature optimum approximately that for net photosynthesis
(Tieszen and Helgager 1977). Both the temperature optimum
and the rate of reduction activity can be varied by growth
temperature and population origin (Billings, et al. 1971;
Tieszen and Helagager 1977; May and Villarreal 1974). In
general, Hill reaction activity seems to increase when
species are grown at low temperatures and is higher in
populations from cool environments. More detailed studies
(Murata and Fork 1975; Armond, Schreiber and Björkman 1978)
confirm that photosystem II activity (the Hill reaction) is
sensitive to growth temperature, but that several other
components of the light harvesting system are also affected. These reactions are all dependent upon the physical structure of chloroplast membranes. As temperatures
are lowered, current membrane models suggest that there is
a transition from liquid-crystalline to more solidified
structure. A reduction in activation energy for membrane-
bound processes frequently accompanies this transition.
Visible disruption of membranes and chloroplast swelling
can be seen at freezing temperatures (Taylor and Craig

1971; Kimball and Salisbury 1973; Forde, Whitehead, and Rowley 1975) and may relate to "lesions" found in key biochemical and physiochemical processes (Steponkus, et al. 1977). In C_4 plants, physical disruption of chloroplast membranes is seen at low temperatures in cold sensitive species, but not in cold tolerant species (Slack, Roughan, and Bassett 1974; Caldwell, Osmond, and Nott 1977). Avoidance of metabolic and photochemical inhibition appears to reside in the membrane constituents. The proportion of unsaturated fatty acids or the amount of lipid material increases at low temperatures (Lyons 1972; de la Roche, Pomeroy, and Andrews 1975; Smolenska and Kuiper 1977). Membrane stability and metabolic function may be maintained as a result of these changes in lipid composition. This mechanism for achieving low temperature stability may simultaneously lead to high temperature instability, but considerably more work is needed on these membrane systems in a variety of plant species.

Stomatal conductance for CO_2 may be altered at low temperatures. In several cases, plants exposed to near freezing temperatures will close their stomates (Chabot 1978; Drake and Raschke 1974; Raschke, Pierce, and Popiela 1976; Marchard and Chabot). High temperature may also induce stomatal closure as in cold climate populations of *Verbascum thapsus* (Williams and Kenys 1976). Sensitivity to temperature and CO_2 increases at low temperatures. Raschke et al. (1976) have found evidence that abscissic acid levels increase at near freezing temperatures leading to a reduction of stomatal conductance. Beyond the increased sensitivity, maximum stomatal conductances appear to be less for plants grown at low versus moderate to high temperatures (Chabot 1978; Larson and Kershaw 1975). These results come from warm climate species and correspond to a reduction in maximum photosynthesis in plants from the colder growth temperatures. Whether tundra species are susceptible to such effects has not been determined.

From the existing evidence it is difficult to produce a coherent picture of how tundra plant species are able to function metabolically at low temperatures. Low temperature clearly represents a stress to plants. Even tundra species eventually show a decline in metabolism at low temperatures. In the case of photosynthesis, several component processes all appear to be affected by temperature. Photosynthesis in temperate and tropical species may be inhibited at low temperatures through disruption in any one of these processes. Tundra species must have relatively greater immunity to low temperatures in all these processes in order to survive.

CONCLUSIONS

Plant ecologists have been interested in arctic species for a long period of time. Study of arctic and alpine species has been a cornerstone of experimental physiolo_ gical ecology. Such interest is still legitimate in that with arctic plants we have a natural experiment, the re_ sults of which will permit a better understanding of how organisms function at low temperatures. However, the results of this experiment are only partially analyzed. It is clear that low temperatures have important effects on cellular metabolic processes and that adaptations at the cellular level are significant to the survival of arctic species. Even with a small sample size of species studied, we can point to the involvement of certain key processes in low temperature adaptation. What is lacking are two things:

1. More detailed studies of what appear to be key cellular processes. These include photochemical activity, membrane structure, carboxylation rates, general enzyme synthesis and degradation, and intergration of metabolic pathways. In most cases the data presented here have come from a few exploratory studies in a small number of species. Some of the studies, such as those dealing with the carboxylase enzymes, need to be repeated to take advantage of updated methodology.

2. A greater degree of integration in studies of temperature influences on plants. We know something about CO_2 exchange in *Dupontia fischeri*, enzyme activity in *Oxyria digyna*, and photochemical processes in *Danthonia caespitosa*. But with such a data base, it is difficult to fit parts of the adaptive puzzle together. The situation would be improved if we were to analyze each of these species with greater thoroughness. Research on low temperature adaptation cannot concentrate on tundra species alone. There is much to be gained from a comparison with plants from warmer climates. At the very least they provide a contrast necessary to appreciate the uniqueness of tundra species. More importantly, among a group of species there are bound to be a variety of ways that various traits can interact to improve low temperatures metabolism.

It is certain that adaptation to low temperatures involves an integration among many processes. The search for a single limiting factor, which has dominated the literature thus far, may be a false paradigm. It is possible that something like membrane structure has general importance in integrating a number of cellular processes. It may be that integration resides most critically in the genome. It is less likely that a simple trait differentiates tundra from temperate species, even at the level of basic metabolism. However, I think that research which is at once comprehensive, coordinated and detailed will be the only way to resolve these possibilities.

REFERENCES

Alexandrov, V.Ya. 1977. Cells, molecules and temperature. Springer-Verlag, Berlin, Germany.

Armond, P.A., U. Schreiber, and O. Björkman. 1978. Photosynthetic acclimation to temperature in the sesert shrub, *Larrea divaricata* II. Light harvesting efficiency and electron transport. *Plant Physiol.* 61:411.

Bauer, H., W. Larcher, and R.B. Walker. 1975. Influence of temperature stress on CO_2-gas exchange. Pages 557-586 in Photosynthesis and productivity in different environments. Cambridge University Press, Cambridge, England.

Billings, W.D.,, and H.A. Mooney. 1968. The ecology of arctic and alpine plants. *Biol. Rev.* 43:481.

Billings, W.D., et al. 1971. Metabolic acclimation to temperature in arctic and alpine ecotypes of *Oxyria digyna*. *Arctic Alp. Res.* 3:277.

Björkman, O., and R.W. Pearcy. 1971. Effect of growth temperature on the temperature despendence of photosynthesis in vivo and on CO_2 fixation by carboxydismutase in vitro in C_3 and C_4 species. *Carnegie Inst. Year Book.* 70:511.

Bliss, L.C. 1971. Arctic and alpine life cycles. *Ann. Rev. Ecol. Syst.* 2:405.

Caldwell, M.M. 1972. Gas exchange of shrubs. Pages 260-270 in Wildland shrubs--their biology and utilization, USDA Forest Service Technical Report INT-1.

Caldwell, M.M., C.B. Osmond, and D.L. Nott. 1977. C_4 pathway photosynthesis at low temperature in cold-tolerant *Atriplex* species. *Plant Physiol.* 60:157.

Chabot, B.F. 1978. Environmental control of photosynthesis
    and growth in *Fragaria vesca*.   *New Phytol.*
    80:87-98.
Chabot, B.F., and W.D. Billings. 1972. Origins and ecology
    of the Sierran alpine flora and vegetation. *Ecol.*
    Monogr. *42*:163.
Chabot, B.F., and W.D. Billings. 1972. Ribulose-1,5-diphos-
    phate carboxylase activity in arctic and alpine pop-
    ulations of *Oxyria digyna*.   *Photosynthetica*. *6*:364.
Darwin, C., and F. Darwin. 1896. The power of movement in
    plants. D. Appleton and Co., New York.
Davis, D.L., and W.B. Gilbert. 1970. Winter hardiness and
    changes in soluble protein fractions of Bermuda grass.
    Crop Sci. *10*:7.
de la Roche, I.A., M.K. Pomeroy, and C.J. Andrews. 1975.
    Changes in fatty acid composition in wheat cultivars
    of contrasting hardiness. *Cryobiology*. *12*:506.
DePuit, E.J., and M.M. Caldwell. 1973. Seasonal patterns of
    net photosynthesis of *Antemesia tridentata*.   *Am. J.*
    Bot. *60*:426.
Drake, B., and K. Raschke. 1974. Prechilling of *Xanthium*
    strumarium L. reduces net photosynthesis and, inde-
    pendently, stomatal conductance, while sensitizing the
    stomata to CO_2. *Plant Physiol*. *53*:808.
Ehleringer, J., O. Björkman, and H.A. Mooney. 1976. Leaf
    pubescence: effects on absorptance and photosynthesis
    in a desert shrub. *Science*. *192*:376.
Forde, B.J., H.C.M. Whitehead, and J.A. Rowley. 1975.
    Effect of light intensity and temperature on photo-
    synthetic rate, leaf starch content and ultrastructure
    of *Paspalum dilatatum*.   *Aust. J. Plant Physiol*.
    2:185.
Forward, D.F. 1960. Effects of temperature on respiration.
    Handbuch der Pflanzephysiologie. *12*:234.
Gerloff, E.D., M.A. Stahmann, and D. Smith. 1967. Soluble
    protein in alfalfa roots as related to cold hardiness.
    Plant Physiol. *42*:895.
Geronimo, J., and H. Beevers. 1964. Effects of aging and
    temperature on respiratory metabolism of green leaves.
    Plant Physiol. *39*:786.
Hall, T.C., et al. 1970. Enzyme changes during acclimation.
    Cryobiology. *6*:263.
Hartgerink, A.P., and J.M. Mays. 1976. Controlled-environ-
    mental studies on net assimilation and water relations
    of *Dryas intergrifolia*.   *Can. J. Bot*. *54*:1884.
Hicklenton, P.R., and W.C. Oechel. 1976. Physiological
    aspects of *Dicranum fuscescens* in the subarctic. I.
    Acclimation and acclimation potential of CO_2 exchange

in reation to habitat, light, and temperature. *Can. J. Bot. 54*:1104.

Huner, N.P.A., and F.D.H. Macdowall. 1976. Effect of cold adaptation of puma rye on properties of RuDPcarboxylase. *Bioch. Biophys. Res. Comm. 73*:411.

Jensen, R.G., and J.T. Bahr. 1977. Ribulose 1,5 bisphosphate carboxylase-oxygenase. *Ann. Rev. Plant Physiol. 28*:379.

Kazaryan, V.O., and A.G. Gasparyan. 1966. Daily changes in the protein nitrogen content of the leaves of alpine plants growing at different altitudes. *Biol. Zh. Arm. 19*:18.

Kershaw, K.A. 1977a. Physiological-environmental interactions in lichens. II. The pattern of net photosynthetic acclimation in *Peltigera canina* (L.) Willd var. *preatextata* (Floerke in Somm) Hue, and *P. polydactyla* (Neck.) Hoffm. *New Phytol. 79*:377.

______. 1977b. Physiological-environmental interactions in lichens III. The rate of net photosynthetic acclimation in *Pelligera canina* (L.) Willd var. *Praetextata* (Floerke in Somm.) Huc, and *P. Polydactyla* (Neck.) Hoffm. *New Phytol. 79*:391.

Kevan, P.G. 1975. Sun-tracking solar furnaces in high arctic flowers: significance for pollination and insects. *Science. 189*:143.

Keys, A.J. et al. 1977. Effect of temperature on photosynthesis and photorespiration of wheat leaves. *J. Exptl. Bot. 28*:525.

Kimball, S.L., and F.B. Salisbury. 1973. Ultrastructural changes of plants exposed to low temperatures. *Am. J. Bot. 60*:1028.

Klikoff, L.G. 1966. Temperature dependence of the oxidative rates of mitochondria in *Danthonia intermedia, Penstemon davidsonii,* and *Sitanion hystrix. Nature. 212*:529.

______. 1968. Temperature dependence of the mitochondria oxidative rates of several plant species of the Sierra Nevada. *Bot. Gaz. 129*:227.

Krasnuk, M., G.A. Jung, and F.H. Witham. 1976. Electrophoretic studies of several dehydrogenases in relation to the cold tolerance of alfalfa. *Cryobiology 13*:375.

Krasnuk, M., F.H. Witham, and G.A. Jung. 1976. Electrophoretic studies of several hydrolytic enzymes in relation to the cold tolerance of alfalfa. *Cryobiology. 13*:225.

Lange, O.L., et al. 1974. The temperature-related photo-
     synthetic capacity of plants under desert conditions.
     I. Seasonal changes of the photosynthetic response to
     temperature. *Oecologia. 17*:97.
Larcher, W. 1973. Limiting temperatures for life functions.
     Pages 195-231 in Temperature and life. Springer-
     Verlag, Berlin, Germany.
Larson, D.W., and K.A. Kershaw. 1975. Studies on lichen-
     dominated systems. XVI. Comparative patterns of net
     CO_2 exchange in *Cetraria nivalis* and *Alectoria och-
     roleuca* collected from a raised-beach ridge. *Can. J.
     Bot. 53*:2884.
Lyons, J.M. 1972. Phase transitions and control of cellular
     metabolism at low temperatures. *Cryobiology. 9*:341.
Mächler, F., and J. Nösberger. 1977. Effect of light inten-
     sity and temperature on apparent photosynthesis of al-
     titudinal ecotypes of *Tribolium repens L. Oecologia.
     31*:73.
_____, et al. 1977. Photosynthetic $^{14}CO_2$ fixation products
     in altitudinal ecotypes of *Trifolium repens L.* with
     different temperature requirements. *Oecologia 31*:79.
Marchand, P., and B.F. Chabot. 1978. Winter water relations
     of treeline plant species on Mt. Washington, New
     Hampshire. *Arctic Alp. Res. 10*:105-116.
May, D.S., and H.M. Villarreal. 1974. Altitudinal differ-
     entiation of the Hill reaction in populations of
     Taraxacum officinale in Colorado. *Photosynthetica.
     8*:73.
McNaughton, S.J. 1972. Enzymic thermal adaptations: the
     evolution of homeostasis in plants. *Am. Natural.
     106*:165.
_____. 1973. Comparative photosynthesis of Quebec and
     California ecotypes of *Typha latifolia. Ecology.
     54*:1260.
Mooney, H.A., and A.T. Harrison. 1970 The influence of
     conditioning temperature on subsequent temperature-
     related photosynthetic capacity in higher plants.
     Pages 411-417 in Prediction and measurement of photo-
     synthetic productivity. Centre for Agricultural Pub-
     lishing and Documentation, Wagengingen, The Nether-
     lands.
Mooney, H.A., and F. Shropshire. 1967. Population varia-
     bility in temperature related photosynthetic accli-
     mation. *Oecol. Plantarum. 2*:1.

Mooney, H.A., O. Bjürkman, and J. Troughton. 1974. Seasonal changes in the leaf characteristics of the desert shrub, *Atriplex hymenelytra*. *Carnegie Inst. Wash. Yearbook.* 73:846.

Moore, R.T., et al. 1973. Seasonal trends in gas exchange characteristics of three mangrove species. *Photosynthetica.* 7:387.

Murata, N., D.C. Fork, and J. Troughton. 1975. Control of photosynthesis by temperature. *Carnegie Inst. Yearbook.* 74:766.

Oechel, W.C. 1976. Seasonal patterns of temperature response of CO_2 flux and acclimation in arctic mosses growing *in situ*. *Photosynthetica.* 10:447.

Parker, J. 1962. Relationships among cold hardiness, water-soluble proteins, anthocyanins, and free sugars in *Hedera helix*. *Plant Physiol.* 37:809.

Pearcy, R.W. 1976. Temperature responses of growth and photosynthetic CO_2 exchange rates in coastal and desert races of *Atriplex lentiformis*. *Oecologia.* 26:245.

Pearcy, R.W., et al. 1971. Photosynthetic performance of two desert species with C_4 photosynthesis in Death Valley, California. *Carnegie Inst. Yearbook.* 70:540.

Pearson, C.J., and L.A. Hunt. 1972. Effects of pretreatment temperature on carbon dioxide exchange in alfalfa. *Can. J. Bot.* 50:1925.

Phillips, P.J., and J.R. McWilliam. 1971. Thermal responses of the primary carboxylating enzymes from C_3 and C_4 plants adapted to contrasting temperature environments. Pages 97-104 in Photosynthesis and photorespiration. Wiley, NY.

Pomeroy, M.K., and C.J. Andrews. 1975. Effect of temperature on respiration of mitochondrial and shoot segments from cold-hardened and non-hardened wheat and rye seedlings. *Plant Physiol.* 56:703.

Raschke, K., M. Pierce, and C.C. Popiela. 1976. Abscisic acid content and stomatal sensitivity to CO_2 in leaves of *Xanthium strumarium* L. after pretreatments in warm and cold growth chambers. *Plant Physiol.* 57:115.

Roberts, D.W.A. 1967. The temperature coefficient of invertase from the leaves of cold-hardened and cold-susceptible wheat plants. *Can. J. Bot.* 45:1347.

———. 1969. Some possible roles for isozymic substitutions during cold hardening. *Internat. Rev. Cytol.* 26:303.

Sawada, S., and S. Miyachi. 1974. Effects of growth tem-
    perature on photosynthetic carbon metabolism in green
    plants. I. Photosynthetic activities of various plants
    acclimatized to varied temperatures. *Plant & Cell
    Physiol. 15*:111.

Sawada, S., H. Matsushima, and S. Miyachi. 1974. Effects of
    growth temperature on photosynthetic carbon metabolism
    in green plants. III. Differences in structure,
    photosynthetic activities and activities of ribulose
    diphosphate carboxylase and glycolate oxidase in
    leaves of wheat grown under varied temperatures. *Plant
    & Cell Physiol. 15*:239.

Schwintzer, C.R. 1971. Energy budgets and temperatures of
    nyctinastic leaves on freezing nights. *Plant Physiol.
    48*:203.

Simon, R.G., et al. 1971. Mitochondrial involvement in cold
    acclimation. *Comp. Biochem. Physiol. 40B*:601.

Slack, C.R., P.G. Roughan, and H.C.M. Bassett. 1974. Selec-
    tive inhibition of mesophyll chloroplast development
    in some C_4-pathway species by low night temperature.
    Planta 118:57.

Slatyer, R.O., and P.J. Ferrar. 1977a. Altitudinal varia-
    tion in the photosynthetic characteristics of snow
    gum, *Eucalyptus pauciflora* Sied. ex. Spreng. II. Ef-
    fects of growth temperature under controlled conditions.
    Aust. J. Plant Physiol. 4:289.

______. 1977b. Altitudinal variation in the photosynthetic
    characteristics of snow gum, *Eucalyptus pauciflora*
    Sieb. ex. Spreng. V. Rate of acclimation to an altered
    growth environment. *Aust. J. Plant Physiol. 4*:595.

Slatyer, R.O., and P.A. Morrow. 1977. Altitudinal variation
    in the photosynthetic characteristics of snow gum, *Eu-
    calyptus pauciflora* Sieb. ex. Spreng. I. Seasonal
    changes under field conditions in the Snowy Mountains
    area of southeastern Australia. *Aust. J. Bot. 25*:1.

Smith, A.P. 1974. Bud temperature in relation to nyctin-
    astic leaf movement in an Andean giant rosette plant.
    Biotropica. 6:263.

Smolenska, G., and P.J. C. Kuiper. 1977. Effect of low
    temperature upon lipid and fatty acid composition of
    roots and leaves of winter rape plants. *Physiol.
    Plant. 41*:29.

Steponkus, P.L., et al. 1977. Effects of cold acclimation
    and freezing on structure and function of chloroplast
    thylakoids. *Cryobiology. 14*:303.

Strain, B.R., K.O. Higginbotham, and J.C. Mulroy. 1976.
    Temperature preconditioning and photosynthetic capa-
    city of *Pinus taeda* L.   *Photosynthetica. 10*:47.

Sysoev, A.F., and T.S. Krasnaya. 1967. Changes in activity of certain oxidative enzymes in wheat seedlings during low temperature hardening. *Doklady Akad. Nauk SSR* *173*:472.

Taylor, A.O., and A.S. Craig. 1971. Plants under climatic stress. II. Low temperature, high light effects on chloroplast ultrastructure. *Plant Physiol.* *47*:719.

Tieszen, L.L. and J.A. Helgager. 1977. Genetic and physiological adaptation in the Hill reaction of *Deschampsia caepitosa.* *Nature.* *:219.*

Tieszen, L.L., and D.C. Sigurdson. 1973. Effect of temperature on carboxylase activity and stability in some Calvin cycle grasses from the Arctic. *Arctic Alp. Res.* *5*:59.

Treharne, K.J., and J.P. Cooper. 1969. Effect of temperature on the acitivity of carboxylases in tropical and temperate *Gramineae.* *J. Exptl. Bot.* *20*:170.

Weintraub, M., H.W.J. Ragetli, and E. Lo. 1972. Mitochondrial content and respiration in leaves with localized virus infections. *Virol.* *50*:841.

Williams III, G.J., and P.R. Kemp. 1976. Temperature relations of photosynthetic response in populations of *Verbascum thapsus L.* *Oecologia.* *25*:47.

Wilson, D.. 1972. Variation in photorespiration in *Lolium.* *J. Exptl. Bot.* *23*:517.

DISCUSSION: WHOLE PLANT GAS EXCHANGE AND ACCLIMATION IN LICHENS

D. W. Larson

Department of Botany and Genetics
University of Guelph
Guelph, Ontario, Canada

Photosynthetic acclimation to temperature in higher plants results in a shift in the optimum temperature for maximum nat photosynthesis in the direction of the altered temperature. The photosynthetic capacity is rarely maintained, however: cold pretreatments greatly reduce maximum photosynthetic rate and increase rates of dark respiration. Warm pretreatments have the opposite effect. In lichens, however, the photosynthetic capacity is maintained despite shifts in temperature optima. The significance of this difference is discussed.

When higher plants are exposed to low temperature a series of effects is seen ranging from altered shoot and root growth, to increase in insoluble carbohydrate levels in leaves, to delayed flower development, to reduced photosynthetic capacity. Most authors have interpreted the latter to be an adaptive response which increases carbon gain under a low energy environment. It is argued here that such altered photosynthetic responses may not be adaptive at all, and may be merely an indication of the interconnection between other changes induced by cold and the rate of photosynthesis. The lichens discussed here do not show effects due to cold treatments, other than acclimation. The process of acclimation may thus be adaptive for these and other lower plants.

INTRODUCTION

The situation in lichens with respect to whole plant
gas exchange and the potential for acclimation is somewhat
different from that found in higher plants. These differences may cause one to question whether or not the process
of acclimation as it is observed in other systems, reflects
a character which imparts a great deal of survival value to
the organisms that exhibit it.

There are at least three important aspects of this
problem to consider:

1. the degree of compensation of absolute rates of
    net photosynthesis exhibited by warm versus cold
    pretreated plants;

2. the ability to induce acclimatory changes without
    any simultaneous effects on cold-hardening, growth,
    or other developmental phenomena; and

3. the increase in survival potential in plants which
    show an ability to acclimate over those which
    appear to be unable to do so.

DEGREE OF COMPENSATION

The "degree of compensation" refers to the ability of
the plant under consideration to maintain the same photosynthetic capacity under differing temperature pretreatments. In most cases where plants of temperate distribution
have been examined, the response of net photosynthesis to
temperature has shown a marked downward shift in the optimum
temperature for maximum photosynthesis when the plants are
cold pretreated (Mooney and West 1964; Strain and Chase
1967; Mooney and Shropshire 1967). This shift is only
apparent, however, when relative rates of photosynthesis are
examined (Figure 1). If absolute rates are considered, most
published work shows a pronounced depression in the maximum
photosynthetic rates (Figure 2). Although some studies
(Billings, et al. 1971; Smith and Hadley 1974) show higher
than expected rates of net photosynthesis in cold pretreated
plants, this can be explained on the basis of marked rate
changes in dark respiration in the experimental material.

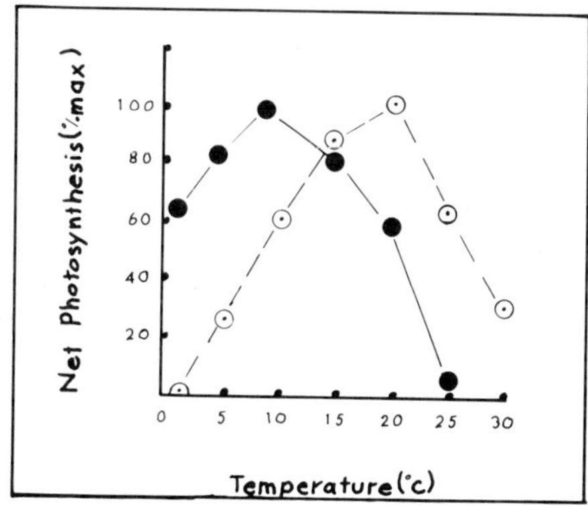

FIGURE 1. *A diagrammatic representation of the effect of cold (●) versus warm (⊙) temperature pretreatments on the pattern of net photosynthesis in higher plants when the rates are expressed as a percentage of the maximum rate under each condition. See Mooney and West (1964) Strain and Chase (1967) for details.*

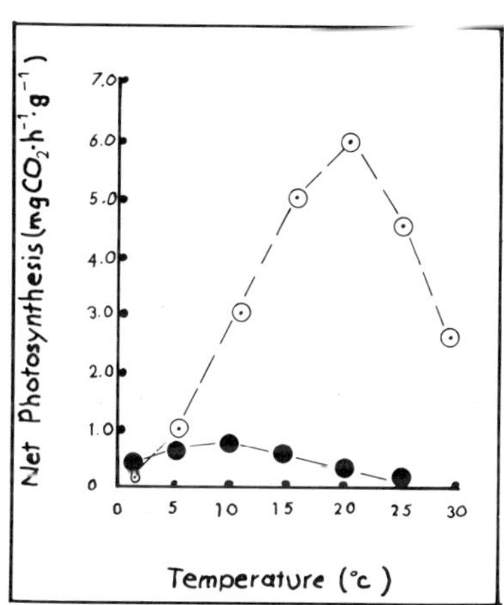

FIGURE 2. *When results such as those shown in Figure 1 are replotted as absolute rates, a marked depression of photosynthesis is often seen. These arate changes cause one to question whether such patterns are adaptive. Cold pretreated plants (●), warm pretreated plants (⊙). See Mooney and West (1964), Strain and Chase (1967) for details.*

The limited data for lichens differs from this pattern. In three subarctic (Larson and Kershaw 1975; Kershaw 1975) and two temperate species (Kershaw 1977), adaptive shifts in optimum temperatures for maximum net photosynthesis have been found under different environmental pretreatments. These changes consistently occur without a significant change in absolute rates of photosynthesis at the different optima, and with no apparent changes in rates of dark respiration (Figure 3). Thus, in these systems it appears that a type of photosynthetic homeostasis is observed which is not found in other plants.

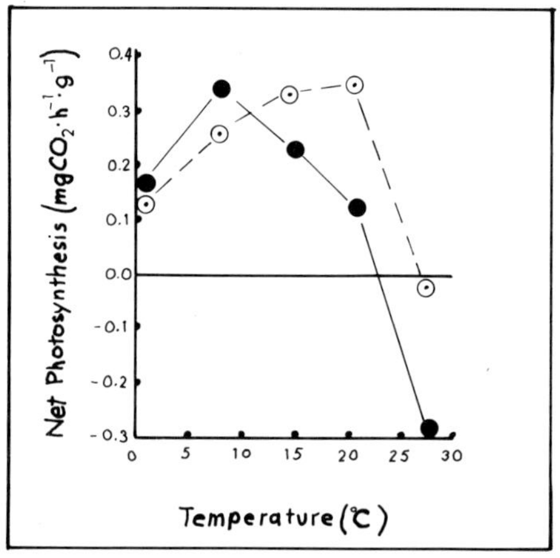

FIGURE 3. *The response of net photosynthesis of cold (●) versus warm (⊙) pretreated plants of the lichen* <u>Alectoria</u> <u>ochroleuca</u>. *Maintenance of absolute rates under the different conditioning temperature is evident (Larson and Kershaw 1975).*

ACCLIMATION OR OTHER PROCESSES?

In higher plants which are undergoing stress-induced acclimation, the normal processes related to growth and development including such phenomena as shoot and root growth, translocation of photosynthates, flowering, senescence, and cold hardening may be occurring simultaneously. Thus, it becomes exceedingly difficult to argue that altered optimal temperatures for maximum photosynthesis are due to the influence of acclimation alone, when similar patterns of physiological change may be induced by other factors singly, or in combination (Sawada and Miyachi 1974). For situations in which acclimation is believed to be operative in the actively photosynthesizing leaf tissue, the influence of other factors on the gas exchange of the whole plant makes it almost impossible to detect changes in the pattern of gas exchange due only to acclimation (DePuit and Caldwell 1973). In lichens, however, little evidence of a seasonal pattern of hardening potential has been found. It appears more likely that (at least for temperate and arctic species) the plants are in a perpetual state of frost resistance. In addition, since the growth and development of these plants is essentially opportunistic, and since phenological changes on a seasonal basis are absent, there is very little possibility of feedback between rates of photosynthesis/respiration and these other factors. To put it very simply, in higher plants the organs that can be most expected to acclimate (the leaves), are constantly modifying the supply of carbohydrate to the rest of the plant, and are themselves influenced by changes in carbohydrate output and by hormonal controls from other areas of the plant. In lichens, however, the leaf-analogue (the thallus) represents the entire plant. Thus the study of whole-plant gas exchange and acclimation in lichens involves a a consideration of the same tissue. Thus, while DePuit and Caldwell (1974) report that seasonal acclimation in *Artemesia tridentata* was obscured after early spring by the effects of growth and development on the pattern of photosynthesis, no such problem is found when lichens are studied (Larson and Kershaw 1975). A similar advantage is found if mosses are treated in a similar way (Oechel 1976).

SURVIVAL VALUE OF ACCLIMATION

Even if the pattern of gas exchange shown in Figures 1
and 2 is accepted as representing acclimation, it is difficult to accept that such changes in absolute rates of net
photosynthesis are terribly beneficial to the plant. It is
not a matter of their gaining extra quantities of carbon
that would otherwise be unavailable if they did not acclimate (as implied by plots of relative rates of net photosynthesis versus temperature), rather it appears to be
simply a matter of limiting the extent of dramatic loss in
carbon fixing capacity under the cold pretreatment conditions. On the basis of the work of Sawada and Miyachi
(1974), it appears as likely that plants showing the pattern
in Figures 1 and 2 are injured, as they are adapted, to the
altered conditions. The suggestion that all patterns of
physiological activity exhibited by plants are adaptive, is
a very weak one. As Oechel (1976) has recently suggested,
it is possible that the lack of correlation between environmental temperature averages, and the commonly found temperature optima in arctic mosses, reflect poorly adapted
organisms. If it is possible that C_3 plants are poorly
adapted to the entire range of temperatures to which they
may be exposed, then perhaps the various reports of acclimation in higher plants reflect such a lack of adaptation.
In no case, has a higher plant been shown in a comparative
carbon balance study, to be dependent on the potential for
acclimation. On what basis then, can it be assumed that
such changes are adaptive?
Lichens on the other hand, appear to represent a
different situation. In most cases to date (albeit the
number of cases is small) photosynthetic homeostasis is
evident. This suggests that under a variety of environmental conditions the plants are able to efficiently adjust
and maximize carbon gain. A second point, is that Larson
and Kershaw (1975) were only able to predict (in general
terms) the distribution of two subarctic lichens when their
differing potentials for photosynthetic acclimation were
considered. In other words, the differing ability to
acclimate appeared to influence the field distribution of
the plants. This suggested that there was immediate survival value in the potential for fully compensating acclimation.
It is possible that this potential is more fully
developed in this group of plants than in most others; the
fact that lichens predominate in arctic and antarctic regions,

and have in some cases (Lange and Kappen 1972) been shown to have optimum temperatures for maximum photosynthesis at 0°C, adds credibility to this argument. Also, the fact that the plants cannot rely on other means of dealing with environmental variation through the storage of photosynthates in roots or stems (as higher plants might), may account for what may be an essential difference between lichens and other plants in their acclimation potential.

CONCLUSIONS

This brief discussion can only serve to point out a number of differences between lichens and higher plants with respect to whole plant gas exchange and the potential for acclimation. These relatively simple plants exhibit an almost completely compensating acclimation mechanism while higher plants do not. Lichens do not show growth or development related effects on net gas exchange patterns, while higher plants have patterns of response to temperature continuously altered by the effects of growth and development as well as acclimation. Additionally, in lichens it would appear the ability to acclimate has indeed an immediate survival value, while in higher plants it is difficult to show that the acclimated plants are benefitting from the altered response surfaces.

REFERENCES

Billings, W.D., et al. 1971. Metabolic acclimation to temperature in arctic and alpine ecotypes of *Oxyria digyna*. *Arctic Alp. Res. 3*:277-298.
DePuit, E.J., and M.M. Caldwell. 1973. Seasonal patterns of net photosynthesis of *Artemesia tridentata*. *Am. J. Bot. 60*:426-435.
Kershaw, K.A. 1975. Studies on lichen-dominated systems. XlV. The comparative ecology of *Alectoria nitidula* and *Cladina alpestris*. *Can. J. Bot. 53*:2608-2613.
Kershaw, K.A. 1977. Physiological-environmental interactions in lichens. 11. The pattern of net photosynthetic acclimation in *Peltigera canina (L.) Willd var. pruetextata* (Floerke in Somm.) *Hue and P. polydactyla* (Neck.) *Hoffm. New Phytol. 79*:387-390.

Lange, O.T., and L. Kappen. 1972. Photosynthesis of lichens
from Antarctica. *Antarctic Terrestrial Biol.*
20:1-19.
Larson, D.W., and K.A. Kershaw. 1975. Studies on lichen-
dominated systems. XVI. Comparative patterns of net CO_2
exchange in *Cetraria nivalis* and *Alectoria ochroleuca*
collected from a raised beach ridge. *Can. J. Bot.*
53:2884-2892.
Mooney, H.A., and F. Shropshire. 1967. Population variability in temperature-related photosynthetic acclimation. *Oecol. Plantarum. 2*:1-13.
Mooney, H.A., and M. West. 1964. Photosynthetic acclimation
of plants of diverse origin. *Am. J. Bot. 51*:825-827.
Oechel, W.C. 1976. Seasonal patterns of temperature response
of CO_2 flux and acclimation in arctic mosses growing *in
situ. Photosynthetica. 10*:447-456.
Sawada, S., and S. Miyachi. 1974. Effects of growth temperature on photosynthetic carbon metabolism of green
plants. 1. Photosynthetic activities of various plants
acclimatized to varied temperatures. *Plant & Cell
Physiol. 15*:111-120.
Smith, E.M., and E.B. Hadley. 1974. Photosynthetic and
respiratory acclimation to temperature in *Ledum gro-
enlandicum* populations. *Arctic Alp. Res. 6*:13-27.
Strain, B.R., and V.C. Chase. 1967. Effect of past and
prevailing temperatures on the carbon dioxide capacities of some woody desert perennials. *Ecology.
47*:1043-1045.

X.   PROSPECTS OF HORMONAL MECHANISMS IN COLD ADAPTATIONS OF ARCTIC PLANTS

George G. Spomer

Department of Biological Sciences
University of Idaho
Moscow, Idaho

No reports of hormones or hormonal mechanisms in arctic plants are found in the literature. Consequently, there is little to review in this regard. But the question of whether hormonal mechanisms might be expected to play a significant role in the adaptations of arctic plants to cold conditions should be considered, and the following review and discussion is directed toward this goal. Justification for investigating or not investigating hormonal mechanisms in arctic plants could then be made upon the basis of at least three lines of evidence from the literature. The first type of evidence is that suggesting that arctic habitats are cold in such a way that coordination in most temperate plants would be seriously disrupted. A second type of evidence would be in the relation between temperature and hormone activity or hormone activity and cold tolerance. And the last type of evidence would be to indicate that arctic plants might differ in their production of hormones under cold conditions when compared with the "norm," i.e., primarily temperate plants.

It might be pointed out, the potential significance of hormones in adaptation of arctic plants goes well beyond any interest in the basic questions of plant ecophysiology. Hence with our ability to manufacture most hormones or analogs of hormones, information about mechanisms could be useful in protecting, managing, or restoring native vegetation of the tundra or in developing cold resistance in agronomic species that might be useful in such regions. What then is indicated in the literature relative to the question?

PLANT TEMPERATURES AND COORDINATION OF GROWTH AND DEVELOPMENT

The general notion that arctic habitats are relatively cold, appears to be based upon correlations with air temperatures, which seldom average more than $10^{\circ}C$ even during the warmest month (Daubenmire 1954). This is misleading because shoot issue temperatures, at least during the daytime, often exceed air temperatures sometimes by as much as $20^{\circ}C$ or more (Stoutjeskijk 1970; Salisbury and Spomer 1964; Warren-Wilson 1957a; 1957b; 1959). Also, low night time tissue temperatures obtain during the growing season in many other habitats that support plants normally excluded from the tundra (e.g., timbered high mountain valleys). But all tundra habitats (alpine as well as arctic) appear to be characterized by a persistant actual or "physiologic" permafrost at relatively shallow depths (Spomer 1964; Salisbury et al. 1968; Bliss 1971; Bliss et al. 1971), and temperatures within the rhizosphere are usually relatively cold, commonly averaging less than $15^{\circ}C$ even during the warmest part of the growing season (Salisbury et al. 1968; Spomer and Salisbury 1968). Therefore, roots of tundra plants must be active under conditions that would usually be expected to impose dormancy or otherwise severely restrict activity in most temperate plants. In essence, we have a condition where the shoots may be experiencing temperatures similar to their counterparts in temperature regions during the growing season, but roots are being exposed to colder temperatures, corresponding more to dormant-season conditions in most temperate regions.

This situation could pose problems in coordination between shoots and roots, and since hormones are important agents in such coordination, one might expect that adaptation by arctic plants to cold soils could involve compensating alterations in hormone synthesis or action. Thus, based upon environmental conditions, it appears that tundra conditions would select against "normal" temperate hormonal patterns.

HORMONAL ACTION AND COLD

Hormones are essentially compounds that function as chemical messengers in coordinating activities throughout multicellular organisms--especially in response to environmental stimuli. Hormones have been found to influence the ecology of many animal species, but their role in plant

ecology is relatively unexplored. Part of this stems from
the difficulty of studying their effects as, unlike most
animal hormones, plant hormones often affect the site of
production as well as other target tissues, and they often
elicit a broad range of overlapping responses. There is
also some suggestion that plant hormones may be stored and
redistributed from tissues other than those in which they
originated. Because of these traits, there has been a great
deal of confusion in attempting to categorize plant hormones
beyond their basic chemical structures (Figure 1). But
there does seem to be some consistency in the general "re-
sponsibilities" of each major type of hormone that is
useful to recognize in considering their possible ecologic
effects. Thus, gibberellins (GA) seem to be primarily
associated with triggering and maintaining a "youthful"
state of activity. Cytokinins appear to function more in
coordinating development and differentiation, while auxins
apparently control many growth reactions to environmental
stimuli (including wounding) through their effects upon cell
wall development--often in concert with ethylene. The
β-inhibitor group, especially abscisic acid (ABA), seems to
be primarily associated with stress avoidance or allevia-
tion. And finally, ethylene appears to control sescence
associated both with normal life cycle development (e.g.,
petal fading and fruit ripening) or severe trauma. With
this brief introduction, we may proceed with the questions
at hand, i.e., does cold affect hormonal activity, or on the
other side of the coin, do hormones affect cold tolerance?
Since so little work has been done with native plants,
effects and possible ecologic significance must be inferred
from experiments with domestic plants.

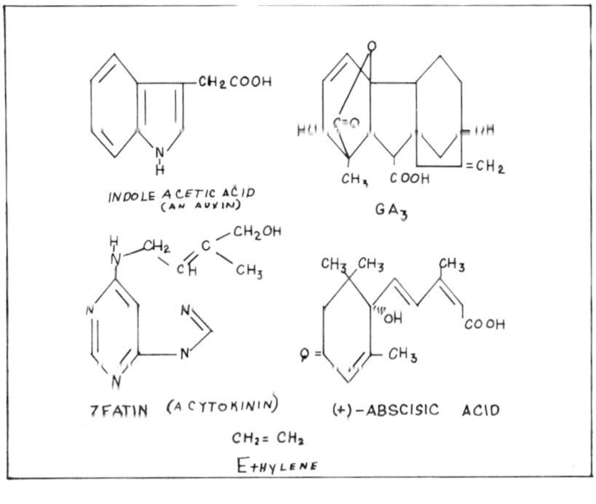

CH₂COOH

INDOLE ACETIC ACID (AN AUXIN)

GA₃

CH₂OH / CH₃

7EATIN (A CYTOKININ)

(+)-ABSCISIC ACID

CH₂= CH₂

E+HYLENE

*FIGURE 1. Examples of the five major types of plant
hormones.*

EFFECTS OF COLD ON HORMONAL ACTIVITY

Hormonal activity may be regulated by tissue temperatures through a combination of both changes in internal concentrations of active forms and changes in the effectiveness at a given concentration of active form. Although these are treated separately for the purpose of discussion, most studies in fact do not clearly differentiate between them. But this should not alter any general conclusions made concerning cold affects upon overall activity. Also because the situation between arctic and temperate habitats do not seem to differ significantly with respect to the dormant period, effects of cold upon hormones regulating vernalization and stratification processes have not been included in the considerations.

There is evidence of temperature effects on hormonal activity. For example, the literature indicates that GA activity usually decreases with decreasing soil temperatures based upon xylem exudate studies of corn (Atkin, Barton, and Robinson 1973). Low soil temperature also produced reductions in cytokinin activity of root exudates (Atkins, Barton, and Robinson 1973), but cool soils may also lead to accumulation of cytokinin activity in the leaves, apparently as a result of reduced cytokinin metabolism (Menhenett and Wareing 1976).

Changes in activity of ether-extractable auxins in response to tissue temperatures on the other hand, seems to vary depending upon the plant species used (Heide 1972). ABA activity is affected by cold, just the opposite to that of GA; i.e., higher levels are detected in exudates at lower temperatures (Atkin, Barton, and Robinson 1973). Ethylene levels do not seem to be directly affected by temperatures except when the cold injures tissues (Wright 1974).

Again, in most reports the mechanism by which cold affects "activity" levels is not clearly delineated, but with respect to auxins, Audus states that changes in auxin synthesis are probably not responsible for the observed temperature effects on this hormone (Audus 1972). On the other hand, low root temperatures are reported to greatly affect basipetal transport of auxin (Audus 1972; Scott and Most 1972; Wilkins and Cane 1970), and we might expect similar retardation for the other hormones where they rely upon active transport. Consequently, some of the observed effects with respect to activity and temperature may be the result of changes in transport rates. This is not likely, however, in instances where temperature effects on auxin and

cytokinin levels in detached *Begonia* leaf regeneration have
been noted (Heide 1965, 1972). It was also concluded from
these experiments, that the differences produced by various
temperatures were not the result of alterations in degradation rates since there were no differences in the rates of
breakdown of labelled exogenous hormones in the tissues.
But there is little direct information about the turnover of
endogenous hormones in plants.

There is also evidence that temperature may affect the
interconversion between forms of hormones, thereby changing
activity levels. Low temperatures, for example, have been
found to increase "bound" forms (glucosides) of GA (Alden
and Hermann 1971; Aung, DeHetogh, and Staby 1969) in some
plants. In one report (Skene and Kerridge 1967), temperature conditions affected the range of cytokinin forms in
grape root exudates as indicated by chromatographic bands.
Yet in this case, such changes did not seem to produce any
variation in the total activity in the exudates.

Based upon the literature, then, temperature does
appear to affect concentrations of hormonal "activity",
albict this may occur by various modes. This seems to imply
that a plant native to the lower latitudes (except for
alpine regions), would experience significant shifts in
activities of GA, cytokinins, and ABA (and possibly auxins)
under normal arctic growing conditions. GA and cytokinins
would generally decline in the tissues (especially in the
roots), while ABA would tend to increase. Auxins could also
decline in the roots due to slowed transport, which might
lead to an accumulation in the shoot tissues. The result,
if such an individual survived, would seem to be a very much
stunted plant in which development was greatly retarded, and
which would be dormant or nearly so for much of the growing
season. This in turn, suggests that adaptation would require some alteration in hormonal mechanisms as affected by
cold conditions, particularly root temperatures.

Some of the affects of temperature upon hormonal concentration, however, might be augmented or amoliorated by
concommitant changes in effectiveness. Again, there is not
much information, but some reports indicate that GA activity
at a given concentration may be reduced at lower temperatures (Goodwin and Carr 1972), and especially with low soil
temperatures (Spomer 1964; Menhenett and Wareing 1976).
Other hormones including auxin (Heide 1965; Jain and Nanda
1972; Philipson, Hillman, and Wilkins 1973; Trewavas 1976).
ABA (Berrie 1968), and coumarin (Khan 1968) apparently also
lose effectiveness at the lower temperatures, but there is
no indication that temperature affects either the action of
cytokinins or ethylene. Unfortunately, it is not possible

to say if the amount of reduction is the same in the effectiveness of the various hormones with a given temperature decrease, so it is difficult to say that the activity of ABA decreases more rapidly than that for GA, which in essence would tend to offset changes produced in concentrations of these two hormones.

HORMONES AND COLD TOLERANCE

Another aspect to consider in the adaptation to cold using hormonal mechanisms, is that levels of hormonal activity may influence cold-tolerance in such a way that "normal" levels might lessen the ability of a plant to survive tundra conditions. If this is the case, then some differences could be expected between arctic and temperate plants in the way they handled hormones.

There is considerable evidence that at least two types of hormones are important in controlling cold-hardiness as indicated by applications of GA_3 (Alden and Hermann 1971; Irving and Lanphear 1968; Rikin et al. 1975) as well as by correlation of endogenous GA levels with the degree of hardiness (Irving and Lanphear 1968). It is interesting to note that the amount of intolerance generated by GA may vary somewhat throughout the growing season (Proebsting and Mills 1974).

GA inhibitors and compounds like ABA that tend to counter the actions of GA have the opposite effect and generally increase cold hardiness. Again this has been verified through applications of these compounds (Alden and Hermann 1971; Irving and Lanphear 1968; Riking et al. 1975; Rikin and Richmond 1976) and by monitoring changes in ABA levels following induction of cold-hardiness (Irving and Lanphear 1968). Furthermore, where cytokinins increase hardiness, it is felt that the response is due to "mobilization" of hardiness promotors (presumably ABA-like compounds) in such tissues (Alden and Hermann 1971).

In light of such effects, it would be tempting to suggest that hormonal mechanisms associated with adaptation are rather simple, *viz.*, cold temperatures reduce GA activity and increase ABA activity resulting in a hardened, albiet stunted plant. Such a possibility should be investigated. But such conditions and, presumably, the resulting hormonal balance, often leads to dormancy or death in many temperate plant species, which suggests that there may be more to it. In addition, the action of hormones in certain

native plants as described below, also indicates that GA and
ABA, respectively, may be acting on separate systems in the
cell so that the actions of one may not totally interfere
with those of the other. All of this then, increases the
justification for more thorough investigations into the
nature of hormonal mechanisms in arctic plants.

HORMONAL MECHANISMS IN NATIVE PLANTS

One of the characteristics of almost all tundra plants
is their dwarfed growth habit. Dwarfing is also evident in
certain agronomic plants, where it is generally attributed
to the lack of GA (Goodwin and Carr 1972). Yet there are
dwarf strains of wheat that do not respond to added GA, and
indeed even the aleuron layer of the grains seem less re-
sponsive in terms of GA induced amylase activity (Gale and
Marshall 1973). GA also fails to elicit growth in certain
steppe cushion plants and alpine cushion plants growing
under field conditions (Spomer 1964) as well as orchard
grass growing in cool soils (Menhenett and Wareing 1976).
Interestingly, alpine cushion plants and orchard grass do
respond with increased growth to GA in warm soil conditions.
GA has also been found ineffective in inducing the main
apical meristem of certain alpine perennial rosette plants,
such as *Geum rossii* (R. Br.) Ser., to bolt (Spomer and
Salisbury 1968).
Such observations might be explained by high levels of
ABA activity, either due to the direct inhibition of stem
growth, or as in the formation of rosettes in the cultivar
Hairy Peruvian of alfalfa (Rikin et al. 1975) or as a
result of formation of GA-glucosides (bound forms), which
are thought to be inactive storage forms of these hormones
(MacMillan 1974). Yet in one study (Spomer and Salisbury
1968), relatively high levels of GA$_3$ applied early in the
season to alpine cushion plants prevented the onset of
dormancy some 60 days later, even though it did not promote
any growth in the same individuals. This suggests that the
GA was not totally inactive.
Because alpine cushion plants and orchard grass do
respond to GA applications in warmer conditions, it is
possible that something other than GA is limiting in cooler
situations. One possibility is that cytokinin levels in
shoot tissues may be reduced due to low soil temperatures.
Cytokinins often originate in the roots, and in many plants
both GA and cytokinins are required to produce growth. But

applications of cytokinins such as benzyladenine (Menhenett
and Wareing 1976) or zeatin (Dawes and Spomer 1976) do not
overcome growth inhibition in cold conditions. Furthermore,
many alpine plants produce elongated floral shoots derived
from axillary buds, which suggest the general presence of GA
but a difference in response between vegetative and reproductive shoots. Such a mechanism would be an advantage in
tundra situations by enhancing floral display and subsequent
pollination as well as aiding seed or fruit dispersal on
annual organs while preserving the low, better protected
perennial vegetative parts of the plant. The observation of
an occasional sessile flower on dormant *Geum* suggests that
the difference is not one of isolated production or differential transport of hormones. Consequently, there are pre-
cedents for expecting differences in hormonal mechanisms as
part of a cold adaptation syndrome.

ADDITIONAL CONSIDERATIONS

Overall, the preceeding evidence implies that arctic
plants may be inherently different in their ability to
control and cope with levels of hormone activities relative
to cold conditions. One must conclude that studies of
hormones and responses to hormones by arctic plants themselves are needed. Furthermore, it would be desirable in
planning such studies, to have a better understanding of the
basic nature of hormonal action. Unfortunately, little is
really known about the biochemistry of plant hormonal
action or the binding sites involved (Jacobsen 1977; Jones
1973; Milborrow 1974; Schneider and Wightman 1974; Kende and
Gardner 1976). Evidence may be found to suggest the primary
action occurs at the transcriptional, translational, or
post-translational stages, but the studies are far from
conclusive. Recently, Trewavas (1976) presented arguments
to suggest that the primary action of hormones was in altering the permeability of membranes, and that what followed
in terms of enzymatic activity, RNA synthesis, etc., was the
result of changes in ions and other compounds in various
parts of the cell due to these permeability changes. (This
could explain, for example, how GA effects upon membrane
properties could influence cold tolerance (Trewavas 1976;
Lyons 1973; Luttge 1974)).

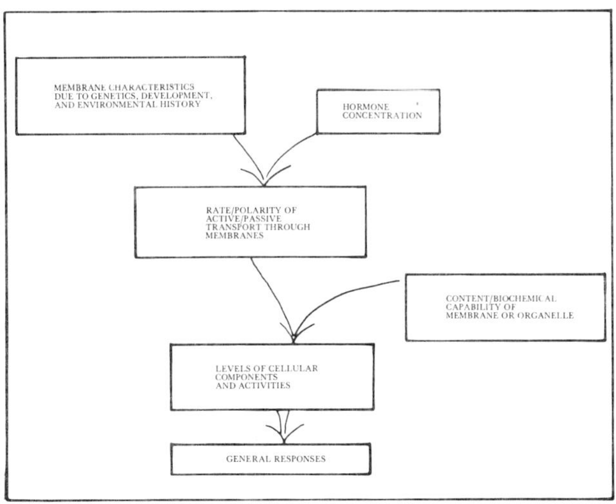

FIGURE 2. Flow chart indicating the various possible major components influencing plant response to hormone levels.

If such a model is correct, "adjustments" may occur due to mutation and selection or acclimation to produce a suitable series of responses relative to hormones at any of three levels (Figure 2). Changes may occur in the biochemistry of the plant to alter the level of hormonal activity. Membrane properties including the number and kinds of hormone binding sites could vary. And finally, the content or potential of the affected membrane cound systems could be altered either in terms of what they might contribute with changes in permeability or in terms of how they responded to alterations in various substances entering. The actual mechanisms could involve one or a combination of these to adjust hormonal activity or regulate response to hormones in a way to increase fitness in the arctic.

While we may be far from knowing what the actual scheme is, it would certainly seem feasible that novel hormonal mechanisms in arctic plants could aid them in adapting to the arctic cold.

REFERENCES

Alden, J., and R.K. Hermann. 1971. Aspects of the cold-hardiness mechanism in plants. *Bot. Rev. 37*:37.

Atkin, R.K., G.E. Barton, and D.K. Robinson. 1973. Effect of root-growing temperature on growth substances in zylem exudate of *Zea mays*. *J. Exptl. Bot. 24*:475.

Audus, L.J. 1972. Plant growth substances. Vol. 1. Chemistry and physiology, 3rd ed. Leonard Hill, London.

Aung, L.H., A.A. DeHetogh, and G. Staby. 1969. Temperature regulation of endogenous gibberellin activity and development of *Tulipa gesneriana L.* *Plant Phusiol.* *44*:403.

Berrie, A.M.M. 1968. The interaction of courmarin and temperature in the germination of lettuce seed. *Physiol. Plant.* *21*:960.

Bliss, L.C. 1971. Arctic and alpine plant life cycles. *Ann. Rev. Ecol. Syst.* *2*:405.

Bliss, L.C., et al. 1973. Arctic tundra ecosystems. *Ann. Rev. Ecol. Syst.* *4*:359.

Daubenmire, R. 1954. Alpine timberlines in the Americas and their interpretation. *Butler Univ. Bot. Studies.* *11*:119.

Dawes, D., and G.G. Spomer. 1976. Univ. of Idaho. Unpublished data.

Gale, M.D., and G.A. Marshall. 1973. Insensitivity to gibberellin in dwarf wheats. *Ann. Bot.* *37*:729.

Goodwin, P.M., and D.J. Carr. 1972. The induction of amylase synthesis in barley aleurone layers by gibberellic acid. I. Response to temperature. *J. Exptl. Bot.* *23*:1.

Heide, O.M. 1965. Interaction of temperature, auxins, and kinins in the regeneration ability of *Begonia* leaf cuttings. *Physiol. Plant.* *18*:891.

_____. 1972. The role of cytokinin in regeneration processes. In Hormonal regulation in plant growth and development. Verlag Chemie, Weinheim, Germany.

Irving, R.M., and F.O. Lanphear. 1968. Regulation of cold hardiness in *Acer negundo.* *Plant Physiol.* *43*:9.

Jacobsen, J.V. 1977. Regulation of ribonucleic acid metabolism by plant hormones. *Ann. Rev. Plant Physiol.* *28*:537.

Jain, M.K., and K.K. Nanda. 1972. Effect of temperature and some antimetabolities on the interaction effects of auxin and nutrition in rooting etiolated stem segments of *Salix tetrasperma.* *Physiol. Plant.* *27*:169.

Jones, R.L. 1973. Gibberellins: their physiological role. *Ann. Rev. Plant Physiol.* *24*:571.

Kende, H., and G. Gardner. 1976. Hormone binding in plants. *Ann. Rev. Plant Physiol.* *27*:267.

Khan, A.A. 1968. Inhibition of gibberellic acid-induced germination by abscisic acid and reversal by cytokinins. *Plant Physiol.* *43*:1463.

Luttge, U. 1975. Co-operation of organs in intact higher plants: a review. In V. Zimmerman and J. Dainty, eds. Membrane transport in plants. Springer-Verlag, New York.

Lyons, J.M. 1973. Chilling injury in plants. *Ann. Rev. Plant Physiol. 24*:445.

MacMillan, J. 1974. Recent aspects of the chemistry and biosynthesis of the gibberellins. In The chemistry and biochemistry of plant hormones. Academic Press, New York.

Menhenett, R., and P.F. Wareing. 1976. Effects of soil temperature on the growth and hormone content of *Dactylis glomerata L.* (Cocksfoot) in controlled environments. *J. Exptl. Bot. 27*:1259.

Milborrow, B.V. 1974. The chemistry and physiology of abscisic acid. *Ann. Rev. Plant Physiol. 25*:259.

Pelton, J.S. 1964. Genetic and morphogenetic studies of angiosperm single-gene dwarfs. *Bot. Rev. 30*:479.

Philipson, J.J., J. R. Hillman, and M.B. Wilkins. 1973. The effects of temperature and IAA concentration on the latent period for IAA-induced rapid growth of *Avena coleoptile* segments. *Planata. 114*:323.

Proebsting, E.L., Jr., and H.H. Mills. 1974. Time of gibberellin application determines hardiness response of 'Bing' cherry buds and wood. *J. Am. Soc. Hort. Sci. 99*:464.

Rinkin, A., and A.E. Richmond. 1976. Amelioration of chilling injuries in cucumber seedlings by abscisic acid. *Physiol. Plant. 38*:95.

Rikin, A., et al. 1975. Hormonal regulation of morphogenesis and cold-resistance. I. Modifications by abscisic acid and by gibberellic acid in alfalfa *(Medicago sativa L.)* seedlings. *J. Exptl. Bot. 26*:175.

Salisbury, F.B., and G.G. Spomer. 1964. Leaf temperatures of alpine plants in the field. *Planta. 60*:479.

Salisbury, F.B., et al. 1968. Analysis of an alpine environment. *Bot. Gaz. 129*:16.

Schneider, E.A., and F. Wightman. 1974. Metabolism of auxin in higher plants. *Ann. Rev. Plant Physiol. 25*:487.

Scott, T.K., and B.H. Most. 1972. The movement of growth hormones in sugar cane. In Hormonal regulation in plant growth and development, Verlag Chemie, Weinheim, Germany.

Skene, K.G.M., and G.H. Kerridge. 1967. Effect of root temperature on cytokinin activity in root exudate of *Vitis vinifera L.* *Plant Physiol. 42*:1131.

Spomer, G.G. 1964. Physiological ecology studies of alpine cushion plants. *Physiol. Plant. 17*:717.

Spomer, G.G., and F.B. Salisbury. 1968. Eco-physiology of *Geum turbinatum* and its implications concerning alpine environments. *Bot. Gaz. 129*:33.

Stoutjeskijk, P. 1970. A note on vegetation temperature
     above the timber line in southern Norway. *Acta. Bot.
     Neerl. 19*:918.
Trewavas, A.J. 1976. Plant growth substances. In Molecular
     aspects of gene expression in plants. Academic Press,
     New York.
Warren-Wilson, J. 1957a. Arctic plant growth. *Adv. Sci.
     13*:383.
_____. 1957b. Observations on the temperatures of arctic
     plants and their environment. *J. Ecol. 45*:499.
_____. 1959. Notes on wind and its effects in arctic-alpine
     vegetation. *J. Ecol. 47*:415.
Wilkins, M.B., and A.R. Cane. 1970. Auxin transport in
     roots. V. Effects of temperature on the movement of IAA
     in *Zea* roots. *J. Bot. 21*:891.
Wright, M. 1974. The effect of chilling on ethylene pro-
     duction, membrane permeability and water loss of
     leaves of *Phaseolus vulgaris.*   *Planta. 120*:63.

XI. RESEARCH SUPPORT FACILITIES ABOVE THE ARCTIC CIRCLE

Gary A. Laursen
John J. Kelley

Assistant Director for Science
Technical Director
Naval Arctic Research Laboratory
Barrow, Alaska

Scientific research and exploration above the Arctic Circle (66°30') have led to intense interest in furthering our understanding of characteristic weather patterns, meteorological phenomena, environmental parameters and the biota, particularly flora and fauna of rare or endangered species and their mechanisms of adaptation to perpetual cold, dry and windswept Arctic habitats. A variety of international interests in the Arctic have led to the development of arctic research stations in seven countries: Canada, Denmark, Finland, Norway, Sweden, USSR, and the United States; all having national boundaries extending above the Arctic Circle. Information is presented on 16 scientific research stations.

INTRODUCTION

During the 19th century, scientific investigations in the Arctic gained notable interest, and exploration often preceded these investigations. Several countries were active in the Arctic; most notably, Denmark (Wood 1967).

Danish scientists have conducted research and exploration in arctic environments in Greenland for over 160 years (Sonesson 1977). Scientific surveys were also undertaken in Iceland during the 18th and 19th centuries by members of the Danish Scientific Society; however, scientific research of significant magnitude in Iceland did not really commence until the early 1900's. Initial research efforts concentrated primarily on the study of agriculture, geology, geography, and meteorology. Studies eventually expanded to include anthropology, archaeology, botany, and zoology.

In 1878 the Danish Scientific Committee on Research in Greenland was established to coordinate, review and follow the research that was proposed and conducted (Peterson 1968). At first, geological investigations were emphasized. During 1888, one of the most significant exploratory travels in Greenland began when Nansen first traversed the Greenland ice sheet (Peterson 1968).

It was not until the early 1900's that other northern countries intensified their interest in the Arctic, when various types of arctic research and exploration gained support. These efforts greaely expanded the knowledge of this region and removed some of the mystery and misconception associated with it. During the early years of arctic exploration, logistic support facilities were few or absent. Research projects had to be self-supporting in the field. Even today, relatively few sites exist that can support multidisciplinary field research efforts.

Currently, most European countries host scientists at research stations or satellite field sites located within various arctic regions. For purposes of this chapter, only arctic and subarctic research support facilities will be considered. This eliminates discussion of many sites maintained at alpine and maritime tundra locations throughout the northern hemisphere. The terms "Arctic" and "subarctic" are not characterized meteorologically or biologically in this discussion. Instead, they are defined by geographical boundaries; arctic regions are above the Arctic Circle. This effort, however, is not exhaustive. The incorporation of additional data are still anticipated from earlier requests made for information.

Although geographically remote from polar latitudes, other nations such as Austria, France, Germany, Great Britain, Italy, Iceland, Ireland, and Switzerland have, and maintain active interest in, scientific investigations in cold regions. However, they do not maintain elaborate field stations of their own where arctic work can be supported. Countries with boundaries, territories, or holdings close to or above the Arctic Circle, also have a long history of Arctic exploration. In most instances, they have all developed research institutions and programs in conjunction with their own national interests and needs.

Recently, the international scientific community has found it advantageous to review and coordinate Arctic research through central planning committees or organizations. An example of this type of coordination is associated with the International Geophysical Year (IGY) of 1957-58 and the International Biological Programme (IBP) in 1969–72. The IBP Tundra Biome study exemplified an international, cooperative, and multidisciplined approach to scientific research.

Some of the more prominent research support facilities operated by nations with strong interests in Arctic research are described.

CANADA

The Canadian Federal Government currently supports and operates two field stations north of the Arctic Circle: Iglooglik (69o20'N) and Inuvik (68o22'N) (Northern Scientific Resource Centres 1977). The construction of three other stations is being planned for Resolute Bay, Whitehorse, and Yellowknife (Lloyd pers. comm. 1978). The National Museum of Science (Ottawa), in conjunction with the Polar Continental Shelf program, also sponsors a station at Bathurst Island.

The Association of Canadian Universities for Northern Studies (ACUNS) currently represents 25 Canadian Universities active in northern studies, research, and training (Fredskild, pers. comm. 1979). Efforts have been made to establish a number of field station sites throughout Canada's northern latitudes and its Arctic Archipelago. Stations presently operational and supported by ACUNS are located on Axel Heiberg Island, at Churchill (Jonkel 1977), Knob Lake (Schefferville), and Rankon Inlet (Hudson's Bay). A site at Frobisher (Lewellen pers. comm. 1978) has also been proposed.

In the past, the Arctic Institute of North America has operated two sizable field efforts at Kluane Lake and Devon Island; the latter was particularly important to the Canadian International Biological Program (IBP) Tundra Biome studies. Devon Island is still used on an occasional basis during spring, summer, and fall, but shifting emphasis and financial restraints have limited the scope of their programs.

Five weather stations are maintained by the Canadian Federal Government and have a satellite support capability. These stations are located at ALERT, Eureka, Isachsen, Moulde Bay (Prince Patrick Island), and Sachs Harbor (Banks Island).

The three principal stations, Igloolik, Inuvik, and
the Devon Island camp are maintained in the eastern, western, and northwestern Canadian Arctic, respectively, to
assist not only government, but university and industrial
scientists as well as the independent scientific community.
The Iglooklik and Inuvik facilities provide an operational
support base throughout the year.

Igloolik, a small island in Foxe Basin northeast of
Melville Peninsula at the entrance to Fury and Hecla Strait,
has a well-protected, deep water harbor with a 2 m tidal
range (Northern Scientific Resource Centres 1977). A
population of 700 (95 percent Eskimo) currently lives on
the island and is supported by a 950 m gravel airstrip with
regular flights. Up to 20 persons can be supported at the
facility. Four temporary field stations at Sarcpa Lake
(Melville Peninsula), Baffin Island (interior), and Baffin
Island, near Penny Icecap, are used and administered from
Igloolik.

The Inuvik station services the western Canadian
Arctic and Mackenzie River Delta area (Northern Scientific
Resource Centres 1977; Western Arctic Scientific Resource
Centres 1977). It is located in Inuvik N.W.T., a community
of about 4,000 people with regular air service. The station is situated 200 km north of the Arctic Circle near the
tree line which demarcates the boreal forest and tundra
biomes. Lakes, mountains, rivers, and marine environments
are characteristic of this area.

Inuvik has a wide, varied research potential and
supports a two-story building that accommodates eight
permanent staff and up to 16 visiting research scientists.
The laboratory is operated by the Northern Coordination and
Research Centre of the Department of Northern Affairs and
Natural Resources. Laboratory utilization is directed by
the Scientific Subcommittee of the Advisory Committee on
Northern Development (Evonuk 1964).

The Churchill Bear Laboratory represents a cooperative
venture with vested interests from private persons, groups,
companies, universities, and Government agencies (Jonkel
1977). It operates primarily under the auspices of the
Manitoba Department of Natural Resources and Transportation
Services. It is also sanctioned by the U.S. Fish and
Wildlife Service, the U.S. Marine Mammal Commission (for
U.S. citizens), the Norwegian Man and the Biosphere (MAB)
Program, the Canadian Wildlife Service, the International
Fund for Animal Welfare Federation (New Brunswick), the
Federal-Provincial Polar Bear Technical Committee, and the

Border Grizzly Technical Committee, under various agreements and terms of grants received by investigators. The primary research thrust is the study of polar bears and their conservation.

The Canadian National Research Council has authorized construction of a new Ice Operations Research Center to be built at the Memorial University at St. Johns, Newfoundland. This center will be devoted to ice-breaking ship research, offshore structures, and safe efficient operations in cold water. It is scheduled to be completed in 1980 at a cost of approximately $50 million.

DENMARK

The Danish Government maintains two research stations in Greenland (Arctic Research Logistics Support Handbook 1977). Southernmost is the arctic station situated near Godhavn (Peterson 1968) on the southern shore of Disko Island off the coast of western Greenland (69°13'N). Access is by air from Sondre Stromfjord or by sea from Copenhagen. The station is attached to the Faculty of Mathematics and Natural Sciences of Copenhagen University. It can accommodate up to 12 guests. Godhavn was established in 1906 by Porsild, a noted botanist and plant systematist.

The operators of the University of Copenhagen Disko Island field research station plan to rebuild the Arctic station commencing 1 July 1979. It will not be in a position to provide the usual logistic support during the summer of 1979, but will be able to support research projects by 1980.

The second and northernmost Danish station is Station Sirus NORD (81°36'N) (Schell 1977). It is located on Greenland's northeastern coast. NORD is maintained by five military personnel whose primary task is to keep the runway operational. Station NORD celebrated its 27th year of Danish occupancy and sovereignty May 15, 1979.

The Danish Government also maintains nine other Greenland sites as unmanned geophysical observatories (Bock and Taagholt; Jensen, et. al. 1976) (UGO). They are lcoated on Carey Island (1975 - 76°38'N), at Daneborg (1975 - 74°18'N), Godhavn, Hall Land (1975 - 81°44'N), Jesup (1974 - 83°38'N). Moltke I and II (1973 - 82°09'N), Narssarssuaq, and NORD. Meteorological observations made are: dew point, soil and air temperature, almospheric pressure, wind speed and direction.

FINLAND

Research in subarctic Finland (Lapland) is primarily
conducted by three universities (Turku, Helsinki, and Oulu)
and eight government research centers. The Finnish government has also established a number of agencies and laboratories (Airaksinen pers. comm. 1977) (Table 1) for supporting various levels of research. Only the Arctic Agricultural Experimental Station, Kevo Subarctic Research
Station, Krunnit Field Station, and the Oulanka Biological
Station in Kuusamo fit the basic criteria for consideration
in this discussion.

The Arctic Circle Agricultural Experiment Station
lies approximately 3 km north of the Arctic Circle and 18
km north of Rovaniemi (66º35'N) on a 975-hectare reserva-
tion (Wood 1967). Fourteen to 16 people work year-round
and up to 12 additional students join the staff during
summer months. Soils, drainage, cultivation, disease,
insect, and a myriad of other agriculturally-oriented
projects are pursued.

The Kevo Subarctic Research Station (69º45'N), established in 1956, is sponsored by the University of Turku,
Department of Botany (Wood 1967). The station's research
emphasis is botanical with prime concern for plant adaptation to ambient thermal and light conditions. Research
interest in zoology and geography in Lapland have had a
decided impact on the station's development. Up to 40
investigators, in addition to the permanent staff, can be
accommodated in the ten-building complex. The station is
located on a 280-hectare site. A growing library and the
international ecotype garden are valuable components of the
station.

The Krummit Station is supported by the University of
Oulu, Department of Zoology, and is located on Finland's
southern islands in the Gulf of Bothnia (Wood 1967).
Brackish water biology constitutes the main research thrust,
and the station can support up to ten people but is well
below the Arctic Circle.

The Oulanka Station in Kuusamo, is jointly sponsored
by the Botany and zoology Departments of the University of
Oulu. Up to 50 students and 10 staff personnel can be
accomodated. A substantial library is also available to
visiting scientists (Wood 1967; Corley 1975; Korhonen
and Wayland No date).

TABLE I. Finland's Research Stations

NAME	ADDRESS
Arctic Circle Agricultural Experiment Station (1938), 66°35'N	727 Apukka 97999 Rovaniemi
Evo Bird Station	Evo
Game and Fish Economic Research Station	Meltaus
Geological Research Institute	96100 Rovaniemi
Geophysical Observatory	99600 Sodankyla
Kevo Subarctic Research Station (1958), 69°45'N, 27°0'E	99980 Utsjoki
Kilpisjarvi Biological Station	99460 Kilpisjarvi
Kolarin (Forest) Research Station	95900 Kolari
Krummit Field Station	
Muddusniemen (Experimental)	99910 Kaamanen
Ore Research Laboratory	Kairatie 56 96100 Rovaniemi 10
Ore Research Laboratory	Kivikatu 6 96400 Rovaniemi 40
Oulanka Biological Station	Kuusamo
Research Institute of Northern Finland	Koskikatu 18 11 96200 Rovaniemi 20
Rovaniemi Forest Research Station	Etelaranta 55 96300 Rovaniemi 30
Sodankylam (Atmospheric) Observatory	99600 Sodankyla
Soderskar Bird Station	Soderskar

(continued on following page)

TABLE I. (continued)

NAME	ADDRESS
Southwestern Archipelago Research Station	*Nauvo*
Varrio Subarctic Research Station	*98820 Varrio*
Water Research Laboratory	*96220 Rovaniemi 20*

ICELAND

In 1919 Iceland's Government authorized the establishment of a scientific research effort. It was not until 1935, however, that a research institute at the University of Iceland was actually established. During 1937, a building was completed and three institutions took up residency. Nearly all scientific research is supported by the government. The Icelandic Museum of Natural History conducts basic research in geology, zoology, and botany. Three small stations are maintained for glaciological and other related research. Basically, the University of Iceland Research Institute, now divided into five institutes, supports all other scientific research. The meteorological office supports an additional 44 synoptic stations, 30 climatological stations and 40 rainfall stations throughout Iceland; its 25 nautical-mile limit along the southeast coast of Greenland, and along air routes to the North American continent and Europe.

NORWAY

Again, as in most northern European countries with interests in Arctic research, university departments and government agencies play a significant role in supporting research programs. There are four major research centers in Norway whose objectives have been strongly oriented toward Arctic research. All are worthy of mention, but do

not fit the concept of field support stations as previously
defined. As late as 1966, the Norsk Polarinstitute in Oslo
was considered by many to be the focal point for Norwegian
arctic research. Its responsibility has been to organize,
support, and conduct arctic investigations.

Bergen has become a center for arctic marine studies.
Trondheim, referred to as Norway's "technical center,"
is the location of the Trollheimen Alpine Station directed
and supported (in part) by the Royal Norwegian Society of
Science and Letters. Marine biological stations are also
found in Trondheim and Agdenes. Tromso, the fourth principal site, supports the Tromso Museum, the Auroral Ob-
servatory, and Weather Control for northern Norway, all of
which have been actively involved in arctic research.

SWEDEN

Sweden's Royal Academy of Sciences, a private organization founded in 1739, is responsible for scientific
research and the operation of six research institutes.
Three observatories in Stockholm, Kiruna (Geophysical), and
Capri (Solar) are maintained. In addition, the Bergius
Botanical Research Station (Stockholm), the Marine Biological Station (Kristineberg), and the Abisko Subarctic
Research Station (Abisko), support a wide variety of arctic
science programs. The Swedish university system plays a
substantial role in promoting and supporting arctic research. The University of Stockholm, for instance, maintains a glaciological field station at Kebnekajsa in Swedish Lapland.

The Swedish subarctic first experienced a strong
impact of economic development from 1898 to 1903. During
this period, the Kiruna-Narvik railroad was built into the
Tornetrask catchment area. This northern tundra site was
not new to human habitation as reindeer domenstication
began in the area approximately 400 hundred years ago.
Economic development brought by the railroad, however, was
new.

The railroad, built primarily for iron ore transport,
frequently brought research scientists into northern Sweden. Dr. Frederik Svenonius (1852-1928), a geologist, had
a great interest in the geology of the northern Scandinav-
ian mountains. His interests stimulated others and the
first research station was constructed at Vassijame, some
30 km west of Abisko. The structure, built in 1903, burned

in 1910. The research station was reconstructed in 1912 at
Abisko. Its present building configuration is pictured in
Figure 1. A road is scheduled to pass through the area and
will service the Abisko Scientific Research Station
(68o21'N) that has been under the direction of the Swedish
Academy of Sciences since 1935 (Royal Swedish Academy of
Sciences 1976). It has since grown to a complex of eight
major buildings.

     The Abisko station (Figure 1) was the site of the
Swedish International Biological Programme conducted from
1970 to 1974. The nature of the research in the Tornetrask
area primarily encompasses the ecology of the tree-line
growth strategies of plants along the chionophitous/chin-
ophobous gradientand the effect of man's impact on northern ecosystems.

*FIGURE 1. Abisko Subarctic Research Station, Sweden.
68o21'N latitude. (Photo compliments of Mats Sonesson)*

UNITED STATES

The only truly arctic lands in the United States are located in the State of Alaska. In Alaska, the area north of the Brooks Mountain Range constitutes a large expanse of arctic tundra and is known as the "North Slope."

This vast, gentle, and northerly dipping coastal plateau features an oligotrophic aquatic habitat covering up to 85 percent of the land's surface of the arctic's low and middle tundra, north of the 10°C July isotherm. Vegetation patterns and species composition repeatedly suggest aquatic influences. The "peat wick" supplies a constant source of surface water during the growing season from a defrosting and receding active layer. Vegetational patterns also reflect influences imposed by geomorphological and physiographical land features that so uniquely characterize Alaskan tundra.

Environmentally, the North Slope of Alaska is a cold desert and lies in the rain shadow of the east-to-west arching Brooks Mountain Range. Precipitation rarely exceeds 150 mm at the northernmost reaches of Point Barrow. Large differences in physical and biological characteristics are manifestations of northern latitude, geological history, climate, low relief landforms, soils, and numerous abiotic and dynamic cycles. Geomorphic processes are dynamic and thus promote substantial soil instability.

Research support for a wide variety of programs is provided by the Naval Arctic Research Laboratory (NARL); the single, largest laboratory of its kind in the world (Frosch 1969; Monson and Sater 1969) (Figure 2).

NARL is the only coastal logistics support base of its kind in the Arctic that the U.S. Federal Government actively maintains throughout the year (Reed 1969; 1971). It is, in fact, the only research support laboratory of its kind throughout the arctic region (above 70°N. latitude) worldwide. Several other northern countries with high latitude arctic tundra within their national boundaries, maintain research stations similar in concept to the NARL facility, but they are smaller and for the most part, operate seasonally.

The laboratory, established in 1947 by the Office of Naval Research (ONR), is owned by the Department of the Navy and operates under a prime contract awarded by ONR to the University of Alaska. Its mission is to provide logistic support, facilities, services, and equipment to the

FIGURE 2.   *The Naval Arctic Research Laboratory as*
seen during the early Spring of 1975.   (Photo by Leslie
Nakashima)

scientific community for accomplishing a variety of basic
and applied research related to arctic problems (Laursen
1978).

Other U.S. Government military services, i.e., Air
Force, Army, Coast Guard, and Navy maintain 42, five, two,
and 10 bases respectively, within Alaska; 24 in Canada; sev-
en in Greenland; and two in Iceland that often lend support
to scientific investigations. None of these military
sites, however, has (as a prime function) a mission to
support scientific research in the Arctic.

The Office of Naval Research has traditionally operated NARL as a national facility rather than one serving
Navy interests alone. Activities have included support of
a wide variety of scientific investigations.

Four large integrated programs have been supported by
NARL in recent years. The U.S. International Biological
Programme (IBP) Tundra Biome Study hosted a series of
innvestigations on the terrestrial and freshwater aquatic
biology of a cold continental environment from 1969 through
1974. The United States represented only one of ten countries working in arctic, alpine, maritime, forest tundras,
and temperate bog sites.

IBP studies evolved into an additional three-year program titled, "Research on Arctic Tundra Environments" (RATE). This terrestrial program consisted of many sub-projects involving several investigators. Research was conducted on a 2,300-acre NARL study area, 65 miles south of Barrow, Alaska near the village of Atkasook on the Meade River. The Meade River field camp is one of the 28 camps NARL uses or maintains throughout the North Slope region for seasonal scientific investigations (Figure 3).

Another large program using NARL logistics support is the Arctic Beaufort Sea studies of the Outer Continental Shelf Environmental Assessment Program (OCSEAP). This program consists of 30 projects and more than 40 investigators addressing environmental problems in the Beaufort and the Chukchi Seas coastal zones, east and west of Point Barrow, Alaska. Arctic studies of the OCS are supported by the Bureau of Land Management (BLM) and are managed by the National Oceanic and Atmospheric Administration (NOAA) through the Fairbanks field office at the University of Alaska.

OCS studies comprise one of the more crucial inter-disciplinary programs currently considering the importance of the Continental Shelf to the United States. Continued emphasis is placed on programs carried out in the shore areas of the barrier island lagoons of the Beaufort Sea.

These shallow-water programs pose difficult logistic problems. In order to provide support, NARL maintains an all-aluminum lightweight warping tug (LWT) (85-foot, 80-ton) which was extensively modified for oceanographic research. The ship, R/V "Alumiak," berths five crew members and four scientists. Unique features of the ship include an ample working deck, shallow draft, and space for two deck-mounted all-weather modular laboratory units. Studies aboard the "Alumiak" concern marine biology, littoral zone fisheries, physical oceanography, and observations of sea birds. The Marine Operations Department also operates a second vessel, the R/V "Natchik."

The Animal Research Facility (ARF) at NARL provides an opportunity to study native species of northern Alaska in a modern and well-equipped laboratory (Laursen and Selby 1978; AIBS Report 1977). This facility has helped to establish and maintain a life science in-house research capability. An important part of the research is concerned with behavioral and physiological mechanisms of adaptation and acclimatization demonstrated by mammals in the arctic environment. Specific areas of interest include metabolic

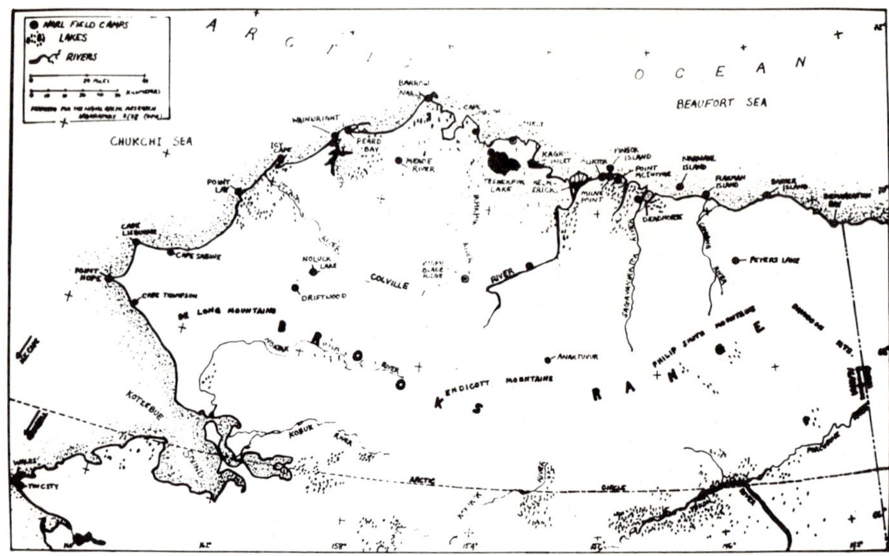

FIGURE 3.   *Naval Arctic Research Laboratory Field Sites and Camps.*

and cardiovascular responses to cold, hematology, body
water and electrolyte regulations, radio telemetry, and
thermoregulatory mechanisms. An understanding of mammalian
physiological mechanisms of cold acclimatization is important in helping mankind cope with living in cold environments.

Most animal species represented at the ARF were originally captured on Alaska's North Slope. Although the
animals are maintained as laboratory specimens, significant
effort has been made to provide them with adequate exterior
living enclosures. They have adjusted well to the laboratory environment with reproductive success noted for
several species. For example, as many as 60 wolves have
been born and raised successfully at the facility; however,
only 20 are presently maintained for research purposes.
Complete clinical records are maintained on all larger
animals which provide researchers with baseline information
on vital statistics, monthly weights, quarterly blood
chemistries, hemotological data, and a detailed history of
the animal's previous role in research.

ARF is comprised of several buildings (15,000 sq. ft.)
situated on a 3.12-acre (1.26 ha) site (Figure 4). There
are 72 exterior cages for housing 183 animals of 13 species. A staff of six, including a full-time resident
veterinarian, maintain the facility. Buildings within the
facility consist of one enclosed high-rise observation unit
that overlooks fox and wolf runs, six buildings dedicated

FIGURE 4. Naval Arctic Research Laboratory, Animal Research Facility (ARF). (Aerial photograph by Arne Hanson)

to indoor animals enclosures, an aviary, an ambient temperature laboratory, and five large freezers for animal food storage. The ARF also contains a food preparation kitchen, cage-washing facility, a clinical lab, surgery suite, metabolic and telemetry laboratories, and a fully-equipped X-ray room. Ample laboratory and office space is available for resident and visiting scientists. The facility also provides veterinary service to the native community.

The ARF research program not only maintains a research veterinarian, but includes a visiting scientist program, research technician, and a postdoctoral program with two participant positions. The facility is open all year. Support and funding is provided by Dr. Arthur B. Callahan, Director of the ONR Biophysics Program Office. During 1978-79, research programs resulted in over 31 technical presentations and publications.

Because of its location, the ARF offers unique opportunities for studies on arctic acclimatized and adapted animals. The ARF is the only facility of its kind in North America and the largest in the world where northern species of mammals are maintained in an arctic environment with its unique temperature and light regimes. These conditions cannot be duplicated in artificial light/dark rooms or temperature chambers.

A new program, "Project Whales," supported by the Department of the Interior's Bureau of Land Management (BLM), commenced at NARL in 1978. This program is designed to investigate impacts of oil and gas field development on principally endangered whale species frequently sighted in or near the lease areas.

NARL houses almost 200 persons year-round. Slightly over one-half of the staff are represented by the Operation and Maintenance subcontractor, ITT/Arctic Services, Inc., a subsidiary of ITT. The remaining staff represents the prime research support contractor staff of the University of Alaska (UA). Additional staff is provided through a variety of contracts, research, and training grants.

The laboratory interacts with the surrounding community not only through employment opportunities but by providing training and information services. A monthly newsletter (NARL NEWS) is distributed. A television series on polar science topics was produced in cooperation with the North Slope Borough School District in 1978, and on-the-job training programs were initiated with support from the U.S. Department of Labor and the Alaska Federation of Natives Comprehensive Employment Training Act (CETA). The primary purpose of the CETA program is to afford opportunities for community residents to gain experience by working in the various laboratory departments.

The library staff consists of one full-time librarian and one part-time employee, both having graduate degrees and experience in library science. Regional emphasis of the NARL library collection is primarily polar (e.g., the Arctic), and Alaska in particular. Subject emphasis is in natural history, snow and ice, geology, oceanography, and anthropology. The library is open daily throughout the year with service as needed. Interlibrary loan service is available. The library maintains a Barrow File, a collection of reports and papers on research conducted by scientists supported by facilities of NARL. The library subscribes to 198 periodicals which cover Alaska, arctic and polar topics, general science, and the physical and

life sciences. There are approximately 20,000 titles in
the general collection which cover all aspects of arctic
science.

In addition, the library has a collection of topographic, hydrographic, and miscellaneous subject maps, with
emphasis on the North Slope and north Alaska coastal regions.

A small reference collection is maintained, including
standard bibliographies of Arctic and Antarctic materials,
special topic bibliographies, and a collection of government reports on microfiche. Miscellaneous materials,
fragile or scarce items, local interest and local history
materials , historical research reports, old photographs,
and special subject collections, constitute special collections.

USSR

Unfortunately, we know very little about Russian
science support facilities in the Siberian Arctic. There
is a report that 96 laboratories exist north of the Arctic
Circle; however, most are reported to be schools in small
villages.

The Russian Academy of Science (Nauk Academia) maintains two Siberian branch research stations at Novosibirsk
and Vladivostok. As many as 11 other stations of varying
size are maintained at Agapa, Ary-Mas, Harp, Kolyma, Maria
Pronchischevo Bay, the Polar Urals, Pronschtschischewa,
Salekhard, Sivaya Maska, Taimyr, and Tareyn. Specific
information was not available at the time of this writing.

ACKNOWLEDGMENT

The authors gratefully acknowledge the financial
support of the Office of Naval Research under contract
number 00014-77-C-003 and the logistical support of the
Naval Arctic Research Laboratory which is operated for the
Office of Naval Research by the University of Alaska.
Appreciation is also extended to the Office of Naval
Research Biophysics Program Office, Code 444 (Grant number
N000-77-C-01-162), and to the numverous agencies and institu-
tions whose research was supported at NARL.

REFERENCES

Abisko Scientific Research Station. 1976. The Royal Swedish
     Academy of Sciences brochure.
AIBS Report. 1977. ONR/AIBS Sponsored Symposium on Arctic
     Environment. *Bio. Sci. 27(6)*:425–416.
AINA Field Station Report. 1978. *Information North*. 4 pp.
Airaksinen, A.R. 1977. Personal Communication.
Arctic Research Logistics Support Handbook. 1977. Prepared
     by AD HOC Working Group on Arctic Logistics IARCC,
     NSF.
Bock, H.C., and J. Taagholt. No date. Greenland and the
     arctic region in the light of defense policies defense
     aims. In People and Defense Series.
Corley, N.T. 1975. Polar and cold regions library resources:
     a directory. Ottawa Northern Libraries Colloquy,
     Rovaniemi, Finland.
Daily News Miner (Fairbanks). 1978. In the arctic. Pages
     10–11 in December 9h Weekender.
Evonuk, E. 1964. Facilities for the study of arctic and
     subarctic biology. In *Fed. Proc. 23*:2298–1201.
Fredskild, B. 1979. Personal Communication.
Frosch, R.A. 1969. The growth of the Naval Arctic Research
     Laboratory. In Proceedings of the U.S. NARL Dedication
     Symposium. APB Manson and J.E. Sater, eds. *Arctic.
     22(3)*:356–364.
Jensen, G., et al. 1976. Unmanned geophysical observatories
     in north Greenland 1972–75. Danish Meteorological
     Institute. ISBN 87 1478 121 9.
Jonkel, C.J. 1977. The Churchill Bear Laboratory.
Korhonen, S., and R. Wayland. No date. Nordkalottens Biblio-
     teksmote. In the Fifth Northern Libraries Colloquy.
     Part I Proceedings, Rovaniemi, Finland.
Laursen, G.A. 1978. Fiscal year 1978 science activity,
     Naval Arctic Research Laboratory, Barrow, Alaska.
Laursen, G.A., and G.M. Selby. 1978. NARL Animal Research
     Facility annual report for FY78. ONR Code 444. Con-
     tract N00014-77-C-0162 Task No. NR 207-117. Technical
     Report No. 2.
Lewellen, B. 1978. Arctic Research, Littleton, CO. Personal
     communication.
Lloyd, T. 1978. ACUNE Executive Director. Personal com-
     munication.
Monson, A.P.B., and J.E. Sater. 1969. Proceedings of the
     U.S. Naval Arctic Research Laboratory. Dedicated
     Synposium, Fairbanks, Alaska of April 9–12, 1969.
     Arctic. 22(3).

Northern Scientific Resource Centres. 1977. By Minister of
    Indian and Northern Affairs. Ottawa, Canada.
Peterson, G.M. 1968. Universitets Arktiske Station, Uni-
    versity of Copenhagen.
Reed, J.C. 1969. The story of the Naval Arctic Research
    Laboratory. In A.P.B. Monson and J.E. Sater, eds.
    Proceedings of the U.S./NARL Dedicated Symposium.
    Arctic. 22(3).
Reed, J.C., and A.G. Ronhovde. 1971. Arctic laboratory: a
    history (1947–1966) of the Naval Arctic Research
    Laboratory at Point Barrow, Alaska AINA. LCCCN 76–
    175103.
Schell, B. 1977. Station Sirus NORD, Denmark's arctic
    outpost. Thule Times 29 April. 8–11 pp.
Sonesson, M. 1977. Abisko Scientific Research Station
    Environment and Research. Paper presented at Fen-
    nosandian Tree-line Conference at Kevo-Abisko, 6–14
    Sept.
Stanka, V. 1958. Institutions of the USSR active in arctic
    research and development, AINA, Washington, DC.
Wood, P.H. 1967. Arctic research in western Europe: a
    directory of institutions. AINA, Washington, DC.

XII. FUTURE TRENDS IN COLD ADAPTATION RESEARCH*

Lawrence S. Underwood
Lawrence L. Tieszen

University of Alaska
Arctic Environmental Information and Data Center
Anchorage, Alaska

Augustana College
Department of Biology
Sioux Falls, South Dakota

A detailed understanding of cold adaptation, especially in the Arctic, has been restricted due to the breadth of the field and lack of communication among researchers. An understanding of cold adaptation requires the synthesis of complex biochemical, physiological, ecological, and behavioral components contributed by researchers who are widely separated both geographically and in their disciplines. Participants in the symposium which precipitated the writing of this volume for example, spanned the continent from northern Alaska to southeastern Canada and from Florida to southern California. The objectives of this volume are: 1) to communicate the latest information available in the field, and 2) to coalesce future research efforts oriented toward the Arctic. The first of these objectives has been met through the preceeding chapters of this volume.

To address future research needs, a workshop was held in conjunction with the symposium. Workshop participants attended the preceeding day's symposium and listened to all paper presentations. While this information was fresh in their minds, participants tackled the often tricky business of recommending future research directions. The results of this workshop was compiled in a report submitted to the Office of Naval Research by the American Institute of Biological Sciences. This chapter reviews the results of that workshop on cold adaptation mechanisms.

*Lawrence S. Underwood summarized research needs for animal studies, and Lawrence L. Tieszen summarized research needs for plant studies.

INTRODUCTION

In a general sense, the need for continued polar
research has been analyzed several times in the past. In
1963 the National Academy of Science, Committee on Polar
Research, recognized that past life science research in
cold dominated regions has been developed opportunistically,
leaving a number of serious gaps in current knowledge. Specifically, in the field of thermal physiology, the report
recognized a number of "colateral studies that should
attract physiologists to the Arctic." In spite of previously intense activity, these suggestions included: 1)
metabolic heat production of individuals under various cold
exposures; 2) thermal economy and exercise; 3) analysis of
insulating devices in naked, hair-covered, and clothed
animals; 4) differentiation of local and temporal variations in temperature of the skin and their integration and
the regulation of body heat; and 5) heterothermal conditions and their influence upon the reaction of tissues and
organisms. In addition, studies of winter hardiness and
rates of biological processes in both plants and cold-
blooded animals were recognized as critical. Since 1963,
some studies have begun to examine these topics in plants,
but animals continue to be virtually unstudied.

In 1970 the National Academy of Science, Committee on
Polar Research, conducted a followup survey. Again a
number of topics were identified that should be "encouraged
and sponsored." These included: 1) physiological mechanisms underlying cold adaptation; 2) low temperature effects
on cells, tissues, and organisms; 3) diving physiology;
4) fish physiology; 5) nutrition, especially vitamin requirements in arctic animals and plants; and 6) comparative
chemotherapy.

In 1972 the Inter-Agency Arctic Research Coordinating
Committee (IARCC) presented a five-year plan for arctic
research. Only three of the member agencies mentioned work
related to cold adaptation. The Department of Health,
Education and Welfare anticipated programs which would,
in part, concentrate on nutritional adaptation of man to
the arctic environment. The National Science Foundation
planned to include some physiological research in its
ecosystem analysis and other extensive studies as a part of
its "Man in the Arctic" program. Interactions between man
and the environment and Inupiat Eskimo nutrition were
stressed. The Department of the Navy, while recognizing
the predominance of physiological research in the early
days of the Naval Arctic Research Laboratory, described no
future plans.

In 1976 the National Academy of Science, Committee on Polar Research, conducted an analysis of the National Science Foundation's arctic programs. Its report mentioned that most recent programs concentrated on ecology and the economics of development in northern Alaska. Relatively little work was being done on physiological topics. Only the Department of Defense (Army) was sponsoring work on cold injury, treatment, and acclimatization in man.

These national surveys concentrated on analyzing the needs of polar science in general. No concentrated efforts have been made to evaluate specific needs for continued cold adaptation until this symposium.

ANIMAL RESEARCH

Questioning Some of the Assumptions of Cold Adaptation Studies in Animals

Past studies of cold adaptation mechanisms have been conducted in the field and in laboratories either on species native to the Arctic, such as brown lemmings, Arctic ground squirrels, Arctic fox, etc., or on such nonarctic species as hamsters and white rats. Additionally, subject animals have usually been adults. Workshop participants felt that rather serious questions can be posed concerning the assumptions on which these studies are based. Is it appropriate to study cold adaptation mechanisms in laboratory species which had evolved in environments where cold tolerance was unnecessary or worse, were they so fundamentally different from "natural" species that they should be considered laboratory artifacts? On the other hand, when conducting laboratory experiments on species recently removed from the wild, how can we separate the effects of cold from the effects of a new and "strange" environment? How much reliance can be placed on measurements of complex physiological mechanisms by primitive field methods? The use of adult subjects poses additional questions. Since natural acclimatization begins at birth, if not before, is it appropriate to concentrate studies on adults, thereby precluding assessment of the importance of growth and development factors? These questions, which cut to the very quick of modern physiological ecology research and may not be limited to cold adaptation studies, should be answered.

Previous studies have often assumed that all species will respond to cold in the same manner. Future studies should recognize that physiological behavior of native and non-native species will likely be different and should utilize a wider variety of experimental species, including native and non-native fishes, birds, and animals. Examination of enzymatic mechanisms are particularly critical in this regard. Relatively few studies have been done comparing cold acclimatization mechanisms in arctic, sub-arctic, and temperate zone species. Research should be conducted to compare the changes seen in arctic vs. temperate hibernators and arctic vs. temperate non-hibernators. Comparisons should also be made of the biochemical, hormonal, and nutrient and energy acquisition mechanisms.

Studies dealing with variations in metabolism also need a general test of assumptions. Intestinal absorption and specific features of active transport mechanisms have been widely studied in rats, hampsters, and a few other laboratory or domestic species. It would be valuable to know if the efficiency of the secum for absorbing small chain fatty acids is greater in arctic mammals than in domesticated laboratory subjects. Is the high metabolic rate of lemmings, other microtine rodents, and shrews reflected in intestinal transport systems in amino acids, monosacharites, etc.? Is the Arctic ground squirrel more efficient in its absorptive capacity during spring arousal or during it's short bout of normothermia? Do the temperature gradients on mucosal and sirrosal surfaces provide clues to the total body heat and temperature tolerance in arctic mammals? Do the thermal properties in the skin of arctic mammals and nonarctic species adapted to cold, differ from those of temperate zone subjects not acclimated to cold.

It would also be instructive to compare responses in a single species over a wide latitude range in order to more fully understand the problems of photoperiod. How are the mechanisms different along geographic gradients in these species? And, how do these changes compare to those seen in species restricted to the Arctic as opposed to those restricted to the temperate zones?

Cold adaptation field studies involve animals naturally acclimatized to winter. Arctic animals adjust not only to seasonal cold, but also to such other potentially important parameters as light conditions, availability of food, etc. Since laboratory studies usually alter only one variable - temperature - many species may not employ the

same mechanisms for laboratory acclimation as they do in
natural settings. For example, white rats acclimated only
to cold often make metabolic adjustments, while those
acclimatized to winter also increase the quality of their
insulation. Future studies should examine the differential
mechanisms by which species acclimate and acclimatize.

Integrated Studies

Questions relating to cold adaptation mechanisms will
not be adequately answered soley by examining individual
needs. Concommittant seasonal, spacial, and thermal adjustments occur in all species, and to a large degree,
influence the direction and magnitude of responses occurring in any given species. Similarly, since interactions
occur between physiological systems that modify and compliment each other, future research in cold adaptation
should simultaneously be conducted at several organizational levels, namely: cellular, species, and ecosystem.
Integration of these studies is important. Arctic animals
could then provide key information for the study of other
basic systems. A major research effort should involve the
development of a set of models which can be used to generate these integrated systems on a seasonal basis. Examples of specific studies which lend themselves to specific integrated approaches follow.

Perturbation Tolerance. Man has become a significant
perturbing element throughout the Arctic. The entire north
polar basin is experiencing large numbers of man-induced
changes, the consequences of which are subtle, diverse, and
potentially serious. Intentional or inadvertent disturbances caused by man tend to amplify the climatic, thermal,
energetic, and nutritive stresses which normally confront
arctic species. When in the Arctic, many of these species
are near the extreme limits of their existence, and the
introduction of additional stress may exceed tolerable
limits. For example, it is now widely recognized that egg
shell formation of predatory birds is affected by the ingestion of insecticides. Insecticides are not currently
utilized extensively in the Arctic; however, many arctic
migrant birds pass through areas in which contaminated
foods are present. Is it possible that ingested insecticides and herbicide chemical contaminants affect egg formation and other physiological processes in arctic species?

Are the physiological effects of these contaminants exacerbated by a cold-dominated environment? Studies of these
and other related phenomenon are currently incomplete and
available pertinent information is inadequate.

   Cellular Mechanisms. Relatively little work has been
done at the cellular level on arctic animals. Yet, such
studies would definitely contribute to large integrated
efforts. For example, it would be quite helpful to integrate studies of nutrient and energy requirements, enzymatic and hormonal mechanisms, and gut physiology and
active transport systems related to arctic animals.
   Neurophysiology is another area that needs study.
Many processes of acclimatization are biochemical, cardiovascular, and behavioral. Alterations induced by cold must
depend on sensory information for their initiation. There
seems to be at least two sources for such information –
warm and cold receptors in the skin, and the correlation of
visual input with previously experienced thermal conditions. Of particular interest, is whether or not thermal
receptors in arctic mammals have properties similar to
those found in nonarctic animals. Furthermore, do thermal
properties in the skin of both arctic animals and nonarctic
animals acclimated to cold, differ from those of temperate
zone, non-acclimated subjects? Can modifications in the
central processing of thermal information be determined or
shown to occur in adapted or acclimatized animals? And,
to what extent do these modifications occur? How do neural
inputs related to environmental temperature govern these
long-term adaptive phenomenon? Are there other nonthermal
environmental cues which govern them? To what extent are
neural patterns of thermal regulatory behavior inherited or
acquired, and what neural input(s) causes their development?

   Temporal Effects. Natural acclimatization to cold
begins at birth, if not before, and factors associated with
prenatal and neonatal growth and development probably play
a major role in the eventual survival and adjustment characteristics in the adult organism. Young animals have an
anatomical and functional plasticity which is useful in
successful adaptation. Natural acclimatization usually
involves a number of environmental variables, and it is
often difficult to precisely define the role of each as
well as the interaction between them. Thus, a considerable

amount of work needs to be done on the ontogeny of cold hardiness and cold resistence. The roles of nutrition, energy requirements and acquisition, and hormonal and enzymatic mechanisms, undoubtedly interact to shape and modify the development of these mechanisms, but how and to what extent?

Many animals native to the Arctic, particularly adults, are subjected to repeated bouts of acclimatization, deacclimatization, and reacclimatization. It seems probable to assume that one bout of acclimatization may influence the response to subsequent bouts. Delineating these carryover effects is a difficult experimental problem, but is nonetheless an important step in characterizing an animal's total adjustment potential. Since most acclimatization studies have used non-native animals, no assessment of growth and development factors has been made. Furthermore, studies are often designed to test only one experimental variable, ignore carryover effects, and do not allow for evaluation of the impacts or interactions of the multiple environmental variables normally found in the arctic environment. Comprehensive experiments on species not native to the Arctic also need to encompass more of these important interacting variables.

Special Single Disciplinary Studies

Nutrient and Energy Acquisition. (See chapters by White and Gessaman this volume for review of background material). Studies of herbivorous, omnivorous, and carnivorous northern animals should be expanded beyond documentation of food habits and food acquisition strategies to give evidence for the existence and selection of specific nutrients and the regulation of their intake. Macronutrients known to be important to northern animals include: nitrogen, phosphorus, calcium, magnesium, sodium, potassium, and sulphur. However, an understanding of the nutrient requirements of animals is based not only on knowledge of diet, nutrition, and general physiology, but also on assessment of the implications of ecosystems affected by the activities of man, and study of the functional aspects of pristine systems.

Selection of habitat by herbivores involves coevolutionary responses of plants and animals. Plant defenses include the utilization of various growth forms, reproductive strategies, special location of storage components,

and the production of secondary compounds which discourage
either specific or generalist herbivores. Herbivores also
evolve mechanisms which allow for exploitation of plants,
even as the plant's defenses evolve. Many such examples
are known from temperate and tropic zones, but the phenomenon is virtually unstudied in the Arctic. Research
should be conducted to provide this information and to
enlarge our understanding of total ecosystem function.

Nutrient acquisition mechanisms in carnivores are
often more specilized than in herbivores and present a
larger spectrum of individual behavior. Similarly, the
problem solved by carnivores in the face of multiple stresses
in hunting conditions; light periods; environmental, thermal, and climatic stresses; and restricted feeding options,
present unique survival circumstances.

A major gap exists in our knowledge of population,
reproduction parameters, and the nutritional status of the
female at conception and during gestation and lactation.
These relationships must be known if we are to understand
how populations respond to changes in food resources. An
understanding of the relative role of energy status as
opposed to the level of nutrient reserves in the female,
will be necessary to answer these questions.

The milk of northern animals has an impressively wide
range of major constitutients, whereas the milk of marine
mammals is more uniformly concentrated and devoid of lac-
tose. The water and protein content in the milk of terrestrial mammals varies greatly. Unifying concepts from
appropriately designed research programs could relate
conception and parturition to the maturation stage of the
neonate at birth, and to the availability and quality of
its food resources, the degree of cold stress, and behavioral patterns associated with predation, avoidance, and
niche exploitation for arctic animals. Adjustment in the
energy and nutrient content of milk in relation to the
balance produced, are undoubtedly variables which are
modified through processes of evolution by a mammalian
species' construction of successful behavorial and growth
patterns. The use of computer modeling in the study of
these phenomenon is especially appropriate and would aid
our understanding of the regulation of gluconeogenesis and
carbohydrate, lipid, protein, and water metabolism in
mammals.

The efficiencies in utilization of metabolizable
energy for maintenance, growth, fat deposition, egg production, and lactation have not been studied in arctic

homeotherms, and are now assumed to be the same as those
for domestic species. Whether this is correct or not is
unknown and should be examined in representative arctic
birds and mammals.

The maximum rates of energy metabolism and thermoregulatory capabilities in arctic homeotherms at rest are
limiting factors at very low temperatures. These rates
have received only cursory attention in one bird species
and in several species of small mammals. These studies
should be expanded to include other birds and to incompass
the study of large mammals acclimatized to both summer and
winter conditions. Young mammals in the Arctic are simultaneously under influences of several environmental stress
sources, and neither their fur, feathers, nor their shivering capability is fully developed. Nonshivering
thermogenesis is a potentially important heat source for
maintaining homeothermia in the young under these conditions. The magnitude of nonshivering thermogenesis in
young arctic mammals should be clearly delineated.

The study of energy utilization for locomotion in
arctic animals (homeotherms and heterotherms) has been
generally neglected. The following areas should be investigated:

1. The metabolic cost of aerial, terrestrial, and aquatic locomotion at cold
   temperatures. (This has been measured
   in only a few arctic species.)

2. The rates of body temperature cooling
   during foraging bouts and the limitation this places on the duration of
   foraging activity.

3. The contribution of heat, produced by
   activity, to the thermoregulatory heat
   requirements of birds and mammals.

4. The seasonal change in metabolic scope
   of arctic fish.

Humans living in the Arctic, particularly native pop
ulations, subsist under nutritional conditions quite different from those living in more temperate zones. On the
basis of available information, we know their diets are
high in protein but may be low in some minerals and vitamins. Many individuals, particularly persons involved in
hunting for food, expend large amounts of energy on a con-

tinuing basis. Subjectively, it appears that these people
are nutritionally well adapted to their environment. Ob-
jectively, however, only limited information is available
to substantiate this impression, since most studies have
been rather superficial. Seasonal nutritional variations
as well as infant and children's diets, are even less well
documented. Conducting nutritional surveys of native
populations is difficult; and technically and sociolo-
gically, the nutritional composition (especially the vitamin and micronutrient components of native diet) com-
plicates evaluation of their nutritional needs.

Basic Physiological Systems. (See chapters by Musacchia, Cooper, and Yousef this volume for review of background information). There are three specific physiological areas which require intensive investigation in arctic
vertebrates -- gastrointestinal, neural, and renal. Intestinal physiology of arctic species and in animals subjected to cold exposure is relatively unstudied. Problems
of increased food intake, energy expenditure, metabolism,
etc., stimulated by environmental cold stress, are generally well recognized. However, despite the knowledge of
"food in-waste out" and "whole body metabolism," it is safe
to say that little is known about how the absorbing surface
functions in cold-exposed subjects and arctic animals. The
absorbtion of nutrients, i.e., glucose, amino acids, fatty
acids, essential ions, etc., is known to be affected by
conditions of stress and by acclimation and acclimitization
to such environmental variables as cold. These latter
events are particularly important relative to meeting nutritional requirements.

All vertebrate groups in the Arctic could contribute
to these studies. In fishes, the effects of temperature
and acclimatization could be readily examined, and functional intestinal absorbtion could be measured *in vitro* sac
preparations or *in vivo* intestinal preparations. The same
experimental approaches could be used in pre- and post-
migratory birds and mammals.

Whereas, temperature effects might be of interest in
fish *per se*, efficiency and competitive absorbtion of
selective compounds and ions might be more practical in
birds. In the latter, calcium absorbtion might well be
investigated because of seasonal egg-laying activity in

many migratory birds. The number of questions is unlimited, but the essential points are listed below:

1. Little is known about gut physiology and active transport systems in arctic animals.

2. Such information is relevant to our understanding of numerous physiological responses related to stress, acclimation, and adaptation phenomenon in the Arctic.

3. Modern laboratory facilities at the Naval Arctic Research Laboratory (NARL) and elsewhere, could bring the investigator closer to the animal in controlled conditions.

In response to seasonal dietary nutritional fluctuations, studies of ingestion and excretion in arctic animals may offer some challenge. The significance of shift in diet quantity and quality of arctic animals imposes continuous stress on renal handling of various elements. Also, diuresis is one of the universal responses exhibited by mammals during cold acclimation. Little is known about renal handling of large intakes of calcium and potassium ions; and the mechanisms underlying disturbances of water metabolism as related to cold diuresis are not well understood. Therefore, kidney physiology in various arctic species as related to the stresses of diet and temperature, should be researched.

Specific Enzymatic Mechanisms. (See chapters by Horwitz and Hettinger, and Roberts this volume for review of background information). Ultimately, an understanding of cold-induced compensatory changes in organisms living at low temperatures will involve delineation of cellular biochemical mechanisms. There appears to be three basic enzymatic strategies that may be used by cold exposed animals - increases in enzyme levels; utilization of enzyme variance (iso-enzymes); and modulation of enzyme activity via alterations of the cellular environment.

Few studies of homeotherms have systematically examined, and relative occurrence of these three strategies has not been determined in cold-induced responses. A

systematic evaluation of the utilization of these strategies should be applied to the following areas:

1. the basis of depression of oxidative enzyme
    activity in some tissues, including liver and
    kidney, during mammalian hibernation as well as
    lack of such a depression in other tissues, such
    as brown fat in the heart;

2. the changes in enzymatic activity during arousal
    from hibernation;

3. the possible occurrence of compensatory meta-
    bolic changes in heteothermic tissues of homeo-
    therms; that is, tissues such as blood, smooth
    muscle of peripherial vasculature, and skeletal
    muscle cells of the extremities; and

4. investigation of nonshivering thermogenic path-
    ways in arctic birds exhibiting a high degree of
    cold resistance, such as redpoles and snow bunt-
    ings.

Moreover, the degree to which the three basis enzymatic mechanisms listed above are involved in the metabolic
adjustments of arctic homeotherms also needs evaluation.
Animals such as the Alaskan king crab and the wood-boring
beetle, obtain their body heat primarily from the environment rather than from their own metabolic activities, and
thus maintain a body temperature similar to ambient temperature. In addition, there are a number of other areas
of a more general nature which require further biochemical
examination:

1. differences between acclimation-induced and
    acclimatization-induced enzyme changes oc-
    curring in homeothermic tissues, e.g., kid-
    ney, liver, and brown fat, as well as in
    heterothermic tissues;

2. effects of seasons on cold-induced enzyme
    changes, and;

3. the time course of enzymatic changes induced
    during prolonged cold exposure in such tis-
    sues as liver, kidney, and brown fat.

Of particular concern is the possibility that upon
exposure to cold, transient changes may occur which result
in the new steady state representing cold acclimation.
Such changes may be overlooked if only the new steady state
is examined.

*Hormonal Mechanisms and Adaptation to Cold in the
Arctic.* (See chapters by Wunder, Fregley, and Heroux this
volume for discussion of background material). Many environmental factors change seasonally in the Arctic, and
these may influence endocrine glands, their control systems, and their interactions. Thus, adaptation to arctic
cold may result not only in adaptation to cold per se, but
also simultaneously to other factors associated with the
seasonal change, e.g., light and darkness.

Since it is well known that hormones affect metabolic,
cardiovascular volume, and electrolyte adjustments to environmental and physiological changes of many types, a
better understanding of the hormonal responses to cold
exposure may allow man to function more effectively at low
temperatures. After a better understanding of the hormonal
responses to cold is achieved, it may be possible to counteract the adverse effects of cold and darkness in polar
regions. It may even be possible to stimulate the deposition and functional control of brown adipose tissue in
adult humans. From an ecological point of view, baseline
values for hormonal secretion rates and other measures of
hormonal function in various arctic species are also needed.
Such measurements might provide insight into the potentially detrimental effects of man's modification of the
natural environment before population levels change or
species become extinct.

Better identification is needed of the environmental
factors that stimulate hormonal response patterns to cold
adaptation in the Arctic. What are the effects of atmospheric factors on hormonal systems? What are the effects
of nutritional states and inputs on hormonal responses and
their feedback patterns? Further identification is also
needed of the interactive effects of hormonal systems now
known to be involved in cold acclimation. It would be
instructive to compare responses in a single species over a
wide latitudinal range in order to further sort out the
problems of photoperiod. Additional knowledge is needed
regarding the effects of acute and chronic exposure to cold
on many hormonal systems.

Hormones interact to produce their effects. An increase or decrease in the rate of secretion of one hormone may affect both the rate of secretion and the cellular activity of another. Hormones can act either synergistically or antagonistically. Our knowledge of hormonal interaction is in its infancy, despite the fact that the interactions constitute the most likely fashion in which hormonally induced changes manifest themselves.

PLANT RESEARCH

Botanical research in the Arctic expanded in new directions and with a renewed intensity around 1970. Prior to that time, research efforts were largely taxonomic in orientation or of a descriptive ecological nature. With the advent of the International Biological Project's Tundra Biome Study (IBP) in the early 1970's, the efforts of a sizeable number of plant physiologists and ecologists were integrated, and their research was oriented toward functional interrelationships within the entire ecological system. The physiological component of these studies documented the ability of certain species to function effectively in the cold climate and during short growing seasons in a soil environment which is often wet, anaerobic, and limited in nutrients. These studies now provide documentation of response patterns in an arctic environment. Things such as seasonal trends in nutrient changes, controls of environmental factors on CO_2 uptake in several species, and rooting patterns among dominant graminoids, can now be described with a certain degree of accuracy.
Little attention, however, has been directed toward explicit understanding of: 1) the adaptations of plants to low temperatures, and 2) the adaptations of plants in arctic ecological systems. Both contexts are in urgent need of attention, especially since present and future industrial developments in the Arctic will modify thermal regimes and perhaps alter the existing "balance" of present systems. The second half of this chapter addresses these issues, presents brief reviews of selected topics, and identifies biochemical, physiological, and ecological research priorities.
In designing a long-term research plan on arctic species, particular attention must be given to the conceptual framework for specific investigations and to the choice of experimental material. It is clear that an

individual plant operates within a complex natural environment in competition or coordination with other components
of the ecosystem. These relationships should be understood
before research on metabolic processes is attempted.
Furthermore, plants are complex and integrated structures.
It is difficult to divorce basic metabolism from cycles of
growth, hormonal regulation, and nutrient acquisition and
utilization. Research must proceed equally on all fronts
and be aimed at understanding the integrated functioning of
tundra plants. While some contributions can, of course, be
made by individual investigators working independently,
there is a great need for cooperation among, and support
for, enough scientists to examine in detail the multi-
faceted behavior of arctic species. For these reasons
integrated projects should be given high priority.

The choice of species to be used in research is also
particularly critical. To date, relatively few arctic
species have been investigated at all, and only a very
small number of these have been examined with any thoroughness. From existing research, different life forms appear
to possess significant differences in their metabolic
behavior. Major life forms include lichens, mosses, graminoid, herbaceous perennial, wood perennial, deciduous and
evergreens. These are based on aboveground traits. Root
systems may be deciduous or perennial and may also influence metabolic patterns. Representatives from all of these
groups should be included in a comparative physiological
survey in order to fully understand the structure of arctic
plant communities and ecosystems. Practical considerations
dictate that will a small number of species can be investigated intensively. At this level, it will be useful to
utilize ecotypes or closely related species which range
between arctic and temperate areas, in order to provide
some perspective on the physiological capacity of arctic
species. In this vein, an important question which extends
beyond concern for the Arctic is: Why are nonarctic species excluded from the Arctic? Answering this question may
directly reveal the unique adaptations of arctic plants.

*Modeling and Ecosystem Approach to Plant Adaptations and
Energy Exchange*

A study of the adaptation of plants to low temperatures requires consideration of both direct and indirect
effects of such temperatures. The direct effects of temperature on plant processes are amendable to laboratory and

field experimentation. Indirect effects, such as the
effects of low air temperatures on soil temperature, permafrost, waterlogging of soil, decomposition, nutrient release, nutrient uptake, growth, and photosynthesis, are
less amendable to direct experimentation. Yet, these
interactions between diverse plant and environmental processes provide the essential stabilizing feedbacks which
allow plants to persist in natural environments. The
necessity of considering a plant in the context of this
complex web of influencing and influenced processes, emphasizes that studies of plant adaptation should maintain
an awareness of the natural ecosystem of the plant and
should integrate the processes of the plant-environment
system.

Modeling can integrate interactions between several
diverse processes and morphological patterns, and can coordinate research on these various aspects. A cost-benefit
analysis approch, often used or implied in ecological and
evolutionary thought, could provide a mechanism of analyzing the relative importance of different physiological and
morphological properties to plant adaptation and survival.
The cost-benefit approach would follow from a quantified,
integrated assessment of these properties. However, most
of the generalities about adaptations of tundra plants are
qualitative, and developed in a non-integrated manner from
observations on diverse organisms. Past emphasis has been
on distinguishing possible adaptation of tundra plants from
those of other large scale vegetation types such as taiga,
desert, or deciduous forest. An 'adaptation' in an evergreen shrub from a dry tundra ridgesite might be generalized to all tundra plants, regardless of site or growth
form. Recent tundra research in the International Biological Project's Tundra Biome Study and the Research on
Arctic Tundra Environments has concentrated on the physiology of plants in relation to micro- or mesotopographic
gradients. This research was not focused specifically on
broad tundra plant adaptation, and its more detailed focus
accentuates the weaknesses of earlier, broad generalities.

Models of tundra plants have been developed for several tundra species, representing several plant growth
forms. Specifically, preliminary models are available for
Dupontia fisheri (a single-shooted graminoid), *Eriophorum
vaginatum* (tussock graminoid), *Ledum procumbens* (evergreen
shrub), *Dryas integrifolia* (evergreen shrub), *Salix pulchra*
and *Betula nana* (deciduous shrubs), and three moss species.
Past research on tundra plant models has concentrated on

model development, not on model validation or generality. Since field research has concentrated on a localized research site, temperature has not been considered a major factor, and its effects on tundra plant processes have not been studied as extensively as water and nutrients.

RECOMMENDATIONS--RESEARCH PRIORITIES

Physical Processes Concerned With Energy Exchange, Effects of Canopies, and the Advantages of Plant Form

Physical processes which affect plant and soil temperatures are fairly well understood and models of these processes are available which can predict plant and environmental temperatures with reasonable accuracy. The models focus attention on a few, very specific, research needs. First, the turbulent exchange of heat and water vapor within tundra plant canopies is not known with sufficient accuracy. Within the range of current values, the effect of the vegetation canopy on permafrost levels can be positive or negative, although field experiments indicate a consistent decrease in soil surface thaw as canopies increase in size. Second, the processes of energy exchange at the moss layer and in the litter must be well characterized. These processes include evaporation which affects the rates of moss growth and litter decomposition. In addition, these surfaces are the boundary between processes in the air and in the soil. Models concerned with energy exchange at this interface have generally assumed saturated, freely evaporating conditions or dry, non-evaporating conditions. The real world is usually somewhere in between, so the effect of surface water content on energy exchange processes is an essential part of this study. Third, there is a heritage of observation and measurements on the effects of plant form on plant temperature. These effects can now be rigorously quantified with the general principles of heat exchange, and should be so defined to assess the cost-benefits of each adaptation.

Plant Processes/Energy Exchange

Models of photosynthesis, respiration and water relations are available for leaves and plant canopies. These models include the processes of both short- and long-wave

radiation, convective and evaporative transfer of energy
and conduction. Research is currently needed on: 1) the
control of stomatal activity in plant leaves and the linking of stomatal activity with rates of transpiration and
photosynthesis; 2) control of CO_2 diffusion and reduction
of CO_2 concentration in leaves; and 3) control and paths of
water uptake by tundra plant roots.

*Secure Experimental Data to Validate Existing Models
and to Extend Their Applicability over a Broader Range of
Tundra.* Preliminary models of plant growth are available
for the species and growth forms mentioned earlier. The
models lack basic information on the influences of temperature, water and inorganic plant nutrients, on plant growth,
and on the control of phenological phases by environmental
and internal factors. Models have largely been developed
from field observations, and postulate physiological
relations to calculate seasonal progression of biomass, the
percent of nitrogen, phosphorus, calcium and TNC in leaves,
stems, and roots. Experimental data are needed to strengthen
the model structures, and to validate the simulation of
plant growth in the field under various perturbations.
Validation should be accomplished, both in the locale for
which the model was originally developed, and throughout
the geographic range of the species. Che construction and
development of models involves several phases: 1) collection of data on processes expected to be important on
the basis for which the model is built; 2) collection of
data on the total plant or system response against which
simulations can be compared, in order to test whether the
model has truly included the important processes and encoded the results correctly; and 3) collection of data in
a variety of geographic and microsite situations so as to
test its generality.

Develop Models of Plant Population Processes. Models
of plant population processes are in a primitive state as
plants reproduce both vegetatively and sexually, and
because arctic plants do not have a heritage of research on
these processes. Research is needed to describe the relationships between initiation and mortality of shrub growing
points or graminoid meristematic regions, as well as the
factors controlling them. Research is also needed on the
initiation and abortion of flowering structures, and the
control of these processes by environmental and internal

factors. Additionally, more studies are necessary on seed
production, viability and dispersal, and seedling establishment, in order to describe the turnover and maintenance
of genetic variability in tundra regions.

*Broaden an Understanding of Photosynthetic and Res-
piratory Processes in all Growth Forms, with an Emphasis on
Nutrient Relationships.* The seasonal surface response of
photosynthesis to light, temperature, and moisture is partially described for several mosses, lichens and vascular
species. Some exploratory information is available on the
effect of nutrient enhancement on photosynthesis. The
rates of, and resistances to, water uptake and loss have
barely been explored. Little information is available on
nutrient uptake rates, controls or pathways. Information
is needed on nutrient availability and gas exchange.

*Document the Extent and Significance of Metabolic
Acclimation.* Seasonal acclimation of photosynthesis and
respiration needs to be studied in an area other than Barrow, Alaska, and cost-benefit analysis performed for patterns of adaptation in these processes. This will require
information on the costs and mechanisms of low-temperature
adaptation of photosynthetic and respiratory gas exchange.

*Describe the Metabolic Costs and Controls of Root and
Rhizome Growth and Activity in Soils Underlain by Permafrost.* Most of the live plant and animal material in
tundra soils is made up of vascular plant roots and rhizomes. This live, belowground material weighs three to
four times as much as aboveground, live plants. The respiration of this belowground material contributes a great
deal of carbon dioxide to the lower atmosphere. Much of
this carbon dioxide is recaptured in photosynthesis.
Decomposition is slow, however, particularly near the
Arctic coast, and organic matter accumulates faster than it
is decomposed.

Far better quantitative data are needed on respiration
rates of roots, rhizomes, and decomposers under field conditions. This would enable models to be more realistic in
relation to soil-atmosphere and carbon dioxide fluxes. The
only data on belowground growth come from studies of root
elongation in a few species of grasses and sedges. At
Barrow, Alaska, such data exist for only three species at
one specific site; this is out of approximately total of
about 150 species. One or two more species have been
studied in relation to belowground growth rates, but even

less intensively. A major effort should be made to study
the effect of photoperiod and light quality on root and
rhizome growth and perinnating bud formation. Preliminary
information indicates these factors may be more important
in stopping root and rhizome growth and inducing dormancy
before soil temperatures begin to drop.

*Develop an Understanding of the Roles of Vegetative
and Sexual Reproduction in Population Processes.* Niche
diversity and differentiation are much greater belowground
in the tundra than aboveground. Each species has unique
root and rhizome characteristics. Populations tend to grow
by vegetative means (tiller and rhizomes). Therefore,
populations in the Barrow, Alaska tundra are often clones
of single genotypes. These clones are several meters wide.
Some species do not readily invade disturbed areas because
of a slower rhizome and tiller growth rate, than that pos-
essed by other species. Inland and/or on sandy soils,
there is some reproduction by seed; but even here, populations are often cloned by vegetative reproduction.
There is almost no quantitative information for modeling on the growth rate of populations by vegetative reproduction. Since this is the principal form of plant reproduction in the wet coastal tundra, it leaves a substantial
gap in predictive modeling.

Studies of Ecosystem Processes

Develop Total System Energy Budgets. Aboveground
physical processes are fairly well known at the ecosystem
level near Barrow, Alaska. Belowground processes are not
so well known. At Meade River and other inland points they
are scarcely known at all, either above- or belowground.
This is particularly true for energy budgets which integrate physical and biological processes in the ecosystem.
Much more information must be obtained on energy
budgets, both above- and belowground, especially for inland
tundra. Also, how energy budgets change from year-to-year,
and from place-to-place, needs to be learned. Much research is needed on the effects of perturbations on the
soil parts of tundra ecosystems, particularly as these
affect the thaw-lake cycle, i.e., drained lakes and accelerated thermokarst erosion along vehicle tracks and
streams.

Extend Community-wide Studies of Carbon Metabolism, Water Use, and Nutrient Cycling. Plant physiological processes are known at the ecosystem level for only the International Biological Project's Tundra Biome Study (IBP) site and in the wet coastal tundra at Barrow, Alaska. Even this information has many gaps especially at the soil nutrient level. Some information is available at a site along the Meade River, approximately 60 miles (100 km) southwest of Barrow, but almost noting is known about the effect of temperature or primary production at the level of photosynthesis and respiration. Some data are available on water gain and loss, but certainly not at the level of the plant community.

Data on community photosynthesis and respiration, even at Barrow, are badly needed. Such data are completely lacking for inland tundras. Since productivity is dependent upon this information, the models are defective at this point for inland tussock tundras which are the main caribou grazing areas.

Some data are available on growth of plants in tundra ecosystems for Barrow and Meade River, but only for a few years and only at one or two sites at each locale. Aboveground growth is far better documented than below-ground growth.

Develop Population Studies to Apply to Problems of Vegetation. Data are needed on growth of clones and populations in all kinds of tundra ecosystems and for most of the dominant plant species. This is particularly relevant for those species invading drained lake basins, impoundments, and vehicle tracks. Rates of revegetation by native species of such perturbed areas also need to be known.

Growth and distribution rates of populations of dominant plant species in both coastal and inland tundras are needed. This is particularly important in changes through time involved in the thaw lake cycle and man-caused perturbations. Changes in population structure and growth are closely tied to changes in depth of thaw, thermokarst erosion, and sand stabilization.

Nutrient Uptake and Utilization

The net effect of the arctic climate is a low annual input of nutrients into the tundra system and an extremely slow rate of nutrient cycling within the system. Hence, the arctic climate results directly and indirectly in the

formation of soils with extremely low nutrient availability. Fertilizer studies in many tundra communities clearly
demonstrate that inorganic nutrients, especially nitrogen
and phosphorus, are among the factors which most strongly
limit plant growth. Tundra plants maximize nutrient acquisition (in part) by maintaining high root-to-shoot
ratios as a result of substantial root production rates and
slow root turnover.

High root-to-shoot ratios have both genetic and environmental determinants. Tundra plants have substantial
rates of root elongation at low temperatures and continue
elongating following freezing. Phosphate uptake by tundra
plants is temperature dependent and shows a temperature
optimum similar to that of temperate plants. The process
functions effectively at low temperature in tundra plants
because the absolute capacity for uptake is high under all
conditions, and because the process is relatively insensitive to temperatures below the optimum. Likewise, tundra
plants respond to nitrogen at low temperature, whereas
temperate plants do not. When grown at low root temperature, plants acclimate and exhibit a higher capacity for
phosphate uptake under all measured conditions. Although
fertilization experiments show growth of tundra plants to
be nitrogen and phosphorus limited, these elements are
present in leaves at substantial concentrations. Plants
restrict growth under conditions of nutrient limitation
rather than produce metabolically inefficient tissues with
nutrient-deficient symptoms. In the short tundra growing
season, nitrogen and phosphorus are translocated into
leaves much more rapidly than carbon. Rapid spring growth
of tundra plants may strain nutrient reserves more severely
than carbon reserves. High nitrogen and phosphorus concentrations are needed in leaves early in the season to
support the high metabolic capabilities characteristic of
tundra plants.

As a result of the nutrient-poor status of the substrate, the plant community has evolved in such a way as to
practice extreme conservation of nutrients by very tight
nutrient cycling--a feature, interestingly, which the
tundra shares with the tropical rain forest, although the
two are different in many ways. The unique nutrient environment of tundra plants has brought about an orchestra-
tion of nutritional adaptations found nowhere else. Following are specific research problems for which there is a
dire need for research activity, and an extraordinary
opportunity for significant advances in understanding mineral
plant nutrition in a situation demanding extraordinary
competence on the part of plants.

RECOMMENDATIONS--RESEARCH PRIORITIES

Establish Controls Over Nutrient Cycling

Nutrients cycle very slowly through the tundra system due to factors resulting directly and indirectly from low temperature. Factors governing nutrient release and movement of nutrients through the soil appear to directly influence rates of primary production. To understand and predict the productivity of natural communities, an understanding of those processes controlling nutrient supply at the plant root surface is necessary. Tundra receives very little nutrient input from the atmosphere or from parent material. Hence, plant production depends upon extremely conservative nutrient recycling within the tundra systems. Within this broad topic the following questions (listed in order of priority) deserve particular attention:

1. Given the low temperature optima and substantial biomass of microorganisms, why are arctic decomposition rates so slow?

2. How do high soil moisture levels enhance plant growth in a water-saturated, partially anaerobic environment? How does water movement influence nutrient movement at the root surface and over greater distances within the soil?

3. What is the role of plants (especially mosses) in nutrient cycling? To what extent do different plant types tie up or change cycling rates of inorganic nutrients?

4. What is the role of animals in changing plant growth and allocation in the Arctic, where herbivory tends to be a periodic rather than a constant selective influence?

Establish the Controls Over Carbon and Nutrient Allocation to Growth.

The growth of tundra plants appears to be simultaneously limited by availability of various mineral nutrients and factors which directly affect carbon gain. Tundra plants do not show nutrient deficiency symptoms and metabolic inefficiencies exhibited in temperate crop plants in nutrient-poor soils. The way in which low nutrient status

controls growth of arctic plants seems fundamentally
different from that of crop plants; yet these controls over
growth and allocation of carbon and nutrients are essentially unexplored. Questions deserving particular attention include:

1. How do carbon and nutrient availability interact
   to control growth rates of arctic plants?

2. What genetic and environmental factors control
   the movement of carbon and nutrients to new
   branches or tillers rather than into the parent
   tiller?

3. What are the mechanisms and costs of rapid shoot
   growth at low temperature exhibited by arctic
   plants?

4. How are the metabolic pathways, which regulate
   carbon and nutrient utilization, controlled? How
   does this control maintain appropriate levels of
   metabolic intermediates?

Establish the Control of Root Growth and Nutrient Uptake

Arctic plants are largely belowground organisms, yet
little is known about the growth, physiology and ecology of
roots and rhizomes. Roots of arctic plants grow in a soil,
which because of its low temperature, poor aeration, and
low nutrient status would prevent growth and function of
temperate plant roots. By understanding the physiological
adaptation of arctic plant roots to their soil environment,
it may be possible to estimate the energy and nutrient
costs of selecting agronomic plants to function in more
extreme soil environments. Topics deserving particular
attention include:

1. What are the genetic and environmental controls
   over the magnitude and seasonal variation in root
   growth and turnover of arctic plants?

2. What metabolic and physiological adjustments in
   carbon and nitrogen metabolism and translocation,
   allow plants to effectively exploit cold am-
   monium-dominated arctic soils?

3.   How effectively do mycorrhizal associations form
      and function at low temperature and low oxygen
      tension?

4.   What physiological and biochemical controls do
      plants exert over nutrient uptake processes to
      permit nutrient absorption from cold, nutrient-
      poor anaerobic soils?

5.   How do the large interspecific and intraspecific
      differences in root morphology affect root func-
      tion?

THERMAL TOLERANCE

It has been established that both temperate and arctic
plants show dramatic changes in resistance to subfreezing
temperatures through their annual cycle. For example,
black locust tree cells will undergo seasonal changes in
freezing tolerance from temperatures of -10°C in their most
tender summer state, to levels in excess of -196°C during
late fall and early spring. In red osier dogwood, cold
acclimation occurs in two distinct phases. The first
involves acclimation from -7°C to -20°C hardiness which is
induced by a phytochrome-mediated photoperiod response, and
a second phase of -20°C to -196°C which is induced by a
single autumn frost.
In contrast, herbaceous species, such as winter wheat,
cold harden in the fall at temperatures in the range of 0°
to +5oC. Such plants can then withstand subzero temperatures ranging from -40°C to -22°C depending on the genetic
constitution of the genotype.
Lichens and mosses of the arctic and alpine regions
also exhibit seasonal changes in cold hardiness. For example, during the most active periods of growth they suc-
cumb to temperatures as high as -2oC. At other times,
these same species will survive immersion in N_2 (-196°C).
It is noteworthy that the tissue moisture content is a
major factor in the ability of these plants to withstand
freezing stress. This phonomenon is a general feature of
lower plants and may suggest differing mechanisms of tolerance as compared to the higher plants that have been
extensively studied in temperate regions.

RECOMMENDATIONS--RESEARCH PRIORITIES

*Characterize Environmental Factors During Stress Periods
and Identify Plant Injuries due to Environmental Extremes*

Identify and quantitate the environmental factors
during freeze stress periods, i.e., high and low extremes
of soil and air temperature, snow cover, wind, and soil
moisture.

Determine the lethal and sublethal injuries to plants
that are directly attributable to extreme environmental
stresses, i.e., midwinter damage due to extreme low temperature minima, unseasonal frost during periods of rapid
growth, winter desiccation, and other easily identifiable
stresses. Injury could be determined *in situ* by visual
assessment and by laboratory tests such as regrowth or
specific assessment of tissue and cell viability.

Document Seasonability of Hardiness

Determine the seasonal fluctuations in hardiness in
the field by collecting plant material and subjecting it to
controlled freeze tests to ascertain their resistance to
freezing. Also, assess the survival in various tissues as
well as the total plant.

Elucidate Controls Over the Development of Tolerance

Identify and quantitate the factors responsible for
the timing and intensity of freezing tolerance, i.e., low
temperature, photoperiod, changes in growth rate, dormancy,
and desiccation. These studies would be largely done in
the laboratory with controlled environmental facilities.
Since these facilities may not be available on site, consideration should be given to having some of these tests
conducted in centers where similar work is being done.
This would require minimal expense and could be completed
in approximately two years. Once the major factors are
identified, on-site facilities could be expanded for further
evaluation of specific factors.

Establish the Physiological Basis for Cold Tolerance

Consideration should be given to more detailed physiological studies on the molecular aspects of cold acclimation and freezing tolerance. This would include studies on the role of hormones, water relations, specific metabolic and enzymatic changes, and structural and functional changes of all organelles and membrane systems.

METABOLIC AND ENZYMATIC ADAPTATION AT LOW TEMPERATURE

Arctic species have evolved to metabolize actively and to grow under temperatures which are continuously below the limits of active growth for a majority of the earth's flora. They are not totally unique in this capacity. A few temperate and tropical-zone species have some ability to survive brief periods of low temperature, but usually cannot sustain this ability for prolonged periods of time. Currently, there is a very small body of information on metabolic processes at low temperature, either in arctic species specifically, or in a much larger group of arctic and non-arctic species. Much of the available research has considered whole-plant gas exchange but very little has dealt with cellular mechanisms.

Several unique features of arctic plants relative to temperate-zone flora can be identified: 1) maintenance of relatively high levels of photosynthesis and respiration at temperatures well below 10°C; 2) ability to differentiate and grow at temperatures close to freezing; and 3) an apparent simultaneous possession of both an active growth capacity and a substantial degree of cold hardiness. The latter has not been fully verified, but rather appears to be true from general observation.

Much of this uniqueness appears to relate to cellular processes which, in turn, seems to be primarily functions of membrane structure and enzymatic activity. The rather large number of membrane-bound reactions, plus an increasing amount of experimental evidence from non-arctic species, suggest that a closer look at membrane structure at low temperatures in arctic plant species is needed.

Photosynthesis and respiratory activity continue in arctic species at temperatures to and below freezing. One part of this ability is a resistance of chloroplast and mitochondrial membranes to phase change at low temperatures. Another part concerns the continued activity of

metabolic enzymes at low temperatures. Amounts of key
enzymes, such as ribulose-1,5-bisphosphate carboxylase, appear to be sustained or increased at low temperatures, in
contrast to non-arctic species where enzyme activity is
often substantially reduced. Existing information on
enzymatic adaptation to low temperature suggests that quantitative changes in amounts of enzymes may be an important
route for maintaining low temperature metabolic capacity.
Additionally, there is limited evidence that different
enzyme forms are produced with altered kinetic and stability properties. Both of these changes need to be confirmed with more rigorous, state-of-the-art methodology.
Protein synthesis capacity is also of obvious importance,
but has received little attention for low temperature
conditions.

Mobilization, translocation, synthesis, and utilization of metabolites is necessary to sustain growth and
differentiation of arctic plants. Very frequently in non-
tundra species, exposure to temperatures below $10^{\circ}C$ retards
these processes, if not producing irreversible lesions in
intermediary metabolism. How tundra species are able to
avoid these damages has not been investigated. These metabolic processes are also necessary to support low temperature hardening which is an active, energy requiring activity.

RECOMMENDATIONS--RESEARCH PRIORITIES

Specific, high-priority research directions which will
permit a better understanding of low temperature metabolism
are listed below. These are aimed at gaining more detailed
information on the unique ability of tundra plant species
to maintain growth and metabolism when subjected to prolonged periods of low temperature.

*Establish the Role of Membrane Structure in Maintenance of
Metabolism*

Specific research is needed to answer the following
questions:

1.   Is membrane fluidity the major property of in-
      terest regarding membrane function?

2. What are the relative roles of qualitative and quantitative changes in lipid fractions in maintaining fluidity? What are the absolute costs and trade-offs involved in changing membrane structure?

3. Are membrane systems (chloroplast, mitochondrial, plasmo lemmo, endoplasmic reticulum) equally susceptible or resistant to low temperature-induced structural changes? If not, why not, and what specific metabolic lesions result?

How and to What Degree, is Protein Synthesis Capacity Maintained at Low Temperatures?

Studies on the temperature responses of the DNA-RNA-protein transcription sequence at low temperature are virtually absent, and extensive research is needed in this area.

Kinetic Properties of Key Metabolic Enzymes--Studies are Needed

Do kinetic properties change substantially in cold-adapted species? Existing studies on the major carbon-fixing enzymes should be repeated, since they were done before the current understanding of the complex requirements of these enzymes.

Research is Necessary on Maintenance of Capacity to Mobilize and Translocate Metabolites at Low Temperature

This research should include how arctic plant species avoid low temperature metabolic lesions that occur in species from temperate latitudes.

HORMONAL MECHANISMS RELATED TO COLD ADAPTATIONS IN ARCTIC PLANTS

Three lines of evidence suggest that hormonal mechanisms may play an important role in the adaptation of arctic plants to cold habitats.

1. various studies indicate some arctic plants are
    often exposed simultaneously to very cold root
    and warm shoot temperatures. Certain plants may
    also be exposed to steep diurnal changes in shoot
    temperatures. Such conditions would be expected
    to present problems in coordinating activities
    (in time as well as space) of various organs and
    tissues in temperate region species that might
    otherwise have access to arctic habitats. Such
    coordination usually involves hormonal signals;

2. a number of studies indicate that low tempera-
    tures can affect hormone content or activity.
    Hormones may also affect the ability of many
    plants to function adequately under cold (i.e.,
    arctic) conditions; and

3. a few studies on the effects of hormone appli-
    cations to alpine plant species (several of
    which also occur in the Arctic) suggest that such
    plants differ in hormonal production and be-
    havior. These differences also appear to be
    adaptive in nature.

Thus, based upon the foregoing type of evidence,
investigation of possible roles of hormones in adaptation
of arctic plants to cold habitats appears to be justified.
Recommendations of how such investigations should proceed
(in the general order of priority) follows.

RECOMMENDATIONS--RESEARCH PRIORITIES

Identify Endogenous Hormone Levels and Forms

At the start of such studies, hormone concentrations
should be followed throughout the growing season in two or
three prominent, primarily endemic species. Again, depending upon the results of application studies, it would undoubtedly be most fruitful to look at ABA and GA levels
first. Such studies should then be expanded to include
other hormones. These studies will require development of
a sensitive, specific method of analysis that can be conveniently applied to large numbers of samples. The recent
advent of radioimmunoassay techniques for plant compounds
promises to provide such a method.

Establish the Role of Hormones in the Control of Allocation to Roots and Shoots; and Vegetative Properties

Identify growth and other responses to exogenously applied hormones under rigorously controlled (laboratory) or monitored field conditions. Abscisic acid levels, known to vary depending upon stress conditions, in concert with GA levels should be given top priority.

ACKNOWLEDGMENTS

The author wishes to thank Ms. Shirley A. Zimmerman for her valuable assistance, patience, and understanding in the preparation of this chapter, as well as the collating and final typing of the entire manuscript.

REFERENCES

NAS. 1970. Polar research, a survey. National Academy of Sciences, Washington, DC. 204 pp.

IARCC. 1972. Five-year coordinated plan for arctic research. Interagency Arctic Research Coordinating Committee. Washington, DC. 150 pp.

N.R.C. 1963. Science in the Arctic Ocean basin. National Academy of Sciences. Washington, DC. 52 pp.

______. 1976. Committee to evaluate National Science Foundation's Arctic program. National Academy of Science, National Research Committee. Washington, DC. Various pagings.

Index